NEUROMETHODS

Series Editor
Wolfgang Walz
University of Saskatchewan
Saskatoon, SK, Canada

For other titles published in this series, go to
www.springer.com/series/7657

Animal Models of Behavioral Analysis

Edited by

Jacob Raber

*Division of Neuroscience and Department of Behavioral Neuroscience,
Oregon Health and Science University, Portland, OR, USA*

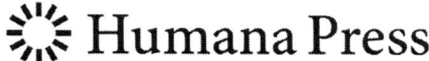

Editor
Jacob Raber, Ph.D.
Department of Behavioral Neuroscience
Oregon Health and Science University
3181 SW. Sam Jackson Park Road
Portland, OR 97239-3098
USA
raberj@ohsu.edu

ISSN 0893-2336
ISBN 978-1-60761-882-9
DOI 10.1007/978-1-60761-883-6
Springer New York Dordrecht Heidelberg London

e-ISSN 1940-6045
e-ISBN 978-1-60761-883-6

Library of Congress Control Number: 2010935599

© Springer Science+Business Media, LLC 2011

All rights reserved. This work may not be translated or copied in whole or in part without the written permission of the publisher (Humana Press, c/o Springer Science+Business Media, LLC, 233 Spring Street, New York, NY 10013, USA), except for brief excerpts in connection with reviews or scholarly analysis. Use in connection with any form of information storage and retrieval, electronic adaptation, computer software, or by similar or dissimilar methodology now known or hereafter developed is forbidden.

The use in this publication of trade names, trademarks, service marks, and similar terms, even if they are not identified as such, is not to be taken as an expression of opinion as to whether or not they are subject to proprietary rights.

While the advice and information in this book are believed to be true and accurate at the date of going to press, neither the authors nor the editors nor the publisher can accept any legal responsibility for any errors or omissions that may be made. The publisher makes no warranty, express or implied, with respect to the material contained herein.

Cover illustration: Collage of selected examples of translational animal models of behavioral analysis for ultimate use in humans

Printed on acid-free paper

Humana Press is part of Springer Science+Business Media (www.springer.com)

In memory of my father, Michael Raber (1919–2007), who taught me how to enjoy life, and in admiration of my children who help me to practice it daily.

Preface to the Series

Under the guidance of its founders Alan Boulton and Glen Baker, the Neuromethods series by Humana Press has been very successful since the first volume appeared in 1985. In about 17 years, 37 volumes have been published. In 2006, Springer Science+Business Media made a renewed commitment to this series. The new program will focus on methods that are either unique to the nervous system and excitable cells or which need special consideration to be applied to the neurosciences. The program will strike a balance between recent and exciting developments like those concerning new animal models of disease, imaging, in vivo methods, and more established techniques. These include immunocytochemistry and electrophysiological technologies. New trainees in neurosciences still need a sound footing in these older methods in order to apply a critical approach to their results. The careful application of methods is probably the most important step in the process of scientific inquiry. In the past, new methodologies led the way in developing new disciplines in the biological and medical sciences. For example, Physiology emerged out of Anatomy in the nineteenth century by harnessing new methods based on the newly discovered phenomenon of electricity. Nowadays, the relationships between disciplines and methods are more complex. Methods are now widely shared between disciplines and research areas. New developments in electronic publishing also make it possible for scientists to download chapters or protocols selectively within a very short time of encountering them. This new approach has been taken into account in the design of individual volumes and chapters in this series.

Wolfgang Walz

Preface

In general, it is hard to compare tests used in the clinic in humans with tests used in animal models. While the term "translational" is being talked about a lot these days, there is little consensus about what translational might stand for. There are actually only a few examples of translational laboratories that critically use and compare models of behavioral analysis in both animal models and humans. Such endeavors are not trivial and are open to a lot of criticism about how well particular behavioral analyses might compare across species. Even if it might be hard to compare all aspects of certain behavioral analyses, there is still great value in pursuing such translational approaches. Based on what is learned in the animal models, tests and treatment strategies might be developed to improve brain function in humans suffering from neurological conditions. In addition, knowledge obtained from human behavioral studies can be used to further improve the animal models of behavioral analysis. Therefore, this book focuses on approaches to translate and compare behavioral tests used in animals with those used in humans. These tests not only increase our understanding of brain function across species but also provide objective performance measures and bridge the gap between behavioral alterations in humans with cognitive disorders and animal models of these conditions. In Chapter 1, the use of eyeblink conditioning in the investigation of neurological conditions at the beginning and end of the life span is discussed. In Chapter 2, the anatomy, physiology, and behavioral analysis of the visual system are reviewed. In Chapter 3, correlates and analysis of motor function and animal models of Parkinson's disease are described. In Chapter 4, spatial learning and memory in animal models and humans are discussed. Spatial learning and memory can be assessed in most animals and humans and are relevant as they are affected by aging and many neurological conditions. Emotional learning and memory, such as measured in fear conditioning, are reviewed in Chapter 6. In Chapter 6, the use of conditioned place preference to assess drug reward in humans and animal models is reviewed. Chapter 7 describes how analysis of social behavior in animal models can be used to study autism. In Chapter 8, the representation of uncertainty in the human and animal brain is discussed. Questions like how and where this occurs in the brain are pertinent to neurological disorders such as schizophrenia that are characterized by a fundamental misrepresentation of uncertainty, arising through increased stochastic noise in neural circuits. Maladaptive styles of coping with uncertainty could be critical for generalized anxiety disorders as well. In Chapter 9, the importance of considering circadian variation in the physiological and behavioral analysis of humans and animals is discussed. In Chapter 10, the behavioral sequelae following Traumatic Brain Injury in humans and animal models are discussed. This is highly relevant as traditionally these sequelae have been used to screen candidate therapeutics for brain-injured patients. In Chapter 11, methods are reviewed to measure psychomotor sensitization in mice. Self-reports of sensitized vigor and energy levels in humans may relate to the more direct measurements of psychomotor sensitization in animals. These studies are important as they allow genetic investigations aimed at determining susceptibility to behavioral sensitization and neuroadaptations related to drug abuse. Chapter 12 discusses behavioral analyses critical to brain injury studies of experimental stroke or cardiac arrest.

These analyses are used to develop therapies to protect the brain during ischemic episodes and to enhance its potential for plasticity and repair after ischemia remains paramount. Finally, in Chapter 13, the use of magnetic resonance imaging (MRI) and in particular diffusion tensor imaging (DTI) to obtain information about cellular microstructure through measurements of water diffusion is discussed. This methodology can be used to characterize cellular morphological changes, for example, those associated with development of the cerebral cortex. Data collected in five species (mouse, rat, ferret, baboon, and human) were compared to determine whether similarities in the trajectory of DTI measurements with development exist in the literature. The ability of DTI to detect changes in neuroanatomy in the normal developing cerebral cortex allows detecting cortical abnormalities associated with various developmental disorders.

I would like to thank all my behavioral neuroscientist colleagues who authored the chapters of this book for their enthusiasm in accepting my invitation and for their outstanding contributions. Your accomplishments made this book possible. My sincere thanks go also to the Humana staff for bringing it all together. Finally, I would like to thank my wife and our children, Daniel, Nathan, Levi, and Eden, for their continuous support and encouragement. My hope is that readers will greatly enjoy this book and that this will help in further developing and appreciating translational animal models of behavioral analysis.

Portland, OR *Jacob Raber, Ph.D.*

Contents

Preface to the Series . . . . *vii*

Preface . . . . *ix*

Contributors . . . . *xiii*

1. Eyeblink Conditioning in Animal Models and Humans 1
   Kevin L. Brown and Diana S. Woodruff-Pak

2. Anatomical, Physiological, and Behavioral Analysis of Rodent Vision 29
   Brett G. Jeffrey, Trevor J. McGill, Tammie L. Haley, Catherine W. Morgans, and Robert M. Duvoisin

3. Correlates and Analysis of Motor Function in Humans and Animal Models of Parkinson's Disease 55
   Alexandra Y. Schang, Beth E. Fisher, Natalie R. Sashkin, Cindy Moore, Lisa B. Dirling, Giselle M. Petzinger, Michael W. Jakowec, and Charles K. Meshul

4. Spatial Learning and Memory in Animal Models and Humans 91
   Gwendolen E. Haley and Jacob Raber

5. Fear Conditioning in Rodents and Humans 111
   Mohammed R. Milad, Sarah Igoe, and Scott P. Orr

6. Conditioned Place Preference in Rodents and Humans 133
   Devin Mueller and Harriet de Wit

7. Social Interactions in the Clinic and the Cage: Toward a More Valid Mouse Model of Autism 153
   Garet P. Lahvis and Lois M. Black

8. The Experimental Manipulation of Uncertainty 193
   Dominik R. Bach, Christopher R. Pryce, and Erich Seifritz

9. Circadian Variation in the Physiology and Behavior of Humans and Nonhuman Primates 217
   Henryk F. Urbanski

10. Traumatic Brain Injury in Animal Models and Humans 237
    Hita Adwanikar, Linda Noble-Haeusslein, and Harvey S. Levin

11. Behavioral Sensitization to Addictive Drugs: Clinical Relevance and Methodological Aspects 267
    Tamara J. Phillips, Raúl Pastor, Angela C. Scibelli, Cheryl Reed, and Ernesto Tarragón

12. Evaluating Behavioral Outcomes from Ischemic
    Brain Injury 307
    *Paco S. Herson, Julie Palmateer, Patricia D. Hurn,
    and A. Courtney DeVries*

13. A Comparative Analysis of Cellular Morphological Differentiation Within
    the Cerebral Cortex Using Diffusion Tensor Imaging 329
    Lindsey A. Leigland and Christopher D. Kroenke

Index 353

Contributors

HITA ADWANIKAR • *Department of Neurological Surgery, University of California, San Francisco, CA, USA*
DOMINIK R. BACH • *Wellcome Trust Centre for Neuroimaging, University College London, London, UK*
LOIS M. BLACK • *Center for Spoken Language Understanding, Oregon Health and Science University, Beaverton, OR, USA*
KEVIN L. BROWN • *Neuroscience Program and Department of Psychology, Temple University, Philadelphia, PA, USA*
A. COURTNEY DEVRIES • *Department of Psychology, The Ohio State University, Columbus, OH, USA*
HARRIET DE WIT • *Department of Psychiatry and Behavioral Neuroscience, The University of Chicago, Chicago, IL, USA*
LISA B. DIRLING • *Portland Veterans Affairs Medical Center, Portland, OR, USA*
ROBERT M. DUVOISIN • *Department of Physiology and Pharmacology, Oregon Health and Science University, Portland, OR, USA*
BETH E. FISHER • *Department of Physical Therapy, University of Southern California, Los Angeles, CA, USA*
GWENDOLEN E. HALEY • *Department of Behavioral Neuroscience, Oregon Health and Science University, Portland, OR, USA*
TAMMIE L. HALEY • *Department of Physiology and Pharmacology, Oregon Health and Science University, Portland, OR, USA*
PACO S. HERSON • *Department of Anesthesiology and Perioperative Medicine, Oregon Health and Science University, Portland, OR, USA*
PATRICIA D. HURN • *Department of Anesthesiology and Perioperative Medicine, Oregon Health and Science University, Portland, OR, USA*
SARAH IGOE • *Department of Psychiatry, Massachusetts General Hospital and Harvard Medical School, Boston, MA, USA*
MICHAEL W. JAKOWEC • *Department of Neurology and Department of Cell and Neurobiology, University of Southern California, Los Angeles, CA, USA*
BRETT G. JEFFREY • *Oregon National Primate Research Center, Oregon Health and Science University, Beaverton, OR, USA; Casey Eye Institute, Oregon Health and Science University, Portland, OR, USA*
CHRISTOPHER D. KROENKE • *Department of Behavioral Neuroscience, Advanced Imaging Research Center, Oregon Health and Science University, Portland, OR, USA; Division of Neuroscience, Oregon National Primate Research Center, Oregon Health and Science University, Portland, OR, USA*
GARET P. LAHVIS • *Department of Behavioral Neuroscience, Oregon Health and Science University, Portland, OR, USA*
LINDSEY A. LEIGLAND • *Department of Behavioral Neuroscience, Advanced Imaging Research Center, Oregon Health and Science University, Portland, OR, USA*

HARVEY S. LEVIN • *Physical Medicine and Rehabilitation Alliance of Baylor College of Medicine and the University of Texas–Houston Medical School, Houston, TX, USA*

TREVOR J. MCGILL • *Casey Eye Institute, Oregon Health and Science University, Portland, OR, USA*

CHARLES K. MESHUL • *Electron Microscopy Facility, Veterans Affairs Medical Center, Portland, OR, USA; Department of Behavioral Neuroscience and Pathology, Oregon Health and Science University, Portland, OR, USA*

MOHAMMED R. MILAD • *Department of Psychiatry, Massachusetts General Hospital and Harvard Medical School, Boston, MA, USA*

CINDY MOORE • *Portland Veterans Affairs Medical Center, Portland, OR, USA*

CATHERINE W. MORGANS • *Casey Eye Institute, Oregon Health and Science University, Portland, OR, USA*

DEVIN MUELLER • *Department of Psychology, University of Wisconsin-Milwaukee, Milwaukee, WI, USA*

LINDA NOBLE-HAEUSSLEIN • *Departments of Neurological Surgery and Department of Physical Therapy and Rehabilitation Science, University of California, San Francisco, CA, USA*

SCOTT P. ORR • *Department of Psychiatry, Massachusetts General Hospital and Harvard Medical School, Boston, MA, USA; Department of Veterans Affairs Medical Center, Manchester, NH, USA*

JULIE PALMATEER • *Department of Anesthesiology and Perioperative Medicine, Oregon Health and Science University, Portland, OR, USA*

RAÚL PASTOR • *Department of Behavioral Neuroscience, Portland Alcohol Research Center, Methamphetamine Abuse Research Center, Oregon Health & Science University, Portland, OR, USA; Area de Psicobiologia, Universitat Jaume I, Castellón, Spain*

GISELLE M. PETZINGER • *Department of Neurology, University of Southern California, Los Angeles, CA, USA*

TAMARA J. PHILLIPS • *Department of Behavioral Neuroscience, Portland Alcohol Research Center, Methamphetamine Abuse Research Center, Oregon Health & Science University, Portland, OR, USA; Portland Veterans Affairs Medical Center, Portland, OR, USA*

CHRISTOPHER R. PRYCE • *Preclinical Laboratory for Translational Research into Affective Disorders, Clinic for Affective Disorders and General Psychiatry, Psychiatric University Hospital Zurich, Zurich, Switzerland*

JACOB RABER • *Department of Behavioral Neuroscience and Neurology, Oregon Health and Science University, Portland, OR, USA; Division of Neuroscience, Oregon National Primate Research Center, Oregon Health and Science University, Portland, OR, USA*

CHERYL REED • *Department of Behavioral Neuroscience, Portland Alcohol Research Center, Methamphetamine Abuse Research Center, Oregon Health & Science University, Portland, OR, USA*

NATALIE R. SASHKIN • *Portland Veterans Affairs Medical Center, Portland, OR, USA*

ALEXANDRA Y. SCHANG • *Department of Physical Therapy, University of Southern California, Los Angeles, CA, USA*

ANGELA C. SCIBELLI • *Department of Behavioral Neuroscience, Portland Alcohol Research Center, Methamphetamine Abuse Research Center, Oregon Health & Science University, Portland, OR, USA*

ERICH SEIFRITZ • *Clinic for Affective Disorders and General Psychiatry, Psychiatric University Hospital Zurich, Zurich, Switzerland*

ERNESTO TARRAGÓN • *Area de Psicobiologia, Universitat Jaume I, Castellón, Spain*
HENRYK F. URBANSKI • *Department of Neuroscience, Oregon National Primate Research Center, Oregon Health and Science University, Beaverton, OR, USA*
DIANA S. WOODRUFF-PAK • *Neuroscience Program and Department of Psychology, Temple University, Philadelphia, PA, USA*

Chapter 1

Eyeblink Conditioning in Animal Models and Humans

Kevin L. Brown and Diana S. Woodruff-Pak

Abstract

The knowledge base on behavioral parameters and neural substrates involved in eyeblink classical conditioning is extensive and continues to expand. The close parallels in behavior and neurobiology in mammalian species including humans make eyeblink conditioning an ideal paradigm for translational studies. Applications presented in this chapter include eyeblink conditioning's utility in the investigation of severe neurological conditions at the beginning and end of the life span: fetal alcohol syndrome (FAS) and Alzheimer's disease (AD). Recent studies in children with FAS demonstrate that delay eyeblink conditioning has a high sensitivity for diagnosis of prenatal exposure to alcohol. Contributing to the understanding of the neural mechanisms in the cerebellum damaged by prenatal alcohol exposure is a developmental rat model tested with eyeblink classical conditioning. Disruption of the hippocampal cholinergic system is associated with impaired delay eyeblink conditioning in normal rabbits. This result led to the hypothesis that the hippocampal cholinergic disruption observed early in AD might cause delay eyeblink conditioning to be impaired. This hypothesis was confirmed, and delay eyeblink conditioning has been demonstrated to have high sensitivity for AD. The hypercholesterolemic rabbit model of AD shows many neuropathologies of human AD and is impaired in delay eyeblink conditioning. Both the rat FAS and the rabbit AD models have demonstrated utility in the elucidation of disease mechanisms and the identification of treatments.

1. Introduction

The neural circuitry underlying classical eyeblink conditioning has been elaborated extensively (1, 2). The essential site of the plasticity for learning in all eyeblink conditioning paradigms resides in the cerebellum, and the hippocampus modulates learning in some paradigms and is essential in others. In eyeblink conditioning paradigms in which the hippocampus is essential, frontal cortical substrates are also engaged (3). Research with normal human adults using positron emission tomography (PET, 4), functional magnetic resonance imaging (fMRI, 5), dual-task paradigms (6), and human patients with neurological disease (7) demonstrates dramatic parallels among mammals in the brain structures engaged in eyeblink classical conditioning.

There are dramatic neurobiological and behavioral parallels in eyeblink conditioning in humans and non-human mammals including mice, rats, cats, rabbits, and monkeys. Parallels in the impairment of eyeblink conditioning in normal aging have been demonstrated in rats, rabbits, cats, and humans. The striking similarities in this behavior and its neural substrates among all mammals make results observed in non-human mammals directly relevant to humans.

The basic classical conditioning paradigm was identified and named by Ivan Pavlov as the "delay" paradigm and involves the presentation of a neutral stimulus, the conditioned stimulus (CS) followed after a time delay by the reflex-eliciting unconditioned stimulus (US). Learning occurs when the organism associates the CS with the US, producing a conditioned response (CR). A widely used variant of the delay paradigm is the "trace" paradigm (also named by Pavlov) in which CS offset precedes a blank trace period when no stimuli are presented. The CS and US must be associated even though they do not overlap. There are a number of additional variants of classical conditioning, but our focus is on delay and trace eyeblink classical conditioning (*see* **Fig. 1.1** for illustrations of various eyeblink conditioning procedures we used in the research described in this chapter). The extensive behavioral characterization of eyeblink classical conditioning over the life span of several mammalian species, including humans, along with knowledge of the basic neural circuitry underlying this form of learning has made it useful in translational research in a number of domains. Our focus for this chapter will be on two applications that occur very early and very late in life: fetal alcohol syndrome (FAS) and Alzheimer's disease (AD). Detailed materials and methods will appear at the end of the chapter, after background information on FAS and AD is provided.

1.1. Fetal Alcohol Syndrome

In addition to the neurobiological parallels among mammalian species, there are dramatic parallels in eyeblink conditioning in development over the life span. Most of the life-span data have been collected in adulthood, where it has been demonstrated that age-related eyeblink conditioning deficits appear in rats, rabbits, and humans in middle age. A basis for drawing parallels between rodent and human developmental models was demonstrated with the description of techniques and results with eyeblink classical conditioning in 4- and 5-month-old infants (8). This approach

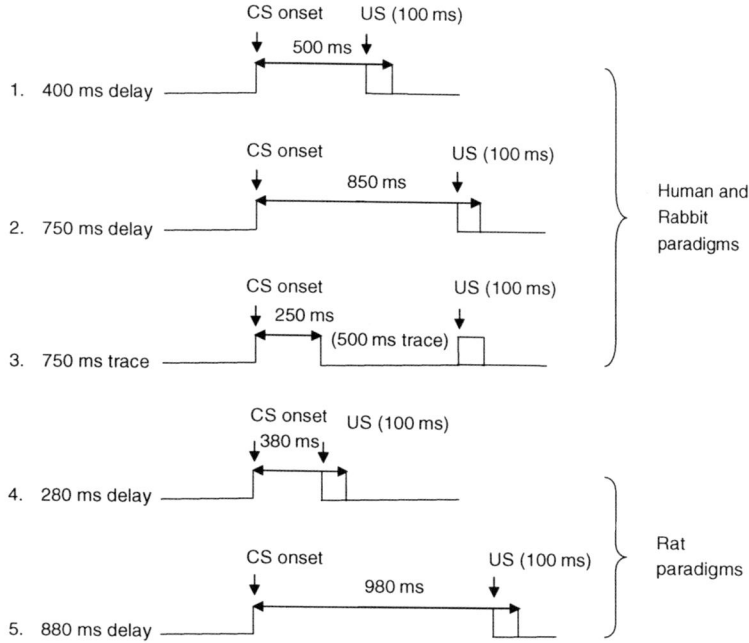

Fig. 1.1. Eyeblink conditioning paradigms typically used in human and rabbit studies of Alzheimer's disease and in rat studies of fetal alcohol spectrum disorders. In the 400-ms delay paradigm (1), a 500-ms conditioned stimulus (CS) precedes, overlaps, and coterminates with a 100-ms airpuff unconditioned stimulus (US). In the 750-ms delay paradigm (2), an 850-ms CS precedes, overlaps, and coterminates with a 100-ms airpuff US. In the 750-ms trace paradigm (3), a 250-ms CS precedes a 100-ms airpuff US. A 500-ms stimulus-free "trace" interval is imposed between the offset of the CS and the onset of the US. The 750-ms delay and trace paradigms are matched for the 750-ms interstimulus interval (ISI) between CS and US onset. In the 280-ms delay paradigm often used in studies with rats (4), a 380-ms CS precedes, overlaps, and coterminates with a 100-ms periocular shock US. The 880-ms delay paradigm (5) is also typically used in rat studies and consists of a 980-ms CS that precedes, overlaps, and coterminates with a periocular shock US. Parameters for auditory CSs used in these studies range between 1–2.8 kHz and 70–85 dB.

has the potential to offer fundamentally important data on the early development of cognitive and neurobehavioral disorders.

1.1.1. Human Studies of Fetal Alcohol Spectrum Disorders

Prenatal alcohol exposure is associated with a variety of physical, behavioral, and neuropsychological impairments and in some cases, fetal death. Fetal alcohol syndrome (FAS), the leading preventable cause of mental retardation in the Western world (9), is characterized by a distinct facial appearance (indistinct philtrum; thin upper lip; short palpebral fissures), CNS dysfunction (microcephaly; IQ deficits), and growth impairments (at or below tenth percentile for height or weight) resulting from heavy prenatal alcohol exposure in humans (10–14).

Fetal alcohol spectrum disorders (FASDs) is a term used to encompass the range of effects resulting from prenatal alcohol exposure (15, 16) and is associated with widespread brain damage and impairments in many domains, including executive functioning/attention, motor coordination, spatial memory, and social functioning (13, 17–21). Incidence of FASDs in the USA is estimated at 0.9–1 of every 100 births (14, 22), with annual costs exceeding $4 billion (23).

1.1.1.1. Anatomical Abnormalities in Fetal Alcohol Spectrum Disorders

Certain brain regions appear to be particularly susceptible to prenatal alcohol exposure, including the corpus callosum, basal ganglia, specific cerebral cortical regions, and – most relevant to this review – the cerebellum (13, 24–26). The cerebellum is associated with motor coordination, including balance, postural adjustment, and associative learning (e.g., eyeblink conditioning). Cerebellar dysmorphology in the form of region-specific hypoplasia and displacement is among the most widely reported anatomical abnormalities in the FASD literature (24, 27–31). Impairments in domains of motor functioning (i.e., balance; coordination) have been cited as well (12, 32, 33). Roebuck et al. (34) examined balance and postural sway in FASD children and controls to assess possible cerebellar-related impairments. The Sensory Organization Test was used to assess the influence of various modes of sensory information on the maintenance of balance. When visual and particularly somatosensory information was compromised or absent, FASD children performed more poorly (as measured by postural stability). However, no group differences emerged when only visual information was compromised, suggesting that cerebellar dysfunction may make FASD children particularly sensitive to altered somatosensory input.

These findings are significant in that they demonstrate the importance of task selection in identifying alcohol-induced impairments. Furthermore, while the deficits appear to be cerebellar dependent, possible effects on peripheral mechanisms cannot be ruled out. To obviate such potential confounds, cerebellar-dependent tasks that limit demands on multi-joint coordination may be ideal (c.f., 35). Toward that end, eyeblink classical conditioning may provide clearer indications of cerebellar dysfunction in FASDs (further details on eyeblink conditioning in the study of FASDs are presented below).

1.1.2. Rat Models of Fetal Alcohol Spectrum Disorders

Various animal models involving early developmental alcohol exposure have produced CNS deficits similar to those observed in human FASD. The period of most rapid brain growth that is often most susceptible to environmental insult (termed the "brain growth spurt") occurs at different relative points in ontogeny across species and must be taken into consideration (36). In rats, the brain growth spurt occurs within the first 2 weeks of postnatal life. This period of brain development in rats is considered "developmentally equivalent" (e.g., in terms of region-specific synaptogenesis and other critical developmental stages) to third trimester human fetuses. The rat model of FASD is useful in that (a) neurological anomalies (e.g., cerebellar damage) in rats exposed to alcohol during critical developmental periods are similar to those observed in human FASD, and (b) because of the extensively documented behavioral capacities of rats, demonstrating functional (behavioral) impairments resulting from early alcohol exposure is easier to demonstrate relative to other species.

The rodent cerebellum, like in humans, is highly sensitive to early developmental alcohol exposure. Studies of alcohol-related cerebellar damage typically focus on Purkinje cells, prenatally generated neurons that receive massive sensory input and represent the sole output of the cerebellar cortex. Early studies revealed Purkinje cell depletion following alcohol exposure spanning prenatal and postnatal periods (GD12–PD4; 37). While alcohol exposure (blood alcohol concentrations (BACs) > 400 mg/dl) restricted to gestation is capable of depleting Purkinje cell populations (38), the neonatal period ("brain growth spurt") appears to be a particularly sensitive period. Marcussen et al. (39) found reduced vermal Purkinje cell numbers at PD10 in subjects exposed postnatally (PD4–9; BACs ~ 200 mg/dl) but not in those exposed prenatally (GD13–18; BACs > 250 mg/dl). Enhanced sensitivity during the neonatal period was further demonstrated by Maier et al. (40), as exposure during all three trimesters equivalent (gestation plus PD4–9) or the "third trimester equivalent" only reduced Purkinje cell populations, with no such impairments observed following alcohol exposure limited to gestation.

1.1.2.1. Binge-Like Exposure Models

Early work was successful in achieving substantial cerebellar microcephaly (small brain relative to body size) and reductions in cerebellar neuronal populations (Purkinje cells; granule cells) at PD10 following postnatal (PD4–10) binge-like exposure to high alcohol doses (BACs > 450 mg/dl; 41). Later studies employing alcohol delivery over this period demonstrated that (a) more moderate doses (4.5 g/kg) are sufficient to produce brain damage, (b) the pattern of exposure is critical, as lower doses (4.5 g/kg) condensed into more frequent deliveries (binge-like) produce greater cerebellar damage than a higher dose (but lower BACs; 6.6 g/kg) gradually delivered over a longer period, and (c) brain damage persists into adulthood (42, 43).

Alcohol was delivered via artificial rearing in some of the early studies (e.g., 42, 43). Artificial rearing entails surgical implantation of feeding tubes into the stomach which justifiably raises concerns over unwanted surgery/stress effects. Another delivery method is exposure to ethanol vapors, which results in comparable BACs and Purkinje cell depletions (44, 45). Both delivery methods include extended periods of maternal separation. Intragastric intubation, a method in which a polyethylene tube is inserted down the esophagus and into the stomach for alcohol delivery (46, 47), avoids these concerns as maternal separation is minimal.

1.1.2.2. Regional Selectivity and Critical Periods in the Neonate

An outcome of neonatal alcohol exposure that is robust across a variety of doses, patterns of dosing, and delivery methods are findings of enhanced regional vulnerability. Deficits are consistently shown in earlier-maturing (i.e., I, II, XI, X) relative to later-maturing cerebellar cortical lobules (i.e., VI, VII; 4, 42, 43, 45, 48). Furthermore, neonatal binge-like alcohol effects upon Purkinje cells depend both on dose and exposure period within the PD4–9 window (the PD4–9 dosing period has often been used in "third trimester" equivalent models). While daily neonatal alcohol exposure to 4.5 g/kg$^+$ (BACs generally > 200 mg/dl) is typically used to obtain Purkinje cell depletion (e.g., 42, 46, 49), slightly lower doses delivered in a binge fashion (i.e., 3.3 g/kg delivered across two feedings or 2.5 g/kg at once – BACs ~ 150–200 mg/dl) are also capable of producing moderate neuronal loss (39, 50).

1.1.2.3. Behavioral Effects

Alcohol-induced brain damage is truly problematic insofar as it compromises quality of life, vis-à-vis behavioral and cognitive impairments. The primary behavioral deficits reported as a result of neonatal exposure in rodents are generally in the spatial and motor domains. Neonatal alcohol exposure also impairs traditionally recognized cerebellar-dependent tasks that place high demands on motor control (i.e., parallel bar and rotating rod; 51–53). Though Purkinje neuronal loss has been correlated with motor performance deficits (53), establishing links between alcohol-induced cerebellar damage and behavior is difficult in these instrumental tasks that require complex multi-joint coordination. Cerebellar-dependent Pavlovian tasks like eyeblink classical conditioning which utilize an easily measurable reflexive response and engage a well-defined neural circuit (54) provide a more promising avenue for investigation of causal relationships between alcohol-induced brain damage and behavior (35, 55).

1.1.2.4. Eyeblink Conditioning in Rats

The extensive behavioral (56) and neurobiological (54) data compiled for the first two decades of research on animal models of eyeblink conditioning has focused on the rabbit model, due in part to their tolerance for restraint. Because rats do not accept restraint as well as rabbits, their testing in eyeblink conditioning requires a freely moving preparation. This was established by Ronald Skelton in the late 1980s using adult Long Evans rats (57). Mark Stanton adapted this freely moving preparation to developing rats. Specifically, the seminal report on this body of work (58) showed that eyeblink conditioning in the rat using a short-delay paradigm (280 ms interstimulus interval – ISI – between CS and US onset – **Fig. 1.1**) develops gradually between postnatal days (PD) 17 and 24. Furthermore, acquisition differences across ages cannot be attributed to differences in CS or US processing (58), increasing US intensity or varying the ISI does not overcome failures of conditioning at PD17 (59), and the learning (conditioned eyeblink responses (CRs)) present in weanlings is cerebellar dependent (60, 61). John Freeman and colleagues have replicated the behavioral emergence of eyeblink conditioning in rats (PD17–24; *see* 62) and, through electrophysiological recording of various components of the putative cerebellar–brainstem "eyeblink" circuit, have offered insights into the underlying mechanisms of the ontogeny of eyeblink conditioning.

1.1.3. Fetal Alcohol Spectrum Disorders Models and Eyeblink Conditioning

The neonatal binge alcohol exposure paradigm results in damage to multiple components of the critical eyeblink conditioning circuit (42, 63, 64). In an effort to translate neurological impairments suffered as a result of neonatal alcohol exposure as directly as possible to functional (behavioral) impairments in cerebellar-mediated tasks, Mark Stanton and Charles Goodlett adapted the neonatal binge alcohol rat model to the freely moving developing rat eyeblink preparation (65). Using the artificial rearing method of binge-like alcohol delivery at doses of 5.25 g/kg/day to Long Evans rats over PD4–9, Stanton and Goodlett (65) reported significant impairments in conditioning in alcohol-exposed rats relative to controls when eyeblink testing occurred over PD23–24. Importantly, these impairments appeared to be associative in nature, as CS and US processing were unimpaired as a result of neonatal alcohol exposure.

Subsequent reports applying the neonatal binge exposure model (5 g/kg/day of alcohol) to the freely moving rat eyeblink preparation have utilized the intragastric intubation method and have advanced the findings of Stanton and Goodlett (65) by (a) further showing that impairments in conditioned responding appear to be associative in nature, as unpaired CS–US training does not yield robust CR generation differences as a function of neonatal treatment (66, 67), (b) associative deficits persist into adulthood (66, 68–71), and (c) CR deficits are correlated with cerebellar neuronal loss (67, 69) and impaired cerebellar electrophysiological activity (68). This rodent model appears to be applicable to the human condition, as children with FAS are also impaired in acquisition of eyeblink classical conditioning (72, 73). Furthermore, recent studies from Mark Stanton's laboratory have revealed deficits in eyeblink conditioned responding in both post-weanling (74) and adult rats (75) following PD4–9 binge exposure to alcohol doses as low as 3–4 g/kg/day. These dose-dependent deficits were revealed when using more complex variants of eyeblink conditioning – e.g., ISI discrimination – relative to the single-cue paradigms typically employed. In ISI discrimination, two distinct CSs (tone and light) are paired with the US at two different CS–US intervals (280 and 880 ms ISI – **Fig. 1.1**).

1.2. Alzheimer's Disease

Severe memory loss is the most prominent clinical symptom of AD, and this memory impairment has long been associated with impairment in acetylcholine neurotransmission. Disrupted cholinergic neurotransmission is not the single cause of memory loss in AD, but this deficit characteristic of AD clearly impairs memory (76). Research in our laboratory testing patients diagnosed with AD began with an idea based on results from the rabbit eyeblink conditioning experiments. We knew from the animal studies that acetylcholine neurotransmission in the medial temporal lobes (specifically, the pathway from the medial septum into the hippocampus) was activated during eyeblink conditioning. Based on the results with the animal model, we hypothesized that patients with a diagnosis of probable AD would be impaired in eyeblink conditioning beyond the deficit due to normal aging, because AD damages the cholinergic system of the human brain (77).

Our hypothesis that eyeblink conditioning would be severely impaired in patients with AD was initially confirmed in a sample of 40 elderly adults, half of whom were diagnosed with probable AD (78). There were very significant differences in eyeblink conditioning between the patients and age-matched, non-demented control subjects. Normal older adults are impaired in eyeblink classical conditioning compared to young adults, but all age groups of normal, non-demented adults, including adults in their 80s and 90s, show clear evidence of associative learning (79). In probable AD, there is very limited eyeblink conditioning in the first session of testing. However, when given a sufficient number of training trials (e.g., 4 or 5 days of 90-trial presentations), patients diagnosed with probable AD acquire conditioned eyeblink responses (called CRs; 80, 81). This slowing of the rate of acquisition occurs in the animal model when antagonists to acetylcholine neurotransmission are introduced (82, 83). There is strong evidence that the site of interference of cholinergic antagonists in rabbits is the hippocampus (84). The facts that patients with AD are slow to acquire CRs and that AD patients do eventually acquire CRs are parallel to results in rabbits injected with antagonists to acetylcholine. Disruption of the hippocampal cholinergic system prolongs acquisition of CRs, but organisms eventually condition.

Solomon et al. (85) reported disruption of eyeblink conditioning in the 400-ms delay paradigm (**Fig. 1.1**) in a sample of probable AD patients in their early 70s – a decade younger than the Woodruff-Pak et al. (78) sample. Woodruff-Pak et al. (86)

carried out an additional replication in the 400-ms delay procedure testing 28 patients with probable AD and 28 healthy age-matched control subjects. Patients diagnosed with cerebrovascular dementia were also tested for this study, and eyeblink conditioning was shown to differentiate some cerebrovascular dementia patients from patients with probable AD. Eyeblink conditioning in adults over the age of 35 years with Down's syndrome and presumably AD was similar to conditioning in probable AD patients (87, 88). Conditioning in patients with neurodegenerative disease that does not affect the critical cerebellar eyeblink conditioning circuitry or disrupt the hippocampal cholinergic system is relatively normal. For example, patients with Huntington's disease (89) and Parkinson's disease (90, 91) have intact eyeblink conditioning.

We combined the data over several studies from 61 probable AD patients and 100 normal age-matched older adults that we had tested in the 400-ms delay eyeblink conditioning procedure. To these data we applied a bivariate logistic regression analysis using percentage of CRs to predict diagnosis (92). With percentage of CRs as the only predictor, the area under the receiver operating characteristic curve (AROC) was 73%. Prediction was improved when age and gender were added to the model. In a multivariable logistic regression model including age, gender, and percentage of CRs, the model accurately distinguished patients with and without probable AD with AROC of 79.8%. Both the bivariate and the multivariable logistic models were statistically significant ($p < 0.0001$). The 400-ms delay eyeblink classical conditioning paradigm has utility in identifying patients with AD.

We have used the delay eyeblink conditioning paradigm on the basis that the hippocampus modulates acquisition of CRs in the cerebellum. When the hippocampal cholinergic system is disrupted, the cerebellum is disrupted and acquisition is delayed. But some have argued that since the trace eyeblink conditioning paradigm is hippocampus dependent, it should be a more sensitive test of AD. Before Clark and Squire (93) demonstrated that the trace had to be 1,000 ms long for the paradigm to be hippocampus dependent in humans, we tested patients with AD in the 750-ms trace paradigm (**Fig. 1.1**) and compared performance to the 400-ms delay paradigm (94). Though AD patients acquired a lower percentage of CRs in the 750-ms trace paradigm relative to controls, AD patients performed better in the 750-ms trace paradigm than in the 400-ms delay paradigm. We are not aware that AD patients have been tested in a long-trace paradigm, which is difficult to implement because the human spontaneous eyeblink rate is around one blink per second. It becomes difficult to determine which responses are CRs and which are spontaneous eyeblinks.

Longitudinal results suggest that 400-ms delay eyeblink conditioning has utility in the early detection of dementia. A 3-year longitudinal study of non-demented adults tested on eyeblink conditioning revealed that three normal subjects testing in the AD range (producing less than 25% CRs) at Time 1 became demented within 2 or 3 years (95). A fourth "normal" subject who scored just above criterion (26% CRs) also developed dementia within 2 years of the initial testing. A fifth subject in this group became demented and died within a year of the initial testing. Thus, of eight non-demented subjects age-matched to probable AD patients who scored on eyeblink conditioning in the AD range, only three were cognitively normal at the end of a 3-year period and 63% became demented within 2–3 years. Age-matched non-demented subjects scoring above criterion remained cognitively intact during the period of the longitudinal investigation. A second longitudinal study followed 20 cognitively normal elderly participants (half good-conditioners; half poor-conditioners) over a 2-year period (96). A neuropsychological test battery administered 2 years after the initial eyeblink conditioning test revealed significantly worse performance in poor conditioners on visuospatial abilities, semantic memory, and language abilities showing early decline in AD. Over a 2-year period, two of the poor conditioners (and none of the good conditioners) failed significantly, and one was diagnosed with probable AD.

1.2.1. Rabbits as Models for Alzheimer's Disease

Rabbits are in the order Lagomorpha and have as separate ancestry from rodents. To assess the evolutionary relationships of Lagomorpha to other taxa, Graur and colleagues (97) used orthologous protein sequences. A total of 91 protein sequences were analyzed, and reconstructions revealed that primates are phylogenetically closer to lagomorphs than are Rodentia. From a phylogenetic standpoint, rabbits are closer than rodents to human primate ancestry.

From the standpoint of the Aβ proteins that are the central pathologic processes associated with neurodegeneration in AD, rabbit models have a definite advantage. The rabbit's closer phylogenetic relation to primates is expressed in the amino acid sequence of Aβ, which is 97% identical to the human sequence (98). They form Aβ plaques. Species such as mouse and rat have a different amino acid sequence for Aβ and do not form these plaques.

As a species for cognitive assessment, rabbits have another advantage. On no other species is there such a large body of parametric data on the classically conditioned eyeblink response as on the rabbit (99). Much of the general literature on classical conditioning is based on data collected in the rabbit. From the early work of Hilgard in the 1930s with human eyeblink conditioning, Gormezano in the 1960s with rabbit eyeblink conditioning, and subsequent work of their colleagues and students, the parallels between rabbit and human eyeblink classical conditioning have been demonstrated empirically for decades.

It was the initial knowledge that eyeblink conditioning was impaired in normal aging in rabbits and humans that made this well-characterized model attractive for the investigation of neurobiological substrates of age-related memory impairment. The model system of eyeblink classical conditioning might be "the Rosetta stone for brain substrates of age-related deficits in learning and memory" (100). Both rabbits and humans show age-associated deficits in conditioning, and these can be easily dissociated from age-associated changes in sensory systems (i.e., differences in sensory thresholds for detection of the conditioned stimulus) or motor systems (differences in reflexive eye blink amplitude). Rabbits and humans also show parallel age-associated changes in the neural substrates critical for eyeblink classical conditioning, the cerebellum, and the hippocampus. Age-associated deficits can be artificially induced in both young rabbit and young human subjects with drugs that affect cholinergic neurotransmission, and age-associated deficits can be reversed in rabbits with drugs approved to treat cognitive impairment in AD (101). The fact that the same classical conditioning paradigm can be tested in both rabbits and humans makes eyeblink conditioning a valuable preclinical test of cognition-enhancing drugs.

One of the strengths of using eyeblink conditioning as a model of age-associated deficits in learning and memory is that both humans and rabbits show similar deficits as they age. Humans begin to show age-associated deficits in eyeblink conditioning at about 40–50 years of age (79, 102). Rabbits begin to show age-associated deficits at approximately 2 years of age (103). Based on declines in reproductive capacity, a 2-year-old rabbit is equivalent to a 40–45-year-old human, which suggests similar onset of age-associated declines in eyeblink conditioning in both humans and rabbits. Furthermore, the results found with rabbits appear to generalize to other non-human species, such as mice, rats, and cats.

The combined advantages of rabbits as a research species are numerous. Rabbits have a closer phylogenetic proximity to primates than do rodents, rabbits have an extensively characterized profile on a measure of learning and memory that closely parallels human performance, and rabbits and humans have a documented profile of age-related parallels in learning impairment. The model system of rabbit eyeblink conditioning has been used extensively to elucidate mechanisms of learning and memory in normal aging. Age is currently the best predictor of AD in humans.

1.2.1.1. The Cholesterol-Fed Rabbit as a Model of Alzheimer's Disease

Vascular risk factors such as hypertension, type 2 diabetes, abnormal lipid levels, and obesity are risk factors common to Alzheimer's and cardiovascular disease. Observations of Aβ-containing senile plaques in the brains of human patients with cardiovascular disease led Sparks and colleagues (104) to examine the heretofore unexplored brains of a long-standing animal model of heart disease, the cholesterol-fed rabbit.

An extensive elaboration of this rabbit model of AD has demonstrated that the cholesterol-fed rabbit is a valid model of AD on a number of levels. At a molecular level, the brain of this animal model has numerous (>12) features similar to the pathology observed in the AD brain including elevated Aβ concentration, neuronal accumulation of Aβ immunoreactivity, extracellular Aβ plaques, reduced levels of acetylcholine, elevated brain cholesterol, apolipoprotein E immunoreactivity, breaches in the blood–brain barrier, microgliosis, and neuronal loss.

Sparks also discovered that increased levels of Aβ in the brains of cholesterol-fed rabbits were dependent on the quality of the water the animals were administered. Animals given tap water accumulated considerably more Aβ in the brain than animals administered distilled water (105). It was in a collaborative attempt to demonstrate cognitive validity of cholesterol-fed rabbits that Sparks discovered that it was trace amounts of copper in the drinking water that were responsible for the higher levels of Aβ in the brain in comparison to the brains of rabbits administered the cholesterol diet and distilled water. In this issue, he elaborates his position that cholesterol in the diet causes accumulation of Aβ, whereas copper introduced in the drinking water impairs elimination of Aβ.

1.2.1.2. Cholesterol-Fed Rabbits, Eyeblink Conditioning, and Therapeutic Interventions

Sparks and Schreurs' (106) demonstration that eyeblink classical conditioning was impaired in cholesterol-fed rabbits, in the light of our own work on eyeblink conditioning in human AD and classical conditioning in normal aging rabbits and humans, was a stimulating discovery that led us to initiate work with AD model rabbits. Our aim was to replicate AD neuropathology in the cholesterol-fed rabbit model, test eyeblink conditioning in this model, and determine if galantamine (Razadyne®) would ameliorate impaired conditioning. Rabbit chow with 2% cholesterol and drinking water with 0.12 ppm copper sulfate were administered for 10 weeks. Control rabbits received normal food and distilled water. Rabbit brains were probed for neuropathology. AD model rabbits had significant neuronal loss in frontal cortex, hippocampus, and cerebellum. Changes in neurons in the hippocampus were consistent with neurofibrillary degeneration and cytoplasmic immunoreactivity for β-amyloid and tau. In a second experiment, cholesterol-fed model rabbits were injected daily with vehicle or 3.0 mg/kg galantamine and tested on 750-ms trace and delay eyeblink conditioning (animals were conditioned first in trace, then in a delay paradigm matched for the ISI between CS and US onset). Galantamine improved eyeblink conditioning significantly over vehicle. We concluded that this animal model of AD has validity from neuropathological to cognitive levels and offers a promising addition to the available animal models of AD. Galantamine ameliorated impaired eyeblink conditioning, extending the validity of the model to treatment modalities (107).

One of the results highlighted in our report was the evident loss of Purkinje neurons and staining for Aβ and tau in the cerebellum of cholesterol-fed rabbits (confirmed by Sparks, unpublished observations). We examined the cerebellum in cholesterol-fed rabbits primarily because the cerebellum is the essential substrate for eyeblink classical conditioning, a behavior on which they were impaired. Although the cerebellum has traditionally been viewed as one of the last structures to be impacted

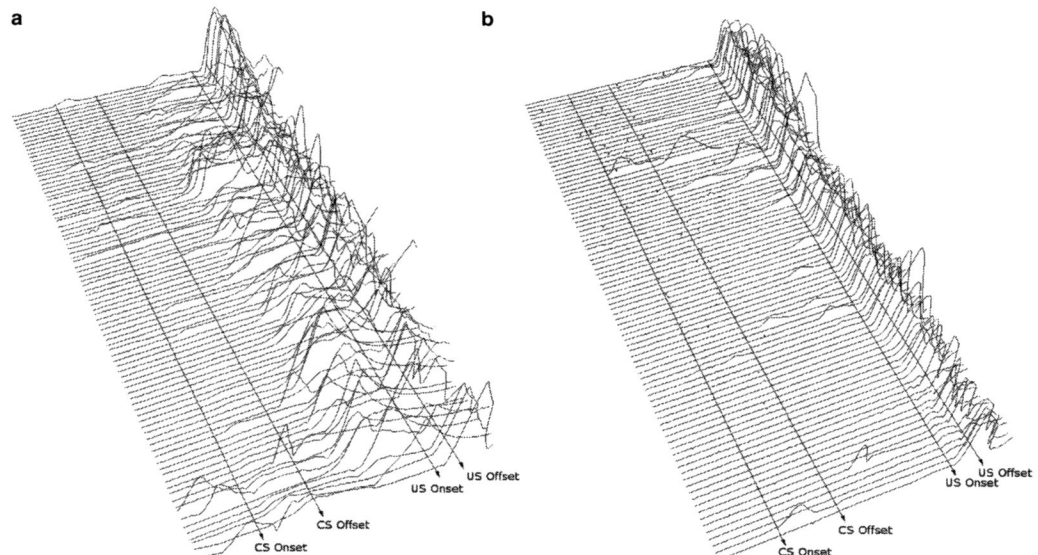

Fig. 1.2. (**a**) and (**b**): Eyelid/nictitating membrane movement in rabbits as recorded from changes in the reflectance from the cornea (of the left eye) detected by infrared. Each *line* represents activity from an individual trial (1–90, with Trial 1 represented at the *bottom* of each figure and Trial 90 represented at the *top* of each figure) from Session 7 of 750-ms trace eyeblink classical conditioning. Total trial length is 1,300 ms. *Lines* are drawn to approximate the onset and offset of both the tone conditioned stimulus (CS) and the airpuff unconditioned stimulus (US). There are 250 ms in the pre-CS period before CS onset, the CS duration is 250 ms, the stimulus-free "trace" interval following CS offset is 500 ms, US duration is 100 ms, and there are 200 ms in the post-US period. A response is scored if eyelid or nictitating membrane movement meets or exceeds 10% of the maximum response generated in that session. Performance is shown for a young male rabbit on a standard diet (control rabbit; **a**) and a young male rabbit following 10-week exposure to a high (2%) cholesterol diet with distilled water supplemented with trace amounts of copper (Alzheimer's model experimental rabbit; **b**). Each subject is representative of their respective groups. The incidence of conditioned responses (CRs) is significantly greater in the control subject (**a**) relative to the Alzheimer's model rabbit (**b**).

by classical AD pathology in humans, evidence has mounted to indicate that the cerebellum is indeed affected by AD (108).

Sparks and Schreurs (106) had tested trace eyeblink conditioning in cholesterol-fed rabbits because medial temporal lobe structures affected in AD are essential for trace conditioning. In their study, rabbits tested in a short-delay paradigm were not impaired. We used the same interval between the tone conditioned stimulus and airpuff unconditioned stimulus in the delay and trace paradigms and found both to be impaired (*see* **Fig. 1.2** for a comparison of a representative control rabbit (a) and a representative cholesterol-fed rabbit (b) trained in the 750-ms trace eyeblink conditioning paradigm). However, both Sparks and Schreurs (106) and our study tested the trace paradigm first. We followed up with a study of cholesterol-fed rabbits to determine if delay eyeblink conditioning would be impaired when tested before trace conditioning (109). Cholesterol-fed rabbits were significantly impaired in the 750-ms delay eyeblink conditioning paradigm (**Fig. 1.1**) when that paradigm was tested first, after 10 weeks of the 2% cholesterol and trace copper diet. Just as humans patients diagnosed with AD are significantly impaired in delay eyeblink conditioning, so are cholesterol-fed rabbits.

2. Materials

2.1. Human Eyeblink Equipment

2.1.1. Apparatus

The eyeblink conditioning apparatus is commercially available from San Diego Instruments and consisted of a portable, automated system with an infrared eyeblink detector and airpuff jet attached to headgear and input to an interface box containing a microprocessor and a miniature air compressor. The interface box is interactive with a computer. The headgear has an adjustable headband holding the airpuff jet and infrared eyeblink monitor. Filtered air is compressed in the interface box and delivered through a jet set approximately 2 cm from the cornea.

The voltage from the infrared device is amplified and differentiated; eyelid data are transferred to the computer for storage and analysis. From the speaker of the interface box, a 1,000-Hz, 80-dB sound pressure level (SPL) tone CS is delivered through headphones. Air for the corneal airpuff US is compressed in the interface box, filtered, and delivered at a pressure of 5–10 psi. The timing of stimulus presentations is controlled by the microprocessor in the interface box.

2.2. Rabbit Eyeblink Equipment

2.2.1. Eyeblink Classical Conditioning

The conditioning apparatus consists of four separate sound-attenuating chambers, permitting four rabbits to be trained simultaneously. A speaker mounted to the wall of each chamber delivers a pure tone (1 KHz, 85 dB SPL) used as the conditioned stimulus (CS). Each box has a fan running throughout the experiment that provides low-decibel background noise. The headpiece, affixed behind the rabbit's ears and under its muzzle, holds a plastic tube to deliver 7–8 psi corneal-directed airpuff unconditioned stimulus (US) and an infrared emitter and detector (San Diego Instruments, San Diego, CA) to measure the rabbit's eyeblink response. This device is positioned approximately 2 cm from the rabbit's eye. Eyelid retractors keep the rabbit's eye open. Output from the infrared detector is stored and analyzed using a PC. This system also controls the timing and presentation of the stimuli. The intertrial interval is randomized and ranges between 20 and 30 s. A single session lasts approximately 45 min and consists of 90-paired CS–US trials.

2.2.2. Food for AD Model Rabbits

Daily rations of the two diets are weighed, packaged in zip-lock bags, individually labeled, and stored in separate bins. All rabbits are fed with 160 g/day of a commercially produced diet (Test Diet 7520) of 2% cholesterol added to Purina Mills High Fiber Diet (experimental diet) or regular Purina Mills High Fiber Diet (control diet). Copper sulfate (0.12 mg/l) is added to distilled drinking water available ad lib for the 40 AD model rabbits, and normal control rabbits have ad lib access to distilled drinking water. Rabbits are treated with the cholesterol/copper or normal control regimen for 10 weeks.

2.3. Rat Eyeblink Equipment

Unlike rabbits, who accept restraint during behavioral testing, conditions must be in place for rodents to move about freely during eyeblink conditioning training. The freely moving rat preparation typically uses electromyography (EMG) to record eyelid muscle movement and periocular shock to elicit the unconditioned blink reflex.

2.3.1. Electrodes

The electrodes that are surgically implanted on the rats are custom-built from Plastics One (Roanoke, VA) – model # MS363/UDEL. Electrodes are composed of a (approximately) 0.7-cm high (0.75 cm diameter) plastic threaded cylindrical base (pedestal) with six female sockets located at the top. These sockets connect to cables used for (1) recording eyelid muscle activity and (2) delivering the shock unconditioned stimulus (US; *see* **Section 2.3.2**). Protruding from the opposite end of the plastic cylindrical component are three functionally distinct components:

(1) Two fine wire, Teflon-insulated EMG wires (these are differential electrodes – one is the "reference" and the other is the "recording" wire). These extend approximately 7 cm from the base of the pedestal.

(2) Two bipolar stimulating electrodes (stainless steel). These two insulated wires – approximately 0.2 mm in diameter – are twisted together and protrude approximately 3 mm from the base of the pedestal. Prior to surgery, the tips of these electrodes are untwisted into a v-shape and 0.1 mm of the insulation is shaved off of each tip.

(3) One ground electrode wire (stainless steel), approximately 0.2–0.3 mm in diameter. This protrudes approximately 4 mm from the base of the pedestal.

This multi-channel electrode set-up is used in recent reports from Mark Stanton's laboratory (e.g., 110) and represents an improvement from previous configurations that used two separate electrodes – one for the recording and ground wire and another for shock delivery.

2.3.2. Rat Eyeblink Conditioning Training Apparatus

The conditioning apparatus outlined here was initially described in experiments using an unrestrained preparation for testing both adult (57) and developing rats (58, 111, 112; available from JSA Designs, Raleigh, NC). It consists of 16 animal chambers (BRS/LVE, Laurel, MD) lined with sound-absorbing foam. Within each chamber animals are kept in stainless steel wire mesh cages measuring 22 × 22 × 26 cm. Each chamber is equipped with a fan which produces "background" noise (< 60 dB), a house light (15 W), and two speakers for delivering the auditory CS. The auditory CS typically used in rat eyeblink studies is a 70–85-dB, 2.8-kHz tone, and activation of a house light (against the dark background) serves as the visual CS. The US is produced by a constant-current, 60-Hz square wave stimulator (World Precision Instruments, #A365R-D, Sarasota, FL) set to deliver (typically) a 2-mA, 100-ms periocular shock to the left eye. Electrodes implanted on the top of the rat's skull (*see* **Section 2.3.1** for electrode design/**Section 3.3.2** for surgery procedure) are connected to wire leads that passed to the peripheral equipment (through a small hole located at the top of the chamber) via a commutator (Airflyte, #CAY-675-6). The commutator is on a swivel and suspended above the chamber. This flexibility allows the rat to move about freely within the mesh cage. The commutator connects to peripheral equipment associated with delivery of the shock and recording of electromyography (EMG) activity from the eyelid muscle. The EMG recording is amplified, rectified, and integrated via peripheral equipment that interfaces with a personal computer that controls delivery of all stimuli.

3. Methods

3.1. Human Eyeblink Classical Conditioning

3.1.1. Procedure

Participants are exposed to at least 72 trials of 400-ms delay eyeblink conditioning. The conditioning paradigm is administered in 10 blocks of 9 trials per block (8 blocks of 9 trials for the healthy control group). The first trial of each block is a CS-alone trial and the remaining trials are paired CS–US trials. The US is a corneal airpuff (5–10 psi, 100-ms duration) and the CS is a tone (80 dB, 1 kHz, typically 500-ms duration). On paired trials CS onset precedes the US by 400 ms and the stimuli coterminate. Participants watch a silent movie ("Milo and Otis") throughout the session to alleviate boredom. Participants are asked to remain alert, attend to the tones, and instructed as follows:

> *Please make yourself comfortable and relax. From time to time you will hear some tones and feel some mild puffs of air in your eye. If you feel like blinking, please do so. Just let your natural reactions take over.*

At that point, the experimenter answers any questions the participant has about the procedure and then initiates the conditioning trials. In both patients diagnosed with AD and in normal control subjects, if the participant's eyeblinks look abnormal – such as reduction in amplitude of the unconditioned response (UR) – the experimenter stops the session and redirects the airpuff nozzle and/or alerts the participant. Bad trials (spike in amplitude prior to CS presentation) and alpha responses occurring 75 ms or less after CS presentation are excluded from analyses.

3.2. Feeding Regimen AD Model Rabbits

On the sixth day after their arrival in the Association and Accreditation of Laboratory Animal Care International (AAALAC)-approved animal facility, rabbits are put on the cholesterol/copper regimen (some control rabbits are fed normal rabbit chow). Daily rations of both types of food are weighed, packaged in zip-lock bags, individually labeled, and stored in separate bins. All rabbits are fed with 160 g/day of a commercially produced diet (Test Diet 7520) of 2% cholesterol added to Purina Mills High Fiber Diet or regular Purina Mills High Fiber Diet. Copper sulfate (0.12 mg/l) is added to distilled drinking water that is available ad lib for the AD model rabbits, and normal control rabbits have ad lib access to distilled drinking water. Rabbits are treated with the cholesterol/copper or normal control regimen for 10 weeks.

3.2.1. Restrainer Training and Rabbit Eyeblink

Restrainer training: Prior to blood collection and injections that require restraint for the rabbits in a Plexiglas restrainer, and especially prior to behavioral training, the rabbits are familiarized and adapted to restrainers. Familiarization training takes place in rabbits' individual cages during the first 5 days. While in the restrainer and at the end of each familiarization session rabbits are rewarded with a treat formulated for rabbits (Kaytee Yogurt Dips). The last 2 days of familiarization take place outside the individual cages, and rabbits are fully restrained.

For the trace eyeblink classical conditioning paradigm, a 750-ms ISI between CS and US onset is typically used in our laboratory. We use this non-optimal interstimulus interval to provide a window of opportunity to demonstrate cognition-enhancing drug effects in rabbits. A 1-kHz, 85-dB SPL tone CS sounds for 250 ms followed by a 500-ms blank period and a 100-ms corneal airpuff US that commences 750 ms after CS onset. The rabbits receive 10 training sessions (5 days/week for 2 weeks).

Changes in the reflectance from the cornea detected by the infrared device are processed and stored in 3-ms bins by the computer. The program records a response when the eyelid or nictitating membrane moves a minimum of 10% of the maximum response generated in that session. A conditioned response (CR) is recorded if the response begins and reaches (at least) the 10% response threshold between 25 and 750 ms after the onset of the CS. An unconditioned response (UR) alone is recorded if the response takes place more than 750 ms after the onset of the CS. A trial is eliminated if nictitating membrane (NM) activity crosses the response threshold within 100 ms prior to the onset of the CS. Additionally, trials are eliminated if pre-CS activity levels exceed 20% of the rabbit's maximum response.

3.3. Rat Procedures

3.3.1. Alcohol Dosing for Neonatal Rats

Alcohol is delivered via intragastric intubation based on methods previously described in Goodlett et al. (46). Long Evans rat pups aged 4–9 days are typically used. The litter (8 pups total) is temporarily removed from the dam during intubations and kept on a commercially available heating pad (GE model #E12107) set to low. With this method, a mixture of a custom milk formula (made according to the procedures outlined in 113) with 95% alcohol (mixed to concentrations typically of 6.8–11.33% alcohol) is delivered to the rat pup via intragastric intubation. Based on the pup's weight (weights are recorded daily), a volume of this solution is taken up into a 1-ml syringe. The needle on the syringe is connected to PE50 tubing. This PE50 tubing is fused with smaller, PE10 tubing. PE10 tubing is lubricated with corn oil (to facilitate passage into the stomach) and gently placed down the esophagus and into the stomach, and the experimenter gently pumps in the solution. To minimize stress, the intubation procedure is completed (typically) within 30 s (to assess stress effects associated with intubation, "intubation control" groups that receive no alcohol or milk formula must be used). Formula is delivered in a volume of 0.02778 ml/g body weight (1/36 of the weight in grams of each pup). Alcohol is delivered in a "binge" fashion such that daily dosings occur 2 h apart, and pups are returned to their mother for the interval between daily doings. On PD4, all pups receive four intubations – two of the alcohol/milk mixture and two of milk alone. The additional calories of the supplementary milk feedings are given during the peak period of intoxication in order to maintain growth rates of alcohol-dosed subjects near those of controls (50). On PD5–9, the dosing regimen is identical with the exception that only one milk-alone intubation (2 h after the second alcohol delivery) is given. On the first day of dosing (typically PD4), a small volume of blood (approximately 20 μl) is collected in a heparinized capillary tube, spun using a centrifuge, and the plasma analyzed for blood alcohol content.

3.3.2. Surgery and Rat Eyeblink

The surgical procedure has been described previously (*see* 112) and is *slightly* modified here to account for the custom electrode (MS363/UDEL; Plastics One) used in recent studies from the laboratory of Mark Stanton (*see* 110).

Steps:
1. Inject the rat with a ketamine/xylazine cocktail (87 mg/kg ketamine; 13 mg/kg xylazine).
2. After thoroughly checking to make sure that the rat is completely anesthetized, shave the top of its head, focusing on the area behind the eyes and in front of the ears.
3. Place the rat on the surgery table (on its stomach). This surface (and all surgery tools) must be sterilized.
4. Apply betadine ointment to the top of the head.
5. Using a sharp scalpel blade, make an incision approximately 2 cm long, starting shortly behind the eyes and going back toward the ears. Make this incision directly at the midline of the skull.
6. Be sure that the top of the skull is completely dry before proceeding (use gauze pads to dry blood if necessary).
7. The following step differs depending on the age of the rat; therefore, two separate lines of directions will follow for this step. After this step the remaining steps are identical across age:
   a. For older rats (at least 40 days of age and older), three separate holes are made on the skull using a drill.

       i. Two holes are made approximately 2 mm anterior to lambda. Each should be approximately 1 cm from midline – one to the right of midline and the other to the left of midline.

     ii. The third hole is made approximately 2 mm posterior to lambda, approximately 1 cm to the right of the midline.

   iii. Screws are carefully inserted into each hole.

  b. For younger rats (younger than 40 days of age), four separate skull holes are made using the tip of a 26-gauge needle.

       i. Two holes are made immediately *posterior* to bregma (each approximately 0.5–0.75 cm from midline; one to the left and the other to the right).

     ii. Two holes are made immediately *anterior* to lambda (each approximately 0.5–0.75 cm from midline; one to the left and the other to the right).

   iii. Two "skull hooks" made of small strips of (malleable) galvanized steel wires (each approximately 1.5 cm long) are configured so that each tip fits securely into the aforementioned holes – one strip secured using the two holes posterior to bregma and the other strip secured using the two holes anterior to lambda.

8. Place the electrode (*see* **Section 2** for details of the electrode) on the exposed skull. The electrode should be placed on the midline, between bregma and lambda (this will fit between the aforementioned screws/wire strips).

9. Two 26-gauge needles are then inserted (on a slight upward angle) directly through the front of the upper eyelid (orbicularis oculi) muscle of the left eye. These should be aimed close to the middle of the eyelid and the two needles should be 2–3 mm apart.

10. Each EMG wire is passed through the tip of one of the needles – taking care that the wires do not cross – and the needles are then gently pulled out from the muscle, leaving the electrodes running through the eyelid muscle.

11. The ground wire is placed so that it is flat against the skull, slightly posterior to lambda and under the skin.

12. The v-tip of the shock delivery electrode (*see* **Section 2**) is placed immediately caudal to the left eye. Most of the length of the bipolar electrodes is running under the skin.

13. Secure the electrode to the skull using a mix of dental acrylic. Gently mix the powder and liquid for 30–60 s before applying to the skull.

14. After the cement has dried, strip the insulation (using fine-pointed forceps) off of the exposed parts of the two EMG wires. Start as far back as possible so that the sections of the EMG wire contacting the orbicularis oculi musculature are completely stripped.

15. Place the rat on a warm surface (electric heating pad; GE model #E12107, maintained at the lowest setting) – and monitor carefully – for approximately 30 min to recover.

3.3.3. Rat Eyeblink Testing

One to two days prior to CS–US eyeblink training, rats are placed in the conditioning apparatus for 1–2 min. At this time, a brief burst of air is blown into the rat's left eye with the integrated EMG signal produced by the blink appearing on an oscilloscope. This "puff-test" provides an indication of recording quality independent of the eyeblink UR elicited by the periocular shock US during training. Depending on the age of the rat being tested, the following day marks the start of CS–US training (post-weanling and juvenile rats; e.g., under 40 days of age) *or* the start of chamber acclimation (adults; e.g., 40 days of age or older). Younger rats begin testing as soon as possible so as to avoid confounding maturation and learning (c.f., 112; note – this is also why multiple daily sessions are used – as many as 3/day – in young rats). Chamber acclimation occurs in the same chambers in which CS–US training will be given and consists of connecting the subject to the recording equipment (i.e., attaching electrode to the recording and stimulating wires) for the duration of a standard conditioning session (approximately 50 min) *without* presentation of the discrete CS (tone or light) or US. Two chamber acclimation sessions are typically given in 1 day (separated by 5 h {±30 min}).

CS–US training consists of 100 trials, typically (in single-cue paradigms) with 90 paired CS–US trials and 10 CS-alone trials (9 CS–US trial and 1 CS-alone trial per block of 10 trials in single-cue conditioning). The ISI between CS and US onset in rat eyeblink conditioning experiments typically ranges between 280 and 880 ms. In delay conditioning, the CS precedes, overlaps, and coterminates with the shock US (in trace conditioning a stimulus-free interval occurs between the offset of the CS and the onset of the US). The shock US is typically 100 ms in duration delivered at an intensity of 1.5–4 mA. The intertrial interval varies randomly around an average of 30 s (range 18–42 s). Training in adults typically consists of 1–2 sessions per day over 5–6 consecutive days. In younger rats (under 40 days of age), 2–3 daily sessions are typically given. As stated above, the condensed training regimen in younger animals is utilized to complete training in as few days as possible so as to avoid confounding learning and maturation. The intersession interval is constant at 5 h (±30 min). In more complex training paradigms (e.g., ISI discrimination), two distinct CSs are used – a tone and a light (*see* **Section 2** for typical tone and light CS parameters). In ISI discrimination, each block of 10 trials consists of 5 tone and 5 light CSs, presented in a pseudorandom order such that no more than 3 consecutive trials of the same CS occurs. Within each block, 4 out of the 5 trials per CS is a paired CS–US trial, and the fifth is a CS-alone trial (8 paired CS–US trials and 2 CS-alone trials total per block of 10 trials).

EMG signals are sampled in 3.5-ms bins during the (typically) 1,400-ms epoch of each trial type. The raw signal is amplified (5 K), rectified, and integrated for analysis. Each trial is divided into five time periods: (1) a 280-ms pre-CS baseline period; (2) a startle response (SR; also termed "alpha response") period reflecting the first 80 ms after CS onset; responses occurring in this period are deemed non-associative "startle" reactions to the CS; (3) total CR period – EMG activity that occurs during either the 200 ms (short-CS trials) or 800 ms (long-CS trials) of CS presentation that precedes US onset; (4) adaptive CR period – EMG activity that occurs during the 200 ms of CS presentation that precedes US onset (for the short CS this is the same as the total CR period); and (5) UR period – EMG activity that occurs from the offset of the US to the end of the trial – for the short CS (280 ms ISI), the UR period is 740 ms; for the long CS (880 ms ISI), the UR period is 140 ms. The recording is interrupted during the 100-ms US presentation to avoid stimulation artifact. The threshold for registering an EMG response is set as 40 arbitrary units above the average baseline amplitude during the pre-CS period (e.g., 57, 58). Trials with excessively high spontaneous blink amplitudes (occurring during the pre-CS period) are not included in analyses.

Acknowledgments

The authors thank Andrey Mavrichev for designing the software used for visualization of eyeblink/nictitating membrane activity and for implementing this software with the rabbit eyeblink classical conditioning data shown in **Fig. 1.2**. Research described in this chapter was supported by grants from the National Institute of Alcohol Abuse and Alcoholism, 1 F31-AA16250 to KLB, 1-R01-AA11945 to Mark Stanton, and National Institute on Aging grants 1 RO1 AG09752, 1 R01 AG021925, 1 R01 AG023742 and grants from the Alzheimer's Association to DSW-P.

References

1. Thompson RF (2005) In search of memory traces. Annu Rev Psychol 56:1–23
2. Thompson RF, Steinmetz JE (2009) The role of the cerebellum in classical conditioning of discrete behavioral responses. Neuroscience 162:732–755
3. Woodruff-Pak DS, Disterhoft JF (2008) Where is the trace in trace conditioning? Trends Neurosci 31:105–112
4. Logan CG, Grafton ST (1995) Functional anatomy of human eyeblink conditioning determined with regional cerebral glucose metabolism and positron-emission tomography. Proc Natl Acad Sci USA 92:7500–7504
5. Cheng DT, Disterhoft JF, Power JM, Ellis DA, Desmond JE (2008) Neural substrates underlying human delay and trace eyeblink conditioning. Proc Natl Acad Sci USA 105:8108–8113
6. Papka M, Ivry RB, Woodruff-Pak DS (1995) Selective disruption of eyeblink classical conditioning by concurrent tapping. Neuroreport 6:1493–1497
7. Timmann D, Drepper J, Frings M, Maschke M, Richter S, Gerwig M, Kolb FP (2010) The human cerebellum contributes to motor, emotional and cognitive associative learning: a review. Cortex 46:845–857
8. Herbert JS, Eckerman CO, Stanton ME (2003) The ontogeny of human learning in delay, long-delay, and trace eyeblink conditioning. Behav Neurosci 117:1196–1210
9. Abel EL, Sokol RJ (1986) Fetal alcohol syndrome is now leading cause of mental retardation. Lancet 2:1222
10. Clarren SK, Smith DW (1978) The fetal alcohol syndrome. N Engl J Med 298:1063–1067
11. Jones KL, Smith DW (1973) Recognition of the fetal alcohol syndrome in early infancy. Lancet 2:999–1001
12. Jones KL, Smith DW, Ulleland CN, Streissguth AP (1973) Pattern of malformation in offspring of chronic alcoholic mothers. Lancet 1:1267–1271
13. Riley EP, McGee CL (2005) Fetal alcohol spectrum disorders: an overview with emphasis on changes in brain and behavior. Exp Biol Med 230:357–365
14. Sampson PD, Streissguth AP, Bookstein FL, Little RE, Clarren SK, Dehaene P, Hanson JW, Graham JM (1997) Incidence of fetal alcohol syndrome and prevalence of alcohol-related neurodevelopmental disorder. Teratology 56:317–326
15. Hoyme HE, May PA, Kalberg WO, Kodituwakku P, Gossage JP, Trujillo PM, Buckley DG, Miller JH, Aragon AS, Khaole N, Viljoen DL, Jones KL, Robinson LK (2005) A practical clinical approach to diagnosis of fetal alcohol spectrum disorders: clarification of the 1996 Institute of Medicine criteria. Pediatrics 115:39–47
16. Manning MA, Hoyme HE (2007) Fetal alcohol spectrum disorders: a practical clinical approach to diagnosis. Neurosci Biobehav Rev 31:230–238
17. Coles CD, Platzman KA, Lynch ME, Freides D (2002) Auditory and visual sustained attention in adolescents prenatally exposed to alcohol. Alcohol Clin Exp Res 26:263–271
18. Hamilton DA, Kodituwakku P, Sutherland RJ, Savage DD (2003) Children with fetal alcohol syndrome are impaired at place learning but not cued-navigation in a virtual Morris water task. Behav Brain Res 143:85–94
19. Mattson SN, Riley EP (1998) Neuropsychological comparison of alcohol-exposed children with or without physical features of fetal alcohol syndrome. Neuropsychology 12:146–153
20. Mattson SN, Goodman AM, Caine C, Delis DC, Riley EP (1999) Executive functioning in children with heavy prenatal alcohol exposure. Alcohol Clin Exp Res 23:1808–1815
21. Schonfeld AM, Mattson SN, Lang AR, Delis DC, Riley EP (2001) Verbal and nonverbal fluency in children with heavy prenatal alcohol exposure. J Stud Alcohol 62:239–246
22. Koren G, Nulman I, Chudley AE, Loocke C (2003) Fetal alcohol spectrum disorder. CMAJ 169:1181–1185
23. Lupton C, Burd L, Harwood R (2004) Cost of fetal alcohol spectrum disorders. Am J Med Genet C Semin Med Genet 127:42–50
24. Mattson SN, Riley EO, Jernigan TL, Ehlers CL, Delis DC, Jones KL, Stern C, Johnson KA, Hesselink JR, Bellugi U (1992) Fetal alcohol syndrome: a case report of neuropsychological, MRI, and EEG assessment of two children. Alcohol Clin Exp Res 16:1001–1003
25. Mattson SN, Schoenfeld AM, Riley EP (2001) Teratogenic effects of alcohol on brain and behaviour. Alcohol Res Health 25:185–191
26. Swayze VW, Johnson VP, Hanson JW, Piven J, Sato Y, Giedd JN, Mosnik D, Andreasen NC (1997) Magnetic resonance imaging of brain anomalies in fetal alcohol syndrome. Pediatrics 99:232–240
27. Archibald SL, Fennema-Notestine C, Gamst A, Riley EP, Mattson SN, Jernigan TL (2001) Brain dysmorphology in individuals

with severe prenatal alcohol exposure. Dev Med Child Neurol 43:148–154
28. Autti-Rämo I, Autti T, Korkman M, Kettunen S, Salonen O, Valanne L (2002) MRI findings in children with school problems who had been exposed prenatally to alcohol. Dev Med Child Neurol 44:98–106
29. Bookstein FL, Streissguth AP, Connor PD, Sampson PD (2006) Damage to the human cerebellum from prenatal alcohol exposure: the anatomy of a simple biometrical explanation. Anat Rec 289B:195–209
30. O'Hare ED, Kan E, Yoshii J, Mattson SN, Riley EP, Thompson PM, Toga AW, Sowell ER (2005) Mapping cerebellar vermal morphology and cognitive correlates in prenatal alcohol exposure. Neuroreport 16:1285–1290
31. Sowell ER, Jernigan TL, Mattson SN, Riley EP, Sobel DF, Jones KL (1996) Abnormal development of the cerebellar vermis in children prenatally exposed to alcohol: size reduction in lobules I–V. Alcohol Clin Exp Res 20:31–34
32. Connor PD, Sampson PD, Streissguth AP, Bookstein FL, Barr HM (2006) Effects of prenatal alcohol exposure on fine motor coordination and balance: a study of two adult samples. Neuropsychologia 44:744–751
33. Mattson SN, Riley EP (1998) A review of the neurobehavioral deficits in children with fetal alcohol syndrome or prenatal exposure to alcohol. Alcohol Clin Exp Res 22:279–294
34. Roebuck TM, Simmons RW, Mattson SN, Riley EP (1998) Prenatal exposure to alcohol affects the ability to maintain postural balance. Alcohol Clin Exp Res 22:252–258
35. Green JT (2004) The effects of ethanol on the developing cerebellum and eyeblink classical conditioning. Cerebellum 3:178–187
36. Dobbing J, Sands J (1979) Comparative aspects of the brain growth spurt. Early Hum Dev 3:79–83
37. Cragg B, Philips S (1985) Natural loss of Purkinje cells during development and increased loss with alcohol. Brain Res 325:151–160
38. Maier SE, West JR (2001) Regional differences in cell loss associated with binge-like alcohol exposure during the first two trimesters equivalent in the rat. Alcohol 23:49–57
39. Marcussen BL, Goodlett CR, Mahoney JC, West JR (1994) Developing rat Purkinje cells are more vulnerable to alcohol-induced depletion during differentiation than during neurogenesis. Alcohol 11:147–156
40. Maier SE, Miller JA, Blackwell JM, West JR (1999) Fetal alcohol exposure and temporal vulnerability: regional differences in cell loss as a function of the timing of binge-like alcohol exposure during brain development. Alcohol Clin Exp Res 23:726–734
41. Pierce DP, Goodlett CR, West JR (1989) Differential neuronal loss following early postnatal alcohol exposure. Teratology 40:113–126
42. Bonthius DJ, West JR (1990) Alcohol-induced neuronal loss in developing rats: increased brain damage with binge exposure. Alcohol Clin Exp Res 14:107–118
43. Bonthius DJ, West JR (1991) Permanent neuronal deficits in rats exposed to alcohol during the brain growth spurt. Teratology 44:147–163
44. Bauer-Moffett C, Altman J (1977) The effect of ethanol chronically administered to preweanling rats on cerebellar development: a morphological study. Brain Res 119:249–268
45. Ryabinin AE, Cole M, Bloom FE, Wilson MC (1995) Exposure of neonatal rats to alcohol by vapor inhalation demonstrates specificity of microcephaly and Purkinje cell loss but not astrogliosis. Alcohol Clin Exp Res 19:784–791
46. Goodlett CR, Peterson SD, Lundahl KR, Pearlman AD (1997) Binge-like alcohol exposure of neonatal rats via intragastric intubation induces both Purkinje cell loss and cortical astrogliosis. Alcohol Clin Exp Res 21:1010–1017
47. Pierce DR, Serbus DC, Light KE (1993) Intragastric intubations of alcohol during postnatal development of rats results in selective cell loss in the cerebellum. Alcohol Clin Exp Res 17:1275–1280
48. Light KE, Belcher SM, Pierce DR (2002) Time course and manner of Purkinje neuron death following a single ethanol exposure on postnatal day 4 in the developing rat. Neuroscience 114:327–337
49. Goodlett CR, Pearlman AD, Lundahl KR (1998) Binge neonatal alcohol intubations induce dose-dependent loss of Purkinje cells. Neurotoxicol Teratol 20:285–292
50. Goodlett CR, Marcussen BL, West JR (1990) A single day of alcohol exposure during the brain growth spurt induces brain weight restriction and cerebellar Purkinje cell loss. Alcohol 7:107–114
51. Goodlett CR, Lundahl KR (1996) Temporal determinants of neonatal alcohol-induced cerebellar damage and motor performance deficits. Pharmacol Biochem Behav 55:531–540

52. Klintsova AY, Cowell RM, Swain RA, Napper RMA, Goodlett CR, Greenough WT (1998) Therapeutic effects of complex motor training on motor performance deficits induced by neonatal binge-like alcohol exposure in rats I: behavioral results. Brain Res 800: 48–61
53. Thomas JD, Goodlett CR, West JR (1998) Alcohol-induced Purkinje cell loss depends on developmental timing of alcohol exposure and correlates with motor performance. Dev Brain Res 105:159–166
54. Christian KM, Thompson RF (2003) Neural substrates of eyeblink conditioning: acquisition and retention. Learn Mem 11:427–455
55. Goodlett CR, Stanton ME, Steinmetz JE (2000) Alcohol-induced damage to the developing brain: functional approaches using classical eyeblink conditioning. In: Woodruff-Pak DS, Steinmetz JE (eds.) Eyeblink classical conditioning, volume 2: animal models. Kluwer Academic Publishers, Boston, pp 135–153
56. Gormezano I, Kehoe EJ, Marshall BS (1983) Twenty years of classical conditioning with the rabbit. Prog Psychobio Physiol Psychol 10:197–275
57. Skelton RW (1988) Bilateral cerebellar lesions disrupt conditioned eyelid responses in unrestrained rats. Behav Neurosci 102:586–590
58. Stanton ME, Freeman JH, Skelton RW (1992) Eyeblink conditioning in the developing rat. Behav Neurosci 106:657–665
59. Freeman JH, Spencer CO, Skelton RW, Stanton ME (1993) Ontogeny of eyeblink conditioning in the rat: effects of US intensity and interstimulus interval on delay conditioning. Psychobiology 21:233–242
60. Freeman JH, Barone S, Stanton ME (1995a) Disruption of cerebellar maturation by an antimitotic agent impairs the ontogeny of eyeblink conditioning in rats. J Neurosci 15:7301–7314
61. Freeman JH, Carter CS, Stanton ME (1995b) Early cerebellar lesions impair eyeblink conditioning in developing rats: differential effects of unilateral lesions on postnatal day 10 or 20. Behav Neurosci 109: 893–902
62. Freeman JH, Nicholson DA (2004) Developmental changes in the neural mechanisms of eyeblink conditioning. Behav Cogn Neurosci Rev 3:3–13
63. Dikranian K, Qin Y-Q, Labruyere J, Nemmers B, Olney JW (2005) Ethanol-induced neuroapoptosis in the developing rodent cerebellum and related brain stem structures. Dev Brain Res 155:1–13
64. Napper RMA, West JR (1995) Permanent neuronal cell loss in the inferior olive of adult rats exposed to alcohol during the brain growth spurt: a stereological investigation. Alcohol Clin Exp Res 19:1321–1326
65. Stanton ME, Goodlett CR (1998) Neonatal ethanol exposure impairs eyeblink conditioning in weanling rats. Alcohol Clin Exp Res 22:270–275
66. Green JT, Rogers RF, Goodlett CR, Steinmetz JE (2000) Impairment in eyeblink classical conditioning in adult rats exposed to ethanol as neonates. Alcohol Clin and Exp Res 24:438–447
67. Tran TD, Jackson HD, Horn KH, Goodlett CR (2005) Vitamin E does not protect against neonatal ethanol-induced cerebellar damage or deficits in eyeblink classical conditioning in rats. Alcohol Clin Exp Res 29:117–129
68. Green JT, Johnson TB, Goodlett CR, Steinmetz JE (2002a) Eyeblink classical conditioning and interpositus nucleus activity are disrupted in adult rats exposed to ethanol as neonates. Learn Mem 9:304–320
69. Green JT, Tran TD, Steinmetz JE, Goodlett CR (2002b) Neonatal ethanol produces cerebellar deep nuclear cell loss and correlated disruption of eyeblink conditioning in adult rats. Brain Res 956:302–311
70. Lindquist DH, Sokoloff G, Steinmetz JE (2007) Ethanol-exposed neonatal rats are impaired as adults in classical eyeblink conditioning at multiple unconditioned stimulus intensities. Brain Res 1150: 155–166
71. Tran TD, Stanton ME, Goodlett CR (2007) Binge-like ethanol exposure during the early postnatal period impairs eyeblink conditioning at short and long CS–US intervals in rats. Dev Psychobiol 49:589–605
72. Coffin JM, Baroody S, Schneider K, O'Neill J (2005) Impaired cerebellar learning in children with prenatal alcohol exposure: a comparative study of eyeblink conditioning in children with ADHD and dyslexia. Cortex 41:389–398
73. Jacobson SW, Stanton ME, Molteno CD, Burden MJ, Fuller DS, Hoyme E, Robinson LK, Khaole N, Jacobson JL (2008) Impaired eyeblink conditioning in children with fetal alcohol syndrome. Alcohol Clin Exp Res 32:365–372
74. Brown KL, Burman MA, Duong HB, Stanton ME (2009) Neonatal binge alcohol exposure produces dose dependent deficits in interstimulus interval discrimination eyeblink conditioning in juvenile rats. Brain Res 1248:162–175

75. Brown KL, Calizo LH, Stanton ME (2008) Dose dependent deficits in dual interstimulus interval classical eyeblink conditioning tasks following neonatal binge alcohol exposure in rats. Alcohol Clin Exp Res 32:277–293
76. Bartus RT (2000) On neurodegenerative diseases, models, and treatment strategies: lessons learned and lessons forgotten a generation following the cholinergic hypothesis. Expt Neurol 163:495–529
77. Woodruff-Pak DS, Finkbiner RG, Katz IR (1989) A model system demonstrating parallels in animal and human aging: extension to Alzheimer's disease. In: Meyer EM, Simpkins JW, Yamamoto J (eds.) Novel approaches to the treatment of Alzheimer's disease. Plenum, New York, NY, pp 355–371
78. Woodruff-Pak DS, Finkbiner RG, Sasse DK (1990) Eyeblink conditioning discriminates Alzheimer's patients from non-demented aged. NeuroReport 1:45–48
79. Woodruff-Pak DS, Jaeger ME (1998) Predictors of eyeblink classical conditioning over the adult age span. Psychol Aging 13:193–205
80. Solomon PR, Brett M, Groccia-Ellison M, Oyler C, Tomasi M, Pendlebury WW (1995) Classical conditioning in patients with Alzheimer's disease: a multiday study. Psychol Aging 10:248–254
81. Woodruff-Pak DS, Romano S, Papka M (1996) Training to criterion in eyeblink classical conditioning in Alzheimer's disease, Down's syndrome with Alzheimer's disease, and healthy elderly. Behav Neurosci 110:22–29
82. Moore JW, Goodell NA, Solomon PR (1976) Central cholinergic blockade by scopolamine and habituation, classical conditioning, and latent inhibition of the rabbit's nictitating membrane response. Physiol Psychol 4:395–399
83. Woodruff-Pak DS, Hinchliffe RM (1997) Scopolamine-or mecamylamine-induced learning impairment: reversed by nefiracetam. Psychopharmacol 131:130–139
84. Solomon PR, Solomon SD, Vander Schaaf E, Perry HE (1983) Altered activity in the hippocampus is more detrimental to classical conditioning than removing the structure. Science 220:329–331
85. Solomon PR, Levine E, Bein T, Pendlebury WW (1991) Disruption of classical conditioning in patients with Alzheimer's disease. Neurobiol Aging 12:283–287
86. Woodruff-Pak DS, Papka M, Romano S, Li Y-T (1996) Eyeblink classical conditioning in Alzheimer's disease and cerebrovascular dementia. Neurobiol Aging 17:505–512
87. Papka M, Simon EW, Woodruff-Pak DS (1994) A one-year longitudinal investigation of eyeblink classical conditioning and cognitive and behavioral tests in adults with Down's syndrome. Aging and Cog 1:89–104
88. Woodruff-Pak DS, Papka M, Simon EW (1994) Eyeblink classical conditioning in Down's syndrome, fragile X syndrome, and normal adults over and under age 35. Neuropsychol 8:14–24
89. Woodruff-Pak DS, Papka M (1996a) Huntington's disease and eyeblink classical conditioning: normal learning but abnormal timing. J Int Neuropsychol Soc 2:323–334
90. Daum I, Schugens MM, Breitenstein C, Topka H, Spieker S (1996) Classical eyeblink conditioning in Parkinson's disease. Mov Disord 11:639–646
91. Sommer M, Grafman J, Clark K, Hallett M (1999) Learning in Parkinson's disease: eyeblink conditioning, declarative learning, and procedural learning. J Neurol Neurosurg Psychiatry 67:27–34
92. Ewers M, Braitman LE, Woodruff-Pak DS (2001) Eyeblink conditioning distinguished Alzheimer's disease in older adults. Soc Neurosci Abstr 27:282
93. Clark RE, Squire LE (1998) Classical conditioning and brain systems: the role of awareness. Science 280:77–81
94. Woodruff-Pak DS, Papka M (1996b) Alzheimer's disease and eyeblink conditioning: 750 ms trace versus 400 ms delay paradigm. Neurobio Aging 17:397–404
95. Ferrante LS, Woodruff-Pak DS (1995) Longitudinal investigation of eyeblink classical conditioning in elderly human subjects. J Gerontol Psy Sci 50B:42–50
96. Downey-Lamb MM, Woodruff-Pak DS (1999) Early detection of cognitive deficits using eyeblink classical conditioning. Alz Rep 2:37–44
97. Graur D, Duret L, Gouy M (1996) Phylogenetic position of the order Lagomorpha (rabbits, hares and allies). Nature 379:333–335
98. Johnstone EM, Chaney MO, Norris FH, Pascual R, Little SP (1991) Conservation of the sequence of the Alzheimer's disease amyloid peptide in dog, polar bear and five other mammals by cross-species polymerase chain reaction analysis. Brain Res Mol Brain Res 10:299–305
99. Woodruff-Pak, DS and Steinmetz, JE (eds.) (2000) Eyeblink classical conditioning: volume 2: animal models. Kluwer Academic Publishers, Boston, MA
100. Thompson RF (1998) Classical conditioning: the Rosetta stone for brain substrates of

age-related deficits in learning and memory. Neurobiol Aging 9:547–8
101. Woodruff-Pak DS, Vogel RW III, Wenk GL (2001) Galantamine: effect on nicotinic receptor binding, acetylcholinesterase inhibition, and learning. PNAS 98:2089–2094
102. Woodruff-Pak DS, Thompson RF (1988) Classical conditioning of the eyeblink response in the delay paradigm in adults aged 18–83 years. Psych and Aging 3: 219–229
103. Woodruff-Pak DS (1988) Aging and classical conditioning: parallel studies in rabbits and humans. Neurobio Aging 9:511–522
104. Sparks DL, Scheff SW, Hunsaker JC III, Liu H, Landers T, Gross DR (1994) Induction of Alzheimer-like β-amyloid immunoreactivity in the brains of rabbits with dietary cholesterol. Exp Neurol 126: 88–94
105. Sparks DL, Lochhead J, Horstman D, Wagoner T, Martin T (2002) Water quality has a pronounced effect on cholesterol-induced accumulation of Alzheimer amyloid β (Aβ) in rabbit brain. JAD 4: 523–529
106. Sparks DL, Schreurs BG (2003) Trace amounts of copper in water induce beta-amyloid plaques and learning deficits in a rabbit model of Alzheimer's disease. Proc Natl Acad Sci USA 100:11065–11069
107. Woodruff-Pak DS, Agelan A, Del Valle L (2007) A rabbit model of AD: valid at neuropathological, cognitive, and therapeutic levels. JAD 11:371–383
108. Coico R, Woodruff-Pak DS (2008) Immunotherapy for Alzheimer's disease: harnessing our knowledge of T cell biology using a cholesterol-fed rabbit model. JAD 15:657–671
109. Woodruff-Pak DS, Agalan L, Del Valle L, Achary M (2008) An animal model of Alzheimer's disease highlighting targets for computational modeling. In: Wang R, Gu F, Shen E (eds.) Advances in cognitive neurodynamics. Springer, Berlin, pp 903–907
110. Brown KL, Stanton ME (2008) Cross-modal transfer of the conditioned eyeblink response during interstimulus interval discrimination training in young rats. Dev Psychobio 50:647–664
111. Stanton ME, Freeman JH (1994) Eyeblink conditioning in the developing rat: an animal model of learning in developmental neurotoxicology. Environ Health Perspec 102:131–139
112. Stanton ME, Freeman JH (2000) Developmental studies of eyeblink conditioning in the rat. In: Woodruff-Pak DS, Steinmetz JE (eds.) Eyeblink classical conditioning, volume 2: animal models. Kluwer Academic Publishers, Amsterdam, pp 105–134
113. West JR, Hamre KM, Pierce DR (1984) Delay in brain growth induced by alcohol in artificially reared rat pups. Alcohol 1: 213–222

Chapter 2

Anatomical, Physiological, and Behavioral Analysis of Rodent Vision

Brett G. Jeffrey, Trevor J. McGill, Tammie L. Haley, Catherine W. Morgans, and Robert M. Duvoisin

Abstract

This chapter provides protocols for the study of rodent vision. An advantage of the visual system is that the physiological and behavioral response to the natural stimulus, light, can be measured. Moreover, the anatomy and circuitry of the system have been the subject of much research. Here, we describe our protocols for the analysis of the distribution of neurotransmitter receptors and signaling molecules in the retina by immunohistochemistry. We also explain in detail how we record the electroretinogram from mice, and we review two behavioral tests of rodent vision. One, the virtual optomotor test, makes use of the optokinetic nystagmus reflex, and thus is simple to perform and does not require training. The other, the visual water maze, is more demanding but provides a true quantitative readout of vision performance.

1. Introduction

As stated by John Dowling in 1987, the retina is an approachable part of the brain (1). The retina has a highly organized structure, being divided into five stratified layers, three cellular and two synaptic (**Fig. 2.1**). The light-sensitive photoreceptors that transduce light into an electrical signal are located in the outermost layer. The cellular and synaptic layers beneath the photoreceptors are collectively known as "neural retina" which is considered part of the central nervous system (CNS). The light response undergoes substantial signal processing within the neural retina before exiting the eye via the optic nerve en route to the

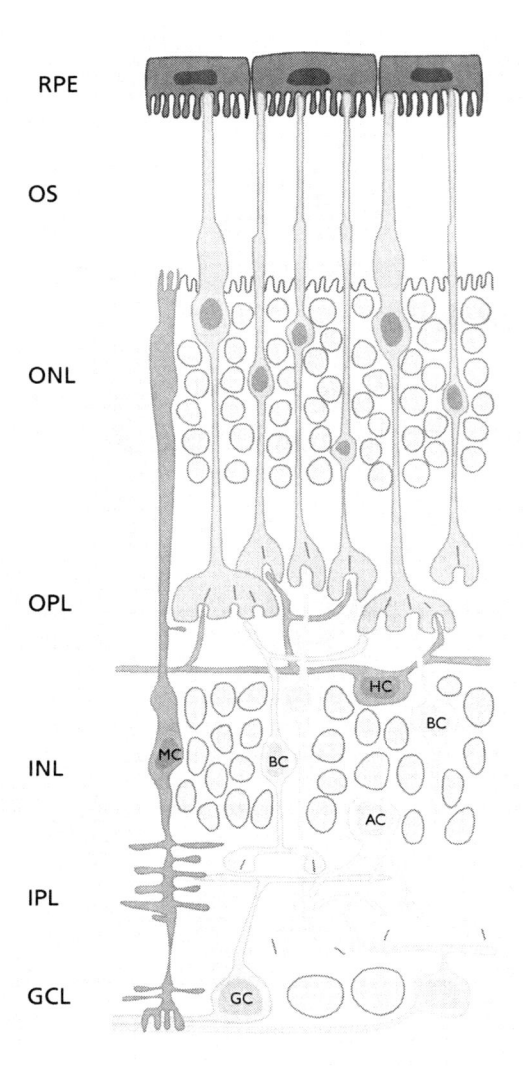

Fig. 2.1. Schematic diagram of the retina. The retina has a laminar structure. Photoreceptors (whose cell bodies are located in the outer nuclear layer; *ONL*) capture photons and transduce light into neural signals in their outer segments (*OS*). Retinal is regenerated and *OS* supported by retina pigment epithelial (RPE) cells. Neural signals are processed in the inner retina composed of two layers of cell bodies (the inner nuclear layer [*INL*] and the ganglion cell layer [*GCL*]). The *INL* comprises bipolar (*BP*), horizontal (*HC*) and amacrine (*AC*) cells. Muller glial cell (*MC*) soma are also located in the *INL*. The *GCL* comprises displaced amacrine and ganglion (*GC*) cells. Processing occurs in two synaptic or plexiform layers (the outer and inner plexiform layers [*OPL* and *IPL*, respectively]).

brain. The neurons within the neural retina are essentially identical to those within the brain: these neurons use most of the same neurotransmitters and neuropeptides and express many of the same receptors, ion channels, and enzymes as those found in the brain. Therefore, the neural retina is a useful system to examine the effects of genetic mutations and pharmacological compounds on CNS function. The anatomy, physiology, and neural circuitry of the retina are well characterized, and the effects of drugs on specific neural pathways as they process light stimuli may be tested.

Rod photoreceptors and the peripheral retina which is rod dominated are well conserved across mammalian species. In contrast, cones and the central retina differ greatly between primates and other mammals. Humans, apes, and old world monkeys have a cone-rich macula and a central cone-only foveola that enables high visual acuity. Other mammals lack this specialized macula although the retinas of some mammals have a visual streak, an area of increased photoreceptor density and therefore, acuity, across the central meridian. Additionally, the cones of human and non-human primates contain one of three opsins; other mammals have one or two cone opsins. Here, we will discuss methods as they apply to rodent vision, particularly the rod system, and we will mention briefly when differences are necessary for studying humans.

In the first part of this chapter, we will provide methods to analyze the distribution of ion channels, neurotransmitter receptors, and signal transduction proteins by immunohistochemistry. In the second part of this chapter, we will present methods to examine the electrophysiological responses of the eye to light stimuli. Finally, we will describe two behavioral tests used to measure vision in rodents: the Visual Water Task is a task used to examine visual acuity, similar to that measured in clinic, and the Virtual Optomotor System, which relies on a visually driven tracking reflex.

2. Materials

2.1. Immunohistochemistry

2.1.1. Aqueous Mounting Medium (AMM)

1. Add 2.4 g Mowiol 4–88 (Hoechst) or Gelvatol 20–30 (Monsanto) to 6 g glycerol in a 50-ml conical tube. Stir to mix.
2. While stirring, add 6 ml distilled water and leave for 2 h at RT on a rocker.
3. Add 12 ml 0.2 M Tris (pH 8.5).
4. Add NaN_3 to a final concentration of 0.02% (optional).

5. Incubate the tube at 50°C for 10 min or until Mowiol is dissolved.
6. Clarify by centrifugation at 5,000 g for 15 min. Store 1-ml aliquots in microcentrifuge tubes at −20°C.
7. Warm tubes to room temperature for use. Opened tubes can be stored at 4°C for approximately 1 month. Discard if any crystalline material is seen in the tube or on the slides.
8. For fluorescence, add 1,4-diazobicyclo-(2.2.2)-octane (DABCO) to 2.5% to reduce fading. This is a free radical scavenger. *Note*: since this is an aqueous medium fluorescence quality can diminish overtime. It is best to image 1 day after tissue is coverslipped.

2.1.2. Paraformaldehyde Fixative (PFA)

1. Heat 10 ml 0.1 M phosphate buffer, pH = 7.4 (PB) to 60°C in a microwave.
2. In a fume hood, for a 4% solution, add 0.4 g paraformaldehyde. *Caution*: paraformaldehyde fumes are toxic.
3. Stir until dissolved. If powder does not dissolve, add a drop of 5 N NaOH.
4. Chill on ice. Filter and adjust pH to 7.4 if necessary. Keep for about 1 week at 4°C.

2.1.3. Antibody Incubation Solution (AIS)

1. 3% (v/v) normal horse serum. (Sometimes less background is obtained by substituting normal goat serum.)
2. 0.5% (v/v) Triton X-100.
3. 0.025% (w/v) NaN_3 in phosphate buffer saline (PBS).

2.2. ERG Recording: Equipment

1. ERG Recording System with Ganzfeld. There are several commercial ERG recording systems that include hardware/software and a Ganzfeld dome, e.g., Espion E^2 (Diagnosys, Lowell, MA), UTAS (LKC Technologies, Gaithersburg, MD), RETI-port (Roland Consult, Brandenburg, Germany). (*See* **Note 1**.)
2. Optometer for measuring flash intensity and background intensity. Commercial optometers/radiometers include the IL1400A (International Light Technologies, Peabody, MA) and the S350/S450 (UDT, San Diego, CA). Both have an integration mode that enables the measurement of light from a brief flash.
3. A heating system to maintain core body temperature of the mouse between 36.0 and 37.5°C while the animal is anesthetized (*see* **Note 2**). Maintaining core body temperature is crucial for both animal health and obtaining maximal ERG amplitude, which drops dramatically with low body temperature. Loss of just a few degrees of body temperature is associated with reduced ERG amplitudes

and prolonged latencies. A simple solution is to place the animal on a warmed sports injury heat pad (2). Commercial systems include a low voltage DC system that enables automated temperature control (Braintree Scientific, Inc., Braintree, MA) and water-circulated heating systems (Roland Consult, Brandenburg, Germany). (*See* **Note 3**.)

4. A system for measuring core body temperature (if a non-automated system is used). We use the Thermalert temperature monitoring system with an RET-3 rectal probe for mice (Braintree Scientific, Inc., Braintree, MA).

5. A system for visualizing the mouse eye in the dark. We use a simple headlamp for most work in the dark. For fine manipulation work such as connecting electrodes we use an OptiVISOR binocular headband with LX-7 lens (2.75 ×) and an attached VisorLIGHT (Donegan Optical Company, Inc., Lenexa, KS). Two layers of deep red film are taped over both light sources and any lighted equipment in the room (e.g., computer monitor and green LED's used as ON lights on most equipment).

2.3. Medical Supply List for ERG Recordings

1. Butterfly needle infusion set (25 ga × 3/4″ with 12-inch tubing; Abbott Laboratories, North Chicago, IL).
2. Ketamine (100 mg/ml; Vedco, St Joseph, MO).
3. Xylazine (20 mg/ml; Lloyd Laboratories, Shenandoah, IA).
4. Saline solution (0.9% NaCl for injection; Hospira, Inc., Lake Forest, IL).
5. Sterile mix bottle (10 ml).
6. Murocel (1.0%; Bausch and Lomb, Tampa, FL).
7. Phenylephrine (2.5%, store at 2–8°C; Bausch and Lomb, Tampa, FL).
8. Tropicamide (1%; Akorn, Buffalo Grove, IL).
9. Proparacaine (1.0%, store at 2–8°C; Akorn, Buffalo Grove, IL).
10. Erythromycin ophthalmic ointment (5 mg/g, E Fougera & Co, Melville, NY).
11. Oxygen tank.
12. General supplies, including insulin syringes (0.5-ml 28 ga), 1-ml tuberculin syringes, fine forceps, tape, KY petroleum jelly, alcohol swabs, cotton-tipped applicators (Q-tips).
13. Miscellaneous equipment: Kitchen cooking scale for weighing mice in the laboratory, aluminum foil for making electrodes shields, tooth brush, and nail file for cleaning electrodes.

3. Methods

3.1. Immunohistochemical Analyses

Our laboratories are interested in localizing molecules involved in retinal synaptic transmission. We perform immunofluorescent labeling of retinal cryosections, since this allows the localization of multiple antigens. Importantly, we have found that short fixation times are essential for preserving antibody staining of synaptic membrane proteins, such as receptors or ion channels. While obtaining lightly fixed human retinal tissue can be problematic, we will describe here our procedure using mice.

3.1.1. Tissue Preparation

1. Mice are euthanized using an overdose of isoflurane or pentobarbital delivered via intraperitoneal injection.

2. Eyes are rapidly enucleated with curved scissors (care must be taken to not damage the optic nerve otherwise the retina will pull away from the eyecup during dissection), and rinsed in cold 0.1 M PB with 2 mM Ca^{2+} and Mg^{2+} added.

3. Eyecups are prepared by cutting just behind the ora serrata followed by removal of the lens. Eye is placed in a Petri dish filled with cold PB (with Ca^{2+} and Mg^{2+}). Hold eye in place with forceps and puncture eye with a number 11 scalpel blade just behind the ora serrata. Roll eye onto optic nerve. Use microscissors to cut around the equator. It may be easier to rotate the eye with forceps as you cut. Gently peel the front portion of the eye away from the eyecup. Roll the lens out of the way. Tilt the eyecup on its side to drain the remaining vitreous. *Note*: Any remaining vitreous will crystallize when frozen and create problems for sectioning. The vitreous freezes at a different rate causing the nerve fiber layer and the ganglion cell layer to be pulled away from the remaining retina. Also the remaining vitreous will fracture during sectioning. This causes the block to cut unevenly. The section will buckle.

4. Eyecups are transferred (via a cut plastic transfer pipette) to the fixative solution. Tissue is fixed for 5–20 min by immersion in 4% PFA. We use glass vials for fixation to avoid plastics that may leach from other containers. PB is used because PBS has a higher osmolarity that can cause tissue damage and poor tissue preservation.

5. The fixed eyecups are washed in PB and then cryoprotected at 4°C by sequential immersion in ice cold 10, 20, and 30% (w/v) sucrose in PB. Use a cut plastic transfer pipette to gently move eyecups into each new solution. Allow each eyecup to sink to the bottom of the vial at each step. (About

10–30 min.) The eyecups can sit overnight in the 30% sucrose at 4°C. The eyecups are then immersed into OCT (Sakura Finetek, Torrance, CA). Gently mix to remove the extra 30% sucrose. You should be able to see the mixing of the sucrose and the OCT. If extra sucrose remains inside the eyecup this will also cause sectioning problems. The sucrose will crumble during sectioning and the retina will be pulled along with the sucrose.

6. The tissue is then embedded in OCT (use plastic molds, such as the cap of a microcentrifuge tube, to help position the eyecups) and quickly frozen by immersion into isopentane/dry ice. Isopentane is placed in a glass beaker. Dry ice is packed around the beaker. Allow for the isopentane to equilibrate (otherwise freeze fracture will occur). The tissue is frozen and equilibrated when bubbles no longer form. The OCT-embedded eyecups can be stored at −80°C, but it is preferable to proceed to the sectioning.

7. The eyecup is removed from the molds and cut at 12–15 μm on a cryostat at −18 to −21°C (2 h equilibration in the cryostat is best). Optimal cutting temperature is determined by the humidity in the cryostat chamber and section thickness. Sections are collected onto Super-Frost glass slides, air-dried (this helps keep the sections attached to the slide during immunostaining and gently dehydrates the tissue for best morphology), and stored at −80°C until used for staining.

3.1.2. Immunostaining of Retinal Sections

1. Eyecup sections are thawed to room temperature (RT) and dried. Create a circle around the section with a PAP pen and allow to dry. Block by incubation at RT for 30–60 min in AIS.

2. Aspirate off solutions to preserve the well created by the PAP pen. Incubate with primary antibody diluted in AIS for either 1–2 h at RT or at 4°C overnight. (Use a humid chamber to prevent the sections from drying. Use only enough antibody solution to cover the section. Otherwise the solution will flow over the well.)

3. After three quick washes (5–10 min each) in PBS, the sections are incubated for 1 h at RT in the appropriate secondary antibody coupled to either Cy3 or Cy2 (Jackson ImmunoResearch Laboratories, West Grove, PA), or to Alexa Fluor 488 or Alexa Fluor 594 (Invitrogen, Carlsbad, CA) diluted 1:500 to 1:2000 in PBS or AIS. (Keep sections protected from light beyond this step.)

4. The slides are washed again three times in PBS (5–10 min each) with one rinse with water (to reduce salt crys-

tal formation) and then coverslipped with AMM. The coverslipped slides are left in the dark overnight to harden before oil immersion lenses are used. (Nail polish is used to seal the edges of the coverslip.)

5. Slides are viewed on a fluorescence or confocal laser scanning microscope using 40× or 60× oil immersion objectives.

3.1.3. Wholemount Retina Staining

1. Following dissection of a mouse retina, four radial cuts are made to flatten the tissue.
2. The flattened retina is placed on a piece of filter paper to prevent the retina from folding or rolling up.
3. The retina on the filter paper is fixed for 1–30 min by immersion in 4% PFA, and then washed in PBS.
4. The tissue is gently removed from the filter paper (this becomes easy after fixation) with a fine paintbrush and is then incubated with primary antibody diluted in AIS for 1–3 days at 4°C on a gently rocking (or orbital) platform.
5. After three 1-h washes in cold PBS, the tissue is incubated overnight at 4°C in the appropriate secondary antibody coupled to either Cy3 or Cy2 (Jackson ImmunoResearch Laboratories, West Grove, PA), or to Alexa Fluor 488 or Alexa Fluor 594 (Invitrogen, Carlsbad, CA) diluted 1:500 to 1:2000 in PBS, again on a rocking platform. (Keep dark.)
6. The tissue is washed again three times for 1 h each in PBS and then carefully placed on a slide, photoreceptor side up. Using a fine paintbrush, the retina is gently flattened and unfolded and then coverslipped with AMM.
7. The coverslipped slides are left in the dark overnight to harden before oil immersion lenses are used. The edges of the coverslip are sealed with nail polish.
8. Slides are viewed on a confocal laser scanning microscope (such as the Olympus FluoView 1000) using 40× or 60× oil immersion objectives.

3.2. Electroretinogram (ERG) Recordings

We record the ERG from genetically altered mice to investigate the roles of different proteins in retinal signaling. Below we detail our particular methods for recording the ERG from the mouse. The associated *Notes* provide more in-depth discussion of methods that vary between laboratories. For a detailed description of ERG components and their cellular origin see Frishman (3). Peachey and Ball (4) and Weymouth and Vingrys (2) provide excellent overviews of the recording, analysis, and interpretation of rodent ERGs

1. Prior to the day of ERG recording, make an anesthetic cocktail by mixing 0.5 ml ketamine (100 mg/ml), 0.25 ml

xylazine (20 mg/ml), and 4.25 ml saline solution (0.9% NaCl for injection; Hospira, Inc., Lake Forest, IL) in a 10-ml sterile mix bottle.

2. Weigh mice and then leave them in the dark to adapt overnight (>12 h).
3. On the day of ERG recording, mice are anesthetized and prepared for recording under dim red light.
4. Anesthetize mice via intraperitoneal injection of anesthetic cocktail (*see* **Notes 4–6**). Dose (ml) = weight (g) × 0.01 (e.g., for a 20-g mouse: dose = 20 × 0.01 = 0.2 ml). Larger doses should be split between two injection sites for better results.
5. Anesthesia is maintained by subsequent injection of a one-third dose (0.07 ml for 20 g mouse) at 30 and 60 min after initial loading dose. For long procedures we taper anesthesia down to 1/4 dose (0.05 ml for 20 g mouse) for injections at 90 and 120 min after the loading dose.
6. As soon as the mouse is sufficiently immobilized, it should be placed on the heating device to minimize any drop in core body temperature.
7. Anesthetize the cornea with a drop of 1.0% proparacaine. After a minute, remove excess fluid with a cotton-tipped applicator ("Q-tip") as excess fluid can reduce ERG amplitude. Mice can aspirate on drops so care should be taken to ensure excess fluid does not roll down snout.
8. Dilate the pupils with a drop of phenylephrine (2.5%) and a drop of tropicamide (1%) and again remove excess fluid.
9. Place a wire loop behind the upper teeth and draw mouse into the nose cone. Start oxygen and fix the wire loop in place so that the head cannot move. (*See* **Fig. 2.2**, *top*, and **Note 7**.)
10. Insert rectal temperature probe.
11. Insert platinum subdermal needle electrode into tail to serve as ground.
12. Insert butterfly needle into flank to allow delivery of anesthetic every 30 min or as required.
13. Tape down temperature probe, subdermal needle, and butterfly needle to heating unit to ensure they do not move during recording.
14. A blanket placed over the mouse will help maintain body temperature.
15. Attach reference and active electrodes (*see* **Fig. 2.2**, *top*, and **Notes 8–10**).

Fig. 2.2. Mouse ERG recording. *Top*: Loop electrode placed over the eye serves as reference. Contact lens electrode placed against the cornea serves as the active electrode. Mouse is stabilized by drawing snout into a nose cone that also delivers oxygen. *Bottom*: Overview of station that holds mouse during recording. Mouse placed on water-circulated heating box. Active corneal electrode held in place by a stand with manipulator at far end. Foil placed over electrode wires is input grounded on the amplifier to reduce 60 Hz interference. After setup is completed, mouse is slid into Ganzfeld dome for ERG recording.

16. To minimize 60 Hz interference, connect the foil shields covering each electrode wire to the amplifier input ground.
17. Slide the mouse forward so that its head is inside the Ganzfeld.

18. Record ERG to a single dim flash (e.g., -3 log sc cd s/m^2) to ensure a good contact has been made.
19. Dark adapt the mouse for a further 10 min before starting to record the ERG (see ERG protocol below).
20. At the end of recording, wash the electrodes and contact lens in warm soapy water to remove old Murocel and any proteins, rinse with alcohol. Cleaning will extend the life of the electrodes.

3.2.1. Recording Conditions

Table 2.1 lists our standard protocol for ERG recordings from the mouse. Total recording time including setup is typically 1.5 h. This protocol is quite extensive and based on our requirements to examine retinal signaling. The full protocol may not be necessary depending on your requirements. For example, screening for retinal degenerations could be done using just the "b wave" and "photopic" parts of the protocol.

Table 2.1
Standard ERG protocol

ERG component	Flash intensity range (log cd s/m^2)[a]	Number of responses averaged	Flash separation (s)	High pass filter (Hz)	Gain
STR	-6.8	60	2	30	10,000
	-4.3	5	3		
b wave	-4.9	10	3	300	2,000
	-1.9	2	10		
a wave	-0.8	1	30	1,000	2,000
	2.1	1	180		
Photopic[b]	-0.8	40	1	300	10,000
	3.7	2	10		

For all ERG recordings: low pass filter = 0.1 Hz, sampling rate = 2.5 kHz, no notch filter.
[a] Each box contains two numbers covering the range of values used for each condition.
[b] Photopic ERGs are recorded 20 min after the onset of a 60-cd/m^2 achromatic background.

Signal averaging is the only way to obtain a useable ERG at low light levels. For example, at the lowest light intensity, we average 60 responses recorded every 2 s in order to obtain a measurement of the scotopic threshold response (STR). Signal averaging increases the signal-to-noise ratio by utilizing the principle that the retinal signal will always be time-locked to the light flash, while non-retinal signals will be random. Therefore, when multiple responses are added, the retinal signal will always superimpose while random signals will tend to cancel each other out. By this process, the signal-to-noise ratio increases proportionally with the square root of the number of responses averaged. For example, increasing the number of responses averaged from 4 to 16 will double the signal-to-noise ratio.

3.2.2. ERG Recording: Useful Tips and Troubleshooting

1. *Main line interference* (50 or 60 Hz) is generated by any equipment using a main line voltage, and this signal is picked up by the electrode wires. Sixty Hertz interference can be minimized either by placing the mouse and recording electrodes in a Faraday cage, or by using the notch filter on the amplifier to completely remove all 60 Hz signal from the ERG. We do not recommend using a notch filter to remove 60 Hz, since we have found significant ERG signal at this frequency in both the ERG *a wave* and oscillatory potentials. Instead, we use a twofold approach in overcoming unwanted 60 Hz contributions to the ERG. The first approach involves running the electrode wire through a grounded shield. We wrap the electrode wires in aluminum foil, which is then connected to the input ground. Using this simple method, we reduce the peak-to-peak amplitude of 60 Hz interference to <10 µV. During recording, main line interference greater than 10 µV is typically caused by either incorrect grounding of the electrode shields, or from poor contact between the active electrode and the cornea. This latter problem can usually be solved by the addition of a small drop of Murocel to the active electrode.

2. *Offset signal averaging* is the second approach we use to minimize 60 Hz interference from ERG recordings. The principle involved is illustrated in **Fig. 2.3a** and **b**. If the time between flashes (inter-flash interval; IFI) is set to 1 s, then the flash will always occur at exactly the same point in the 60 Hz cycle. This is because the main line frequency of 60 Hz is not random and by definition there are exactly 60 cycles in 1 s. Offset sampling involves adding a small-time increment to the 1 s IFI such that each flash will occur at a different point in the 60 Hz cycle. **Figure 2.3a** shows five different starting points in a 60-Hz cycle. The first flash will start sampling from point 1, the second flash will start sampling from point 2, etc. **Figure 2.3b** shows the first 20 ms of each 60 Hz waveform from each starting point in **Fig. 2.3a**. The average of these five waveforms is zero (**Fig. 2.3b**, *horizontal line*).

   The offset signal averaging method demonstrated in **Fig. 2.3a** and **b** was achieved by setting the inter-flash interval to 1.00333 s. The addition of the 3.33 ms shifts the start of sampling by 1/5 of the 60 Hz cycle (i.e., 1/5 × 16.667 ms). Averaging any number of flashes that is a

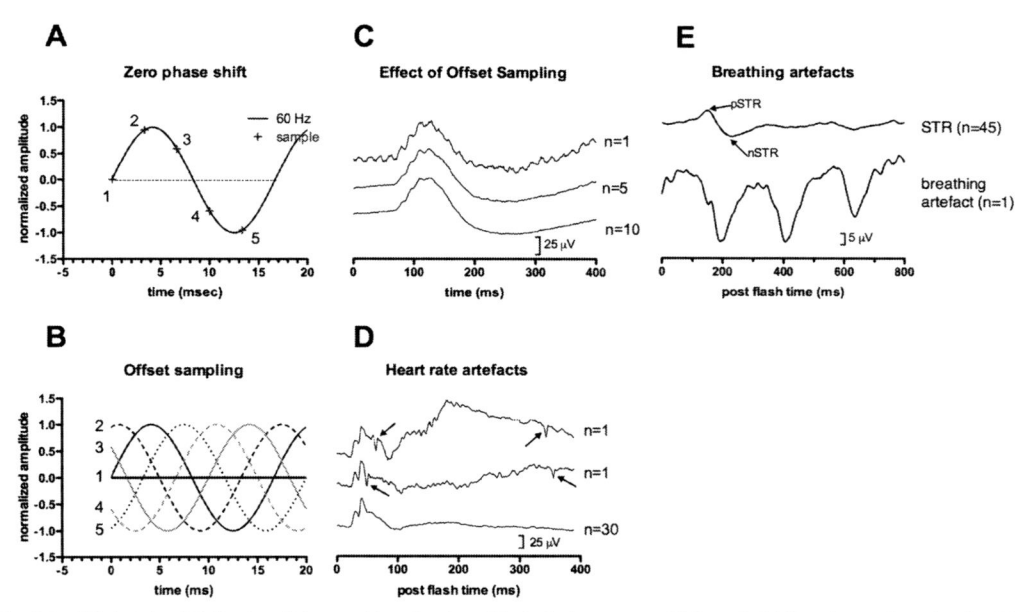

Fig. 2.3. Minimizing 60 Hz, breathing and heart rate artifacts in the mouse ERG. **a** *Solid line* shows 1 cycle of a 60-Hz waveform which lasts 16.667 ms. The number points show starting locations of sampling in five sequential epochs. **b** Each trace shows the time course of the 60 Hz cycle for the 20 ms following each starting point. The *horizontal line* shows the average of the five waveforms. **c** Scotopic ERG response to a dim flash (−3.2 log sc cd s/m^2) in response to a single flash (*top*) or for the average of 5 or 10 responses. **d** Photopic ERG response to a flash (1.2 log ph cd s/m^2) presented against a achromatic background (60 cd/m^2). *Top* two traces show responses to a single flash and *arrows* show intrusion of the heart beat into the ERG response. *Bottom* trace shows the average of 30 responses and the heart beat, which is not synchronous with the flash, can no longer be seen. **e** The STR obtained from the average of 45 ERG responses recorded to a very dim flash (−6.1 log sc cd s/m^2). Note that the peak-to-peak amplitude is only about 10 μV. Below the average is a single response over the 80 ms following this flash. The large oscillations are breathing artifacts in which the body movements of the mouse are causing the eye to move relative to the active electrode and are thus recorded.

multiple of 5 will result in a zeroing of the main line 60 Hz interference (*see* **Note 11**). **Figure 2.3c** shows the effectiveness of offset signal averaging. The ERG response to a single dim flash (*top trace*) has prominent 60 Hz interference. The *bottom two traces* show that this non-physiological effect is no longer present after offset signal averaging of 5 or 10 traces. Importantly, any physiological 60 Hz component of the ERG will not be eliminated since this signal will always be synchronous to the flash.

3. *Heartbeat artifacts*: It is not uncommon for the heart beat to appear on ERG traces (**Fig. 2.3d**; *top two traces*). However, since the heart beat is not synchronous with the flash, evidence of the heart beat is undetectable in the averaged response (**Fig. 2.3d**, *bottom trace*). If the heartbeat artifact is large enough to be a problem, moving the electrode position may be required, although this is rarely necessary.

4. *Breathing artifacts:* The *bottom trace* in **Fig. 2.3e** shows an example of a breathing artifact, which is commonly observed in mouse ERG responses. The breathing artifact is caused by excessive body movement, which in turn moves the eye back and forth across the corneal electrode. We have found that breathing artifacts are minimized with supplemental oxygen and by stabilizing the head. In order to stabilize the head, we place a wire loop behind the upper teeth and draw the mouse into a nose cone, which also delivers oxygen (**Fig. 2.2a**). After being drawn into the nose cone, the wire loop is tied off such that the head is held firmly in place. A large breathing artifact may also be seen when anesthesia is too light. As for the 60 Hz and heart rate artifacts, sufficient averaging will zero the breathing artifact and enable the ERG signal to be obtained. The *top trace* in **Fig. 2.1e** shows the STR obtained for an extremely dim flash by averaging 45 responses each of which had the breathing artifact seen in the bottom response.

5. *Cataracts:* Mice develop cataracts quite soon after injection with the anesthetic cocktail used for ERG recording. However, these cataracts do not pose a problem to ERG recording since they act as a natural light diffuser and very large ERG amplitudes are still obtained even with apparently dense cataracts. Cataract formation is minimized by keeping the animal warm and the cornea lubricated.

6. *Small ERG amplitudes:* Smaller than expected ERG amplitudes can be attributed to poor positioning of the electrode against the cornea (i.e., electrode is not centered), poor electrode contact with the cornea, or poor positioning of the inactive electrode. For example, with our electrode configuration, smaller amplitudes are obtained if the reference loop electrode is not placed fully behind the eye. Additionally, adding too much Murocel to the contact lens will also reduce ERG amplitude as it shorts the electrical contact between active and loop electrodes.

3.2.3. ERG Notes

1. We use a custom-made recording system based on ERGTool software (kindly supplied by Dr. Richard Weleber, Casey Eye Institute, Portland, OR), installed on a G4 Macintosh Computer running OS9. The system also requires a custom-made interface box and National Instrument 16 channel A/D boards. A new version for the Intel chip-based Macintosh Computers running OSX is currently under development. We use a number of light sources mounted together in one box and channeled to the opening on a 40-cm Ganzfeld via a light tunnel. High-intensity flash stimulation is provided by photoflash units

(2405CX and a modified 1205CX power supplies with 205 flash units: Speedotron, Chicago, IL). Low-intensity flash stimulation is provided by a Grass PS22 (Astro-Med, West Warwick, RI). Flash intensity is adjusted by the use of metallic neutral density filters, and the wavelength is modified with glass color filters (Melles Griot, Optics Group, Rochester, NY).

2. The use of xylazine as an anesthetic has the advantage that it prevents the eye moving. Two unwanted side effects include suppressed respiration and a dramatic reduction in core body temperature. We supply a low flow of oxygen through a nose cone, and the mouse is placed on a water-circulated heating source during recording (**Fig. 2.2**).

3. We built a water-circulated heater using a 3.5″ × 6″ electronics box from Radioshack. A cutout, slightly larger than the size of a mouse, was made in the plastic top of the box. The metal plate that comes with this electronics box was then glued to the underside of the plastic top using marine grade epoxy available from hardware stores. Two 1/2-inch copper connectors, which carry the heated water, were glued into one end of the box. Finally the modified plastic/metal top was glued to the box using the marine grade epoxy. Water is circulated through the box via 1/2-inch plastic tubing that carries the water from a laboratory water heater (**Fig. 2.2**).

4. Anesthetic dose (ml) = [weight (mg) × dose (mg/g)]/concentration (mg/ml). For a mouse we use a ketamine dose of 0.1 mg/g. The ketamine concentration of the cocktail = 10 mg/ml. Therefore, for a 20-g mouse:
   a. Dose (ml) = [20 g × 0.1 mg/g]/10 mg/ml = 0.2 ml.
   b. The cocktail is mixed at ketamine to xylazine ratio of 10:1; therefore, the loading dose for xylazine is 0.01 mg/g.

   An anesthetic protocol is not something that can be set in stone. Even in animals that are from the same litter, anesthetic requirements can differ greatly between animals depending on many factors such as fat content and excitability. Mice with a lot of adipose tissue or that are highly agitated/excited at initial injection may need more initial anesthetic to reach a suitable anesthetic plane for ERG recording. For some mutant mice, we lower the anesthetic dose because of increased mortality during ERG recording. Alternatively, in one strain, we had to increase the anesthetic dose in order to achieve 30 min of sedation between re-doses.

5. Eye movements and the presence of breathing artifacts on ERG traces and/or direct visualization of whisker movements are all used to assess the depth of anesthesia during testing. As the mouse passes from the alert state to the required anesthetic plane, whisker movements slow down and eventually stop and the breathing artifact disappears from the ERG response. Conversely, during ERG recording, the appearances of eye movements, breathing artifacts, or whisker movements indicate that a re-dose of anesthetic is required.

6. Anesthetic protocols vary between laboratories, each of which finds a protocol that works for them. Most laboratories use a ketamine:xylazine (K:X) loading dose followed by a re-dose at varying time intervals, e.g., Hetling and Pepperberg (5) (K:X = 0.15:0.01 mg/g; re-dose 1/8 to 1/4 at 45 min); Saszik et al. (6) (K:X = 0.07:0.007 mg/g; re-dose K:X = 0.072:0.005 mg/g at 45 min); and Peachey (4) (K:X = 0.08:0.016 mg/g; given as 75% initial dose then remaining 25% 10 min later, re-dose K:X = 20% of initial dose after 30–40 min). Woodward (7) used isoflurane for mouse ERG recordings as they found an unacceptably high mortality rate with ketamine and xylazine in their mouse mutants.

7. Oxygen delivery and head stabilization help minimize breathing artifacts and lower anesthetic deaths. We fashioned a nose cone from a cutoff 3 ml syringe: The syringe is pushed into one end of a 3-way valve. A cutoff 3/16-inch nasal canula is inserted into the "T" of the valve and carries oxygen from the tank. A suture bent into a hook is placed under the upper teeth of the mouse and serves to pull the mouse into the nose cone and hold the head steady. During setup, the suture line is fed through the syringe and 3-way valve. Once the mouse is pulled firmly into nose cone, the suture line is held tight by wrapping it around a small cleat attached to the front of the heating box.

8. We use a custom-made contact lens electrode placed on the cornea for the active electrode and use a loop placed over the proptosed eye as the reference (**Fig. 2.2**). The contact lens is formed from heated Aclar plastic. A hole is pierced through the center of the contact lens with a heated 28-ga needle. Platinum electrode wire, with one end formed into an "L" shape, is threaded through the center of the contact lens. The contact lens can move freely along the electrode wire. For ERG recording, the L-shaped end of the electrode wire is pushed up against the cornea. The contact lens with a very small amount of Murocel is then moved down the wire with a pair of fine forceps until the

lens covers the cornea. The contact lens helps prevent drying of the Murocel, thereby producing a more stable ERG for a longer period.

9. The reference loop electrode is placed over the proptosed eye. Once over the eye, the loop is pulled back gently to ensure the loop is in good contact with the front of the eye (**Fig. 2.2**, *top*). The electrode wire is run over the head between the ears and held in position with Velcro straps (**Fig. 2.2**, *bottom*). Both active and reference electrodes are made from approximately 2–3-inch long sections of platinum wire (0.01 inch diameter; World Precision Instruments, Sarasota, FL). One end of each electrode wire is wrapped tightly around blunted platinum subdermal needle electrode wires (Astro-Med Grass, West Warwick, RI).

10. Electrode configurations vary widely across laboratories. An L-shaped wire placed against cornea is commonly used for the active electrode. However, the material used varies, with some preferring Ag/AgCl, stainless steel, or precious metals such as gold or platinum. Another option is a DTL fiber (cotton fiber-embedded Ag/AgCl particles) held in place with a contact lens. The advantage of Ag/AgCl electrodes is that they produce a very stable response with little drift and, therefore, the high pass filter can be set for direct current recording (i.e., 0 Hz cutoff). The disadvantage of Ag/AgCl electrodes is that they produce very large flash artifacts in response to bright flashes. We use platinum for all our electrodes since it provides a stable ERG response with no flash artifact. For our ERG recording we set the high pass filter to 0.1 Hz. The choice of reference location varies considerably between laboratories. Reference electrodes used include a DTL fiber under an opaque contact lens on the eye contralateral to the one tested, a subdermal needle electrode inserted in the cheek beneath the eye tested, and metal electrodes placed in the mouth. The ERG is a measure of the change in voltage across the retina in response to a light stimulus. For this reason, we use the loop placed over the eye as reference because it enables electrodes to be essentially placed on either side of the retina and allows the most direct in vivo method of measuring the voltage across the retina.

11. Our system allows microsecond control of timing. In other systems, offset averaging can still be used even if there is less control of the time between flashes. For example using an offset of 2 ms and averaging multiples of 8 responses (total = 16 ms instead of 16.667 ms) will still reduce 60 Hz interference.

3.3. Measuring Vision in Rodents

Laboratory rodents may seem like a peculiar choice as animal models for studying the human nervous system, particularly ones for testing visual function because rodents have relatively small eyes, have a small visual cortex, and have a predominately rod-based retina. However, upon closer comparison, the rodent and primate visual systems are quite similar and many of the same visual functions can be measured in both species. Although, given the architecture of the human macula and the comparable size difference between human and rodent brains, the human visual system outperforms that of the rodent. For example, average human visual acuity is approximately 30 cycles/degree (cpd) (20/20), whereas the average visual acuities of rats and mice are ~1.0 cpd (20/600) and 0.6 cpd (20/1000), respectively. Although visual acuity is lower in rodents compared to humans, rodents do have quantifiable vision, which can be quite detailed at close distances (8). Furthermore, it has recently been shown that complex visual capabilities such as motion coherences, which were thought to not exist in rodents, are not only quantifiable, but in fact, rodents are quite good at them (9).

Since the 1930s, a number of methods have been devised to quantify rodent vision. Lashley's jumping stand may have been the first method used to quantify rat vision (10, 11), and it is still used to a limited extent. Y-mazes (11), conditioned aversion (12), and operant tasks (13, 14) have all been employed with some success, but in general, these methods require a considerable amount of time to train and test rats, which probably accounts for their limited popularity. In addition, the harsh negative consequences of an incorrect response, in particular the ones used with Lashley's jumping stand, are not optimal in training and testing animals behaviorally. Some experimenters have also used a modification of the Morris water task in which rats learn to swim to a platform that is raised above the water's surface, guiding themselves with the use of visual cues (15). However, viewing distances are hard to control in this situation, making quantitative measurements nearly impossible. Since then, two methods have emerged that are designed to minimize the constraints and limitations of previous methods and allow for quantitative measurements.

3.3.1. The Visual Water Task

The Visual Water Task (VWT; **Fig. 2.4**) is an apparatus used to evaluate visual perception of rodents in a psychophysical manner. Animals are trained to discriminate between two visual stimuli and are positively reinforced for each correct response. The apparatus consists of a trapezoidal-shaped tank containing water, with two computer monitors facing through a clear glass wall into the wide end of the pool. Visual stimuli are generated and projected on the screens using a computer program (Vista©; CerebralMechanics). The choice point, defined by a 46-cm long midline divider,

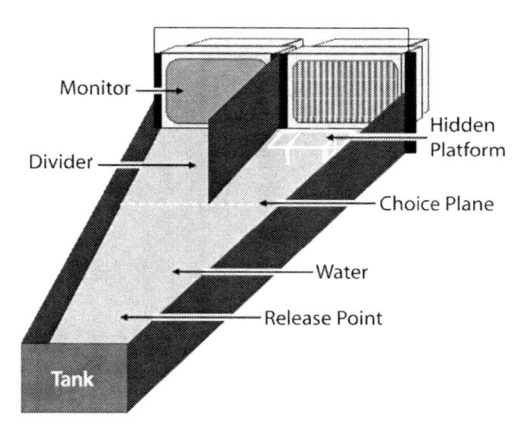

Fig. 2.4. The Visual Water Task. A task used to evaluate the visual perception of rodents (8, 9).

which extends into the pool from between the monitors, creates a Y-maze with a stem and two arms. A moveable, transparent Plexiglas escape platform (37 cm L × 13 cm W × 14 cm H) is always submerged directly below whichever monitor displays the grating. A LRLLRLRR sequence, a pattern the animals cannot memorize, is used for the location of the gratings. Animals are released into the pool from the wall opposite the monitors, and the end of the divider within the pool sets a choice point for the rats that is as close as they can get to the visual stimuli without entering one of the two arms. The length of the divider, therefore, sets the effective spatial frequency of the visual stimuli. The animals usually stop at the end of the barrier and inspect both screens before choosing a side. If the animals swim to the platform below the positive stimulus without entering the arm with the monitor showing the negative stimulus, the trial is considered correct; if they swim into the arm of the maze that contains the negative stimulus, the trial is recorded as an error.

The most common measure of vision is acuity, measured clinically with the use of a Snellen chart. This technique measures the ability to resolve two high contrast items as distinct; two parts of the same letter on a Snellen chart. Discrimination between sine wave gratings and a gray screen of the same mean luminance in the VWT is akin to Snellen acuity and allows researchers to measure visual acuity in rodents in a similar manner to those measured clinically, helping validate experimental results. Animals are first trained to discriminate between a low spatial frequency (~0.1 cpd), vertical sine wave grating (+ stimulus; 100% contrast), and uniform gray of the same mean luminance (36.2 cd/m^2 at the choice point). The animals are tested in groups of 5 or 6, with 15–20 interleaved trials each, with each session lasting 45–60 min. No more than two sessions, separated by at least 1 h, are performed in

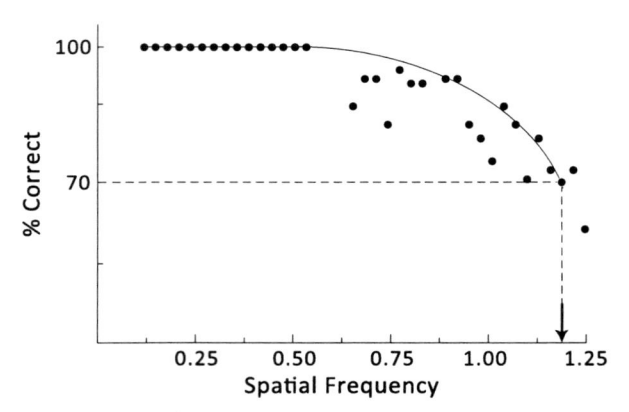

Fig. 2.5. An example of a frequency-of-seeing curve. Frequency-of-seeing curves are used to estimate the visual acuity based upon a 70% correct criterion.

a single day. All trials are run with the room lights off. Once animals achieve near-perfect performance (90% or better over at least 40 consecutive trials), the animals are then tested for visual acuity. A flexible method-of-limits procedure is used in which incremental changes in the spatial frequency of the sine wave grating are made until choice accuracy falls below 70% (**Fig. 2.5**). Accuracy for a given frequency is measured in blocks of ten trials when near threshold, and shorter blocks at the low spatial frequencies, thereby minimizing the number of trials far away from threshold. A preliminary grating threshold is established when animals fail to achieve 70% accuracy at a spatial frequency. In order to assess the validity of this estimate, the spatial frequency of the grating is increased by about 0.1 cpd, and the experimental procedures described above are repeated until a stable pattern of performance is established. The performance at each spatial frequency is averaged for each animal and a frequency-of-seeing curve is constructed. The point at which the curve intersects 70% accuracy is recorded as the visual acuity (**Fig. 2.5**).

Contrast sensitivity is assessed using similar procedures, except the minimal contrast required to differentiate between the screens at different spatial frequencies is measured (**Fig. 2.6**). Contrast thresholds are typically measured after grating acuity is assessed, so minimal re-training is required. Seven spatial frequencies are normally tested; 0.059, 0.119, 0.208, 0.297, 0.505, 0.712, and 0.890 cpd. At each spatial frequency, trials are initiated at 100% contrast and the contrast is decreased systematically until performance falls below 70% accuracy. The contrast threshold is measured independently at least three times, after which final values are computed as above from frequency-of-seeing curves of the combined data. Measuring responses to different types of visual stimuli can test a range of visual functions, beyond visual acuity and contrast sensitivity. These include oblique gratings, color gratings, moving gratings, and dot motion among

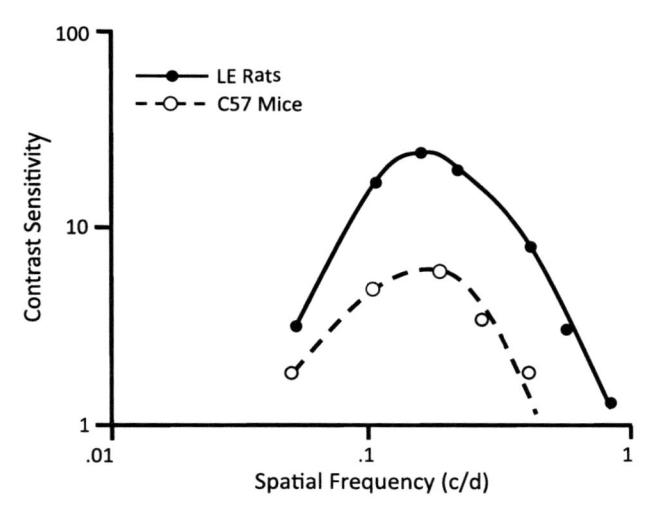

Fig. 2.6. Contrast sensitivity curves generated using the Visual Water Task.

others (8, 9, 16). Testing each of these visual functions follows the same general procedure, only the visual stimuli are systematically changed.

A significant limitation of the VWT is that the time invested in training and testing animals is on the order of weeks, which limits longitudinal studies to a maximum measurement frequency of once per month. On the other hand, once the animals are well trained, they do not require a full re-training before subsequent testing, but rather, a day or two of swimming trials are sufficient to restore optimal performance. Visual thresholds are generated by compiling data collected over the course of 1–2 weeks; therefore, daily measurements of vision are not possible in the VWT. As a consequence, the earliest age at which visual thresholds can be measured from rodents in the VWT is approximately P30, about 2 weeks after the day of eye opening.

3.3.2. The Virtual Optomotor System

The Virtual Optomotor System (**Fig. 2.7**) is primarily used to evaluate spatial frequency thresholds (optomotor acuity) and contrast sensitivity. The apparatus consists of four computer monitors positioned around a square testing arena. An unrestrained rat is placed on a platform in the center of the arena, and a sine wave grating drawn on a virtual cylinder is projected on the monitors in 3D coordinate space (OptoMotry©; CerebralMechanics). A video camera provides real-time video feedback from above, and the position of the head on each frame is used to continually center the hub of the cylinder at the rat's viewing position. On each trial the cylinder is rotated at a constant speed (12°/s) and the experimenter judges whether the rat makes tracking movements with its head and neck to follow the drifting grating. The spatial frequency threshold, the point at which animals are no longer able to

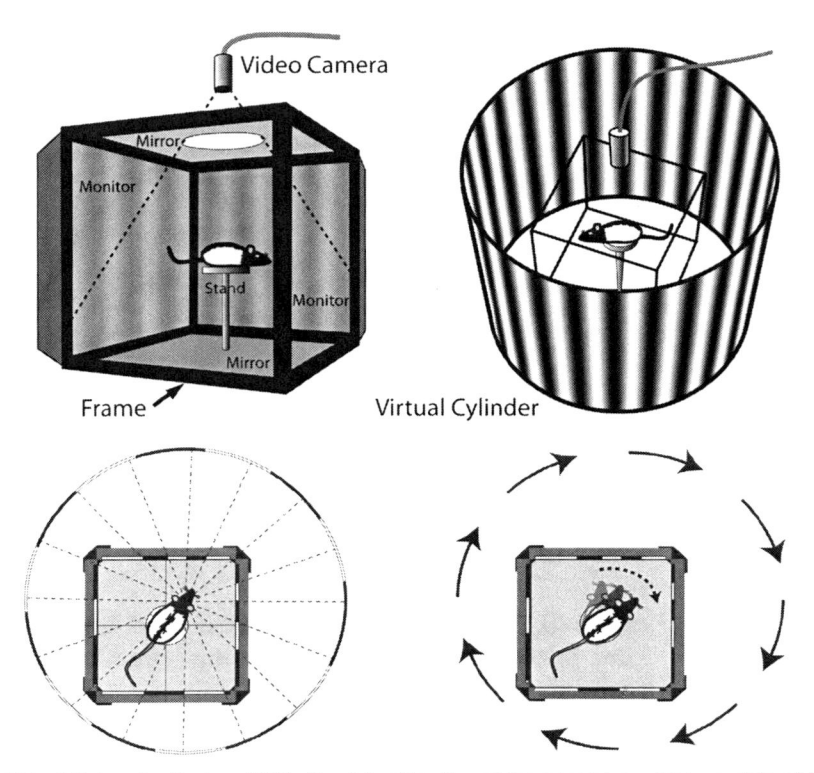

Fig. 2.7. The Virtual Optomotor System (VOS). *Top left*: side view of the apparatus, which consists of four computer monitors facing inward (two removed for ease of viewing) into an arena. An animal is placed on a platform located in the middle of the arena and a video camera viewing from above provides real-time feedback of animals' behavior. *Top right*: the virtual cylinder displaying sine wave gratings with the testing arena and animals highlighted within. *Bottom left*: diagram depicts the animals viewing angle of the visual stimuli. The sine wave gratings are displayed as different sizes on the monitors in order to keep the virtual cylinder homogeneous throughout the testing area. *Bottom right*: arrows show the direction of rotation of the virtual cylinder and the corresponding movement of an animal performing the optokinetic tracking behavior (17, 21).

track, is obtained by incrementally increasing the spatial frequency of the grating at 100% contrast. Contrast sensitivity thresholds are measured at up to eight different spatial frequencies by systematically decreasing the contrast until no tracking is observed. Thresholds through each eye are measured separately by simply reversing the rotation of the cylinder (17).

One of the major advantages for using the VOS is that animals require no reinforcement training prior to being tested. This greatly decreases the time required to generate a visual threshold, and therefore allows for much larger group sizes to be examined. In addition, the ability to generate thresholds quickly, in some cases as quickly as 5–10 min, allows for thresholds to be assessed on a daily basis. This is of particular importance in developmental studies as animals can be tested from the day of eye opening and for very fast retinal degenerations where visual decline can occur sooner than it can be measured with other tasks. Another advantage arises from a combination of inherent properties of the task itself and the structure of the rodent visual system. That is, when the virtual cylinder that displays the sine wave grating is rotated in the clockwise direction, only the left eye (tracking in the temporal to nasal direction) responds to the movement, resulting in an elicited optomotor behavior moving in the same direction as the grating movement. Conversely, if the cylinder is rotated in the counter-clockwise direction, only the right eye (again temporal to nasal) responds to the stimuli. This phenomenon is particularly advantageous as therapies for retinal disease are often performed monocularly, maintaining the contralateral eye as a control. Therefore, employing the VOS allows for within animal controls to be used in the experimental design, a powerful method in research. Finally, the VOS is also designed to provide a "blind" psychophysical testing methodology, which can be used to eliminate operator bias. This is of particular importance when evaluating the efficacy of an experimental therapy.

3.3.3. Differences Between the Virtual Optomotor System and the Visual Water Task

There are many different aspects of vision such as visual acuity, visual motion, color vision, and form vision. The VOS and the VWT are tasks designed to evaluate two different forms of visual thresholds, reflexive optomotor responses and perceptual visual acuity, respectively. Because the tasks evaluate different components of vision, the sensory and motor circuits each task relies upon are likely to be different. For example, the VWT is normally dependent on the visual cortex, and surgical removal of the visual cortex (V1) results in significantly lower visual thresholds (unpublished observations). The VOS, however, is normally mediated by subcortical circuits and removal of the cortex does not affect the measured thresholds (16). Finally, the visuospatial thresholds of naïve adult rodents are lower when measured in the VOS than when measured in the VWT, which is likely a reflection of the VOS relying on subcortical circuits and the VWT relying on the visual cortex. Therefore, these and other factors should be taken into consideration in choosing an appropriate method of testing vision.

3.3.4. Applicability of the Tasks

The VWT has been used for a number of studies examining rodent vision. The first published study using the VWT examined the visual acuity of Long Evans rats and C57BL/6 mice (8). This study revealed that behaviorally measured visual acuity in rats is approximately double that of the mouse, although both species could be tested repeatedly and reliably. These studies were then extended into other rodents including albino rats and other pigmented rat strains showing that differences within the same species can also be detected (18). The VWT has also been used to examine different forms of vision such as motion acuity (9), dot motion coherence (9), and acuity using oblique or horizontal gratings. Since then, the use of the VWT has been extended to include studies involving damage to the visual system. For example, ablation lesions of the visual cortex and visual cortex stroke result in a significant drop in visual acuity from 1 cpd down to ~0.7 cpd (unpublished observations). In 2004, McGill et al. (19) performed a longitudinal characterization of visual acuity (measured once every 30 days) of rats that undergo photoreceptor degeneration. The authors showed that from 30 days of age until 11 months, the rats progressively declined from near-normal vision, to the inability to discriminate between a white and a black screen. This was the first longitudinal quantification of spatial vision in a model of retinal disease. This study was followed closely by cell-based therapies used to prevent the degeneration of vision in the RCS rat (20), where both a human-derived RPE cell line (ARPE19) and human Schwann cells were shown to significantly limit the progression of visual deterioration.

The VOS is newer than the VWT and therefore has been used to a lesser extent. However, over the last few years the task has been used to quantify visual thresholds in normal mice (17), developing mice (21), monocularly deprived mice (22), in normal and experimentally enhanced rats (16, 17), and in retinal degenerative rats (23) and rats receiving a potential neuroprotective treatment (24). The VOS has also been used to examine vision in mice without functional rods or cones (25), noerg-1 (26), and nob4 mice (27) and TRPM1-deficient mice (28). Finally, the VOS has also been used to evaluate the role of ON bipolar cells that were engineered to be photosensitive (29).

In summary, the Visual Water Task and the Virtual Optomotor System allow for detailed examination and quantification of visual thresholds in both normal and diseased rodent models. These behavioral tasks allow for therapeutic interventions to be evaluated using multiple approaches to testing visual function. Finally, each task measures visual thresholds in a manner similar to those used clinically, helping validate the use of these tasks for vision research in rodents.

Both these tasks and a comprehensive list of scientific publications using these tasks can be found at: www.cerebralmechanics.com.

References

1. Dowling JE (1987) The retina: an approachable part of the brain. The Belknap Press of Harvard University Press, London
2. Weymouth A, Vingrys A (2008) Rodent electroretinography: methods for extraction and interpretation of rod and cone responses. Prog Retin Eye Res 27:1–44
3. Frishman LJ (2006) Electrogenesis of the ERG. In: Ryan SJ (ed.) Retina. Elsevier Mosby, Philadelphia, PA, pp 103–113

4. Peachey NS, Ball SL (2003) Electrophysiological analysis of visual function in mutant mice. Doc Ophthalmol 107:13–36
5. Hetling JR, Pepperberg DR (1999) Sensitivity and kinetics of mouse rod flash responses determined in vivo from paired-flash electroretinograms. J Physiol 516:593–609 (Pt 2)
6. Saszik SM (2002) The scotopic threshold response of the dark-adapted electroretinogram of the mouse. J Physiol (Lond) 543:899–916
7. Woodward WR, Choi D, Grose J, Malmin B, Hurst S, Pang J, Weleber RG, Pillers DA (2007) Isoflurane is an effective alternative to ketamine/xylazine/acepromazine as an anesthetic agent for the mouse electroretinogram. Doc Ophthalmol 115:187–201
8. Prusky GT, West PW, Douglas RM (2000) Behavioral assessment of visual acuity in mice and rats. Vision Res 40:2201–2209
9. Douglas RM, Neve A, Quittenbaum JP, Alam NM, Prusky GT (2006) Perception of visual motion coherence by rats and mice. Vision Res 46:2842–2847
10. Lashley KS (1930) The mechanism of vision: I. A method for rapid analysis of pattern vision in the rat. J Gen Psych 37:453–460
11. Seymoure P, Juraska JM (1997) Vernier and grating acuity in adult hooded rats: the influence of sex. Behav Neurosci 111:792–800
12. Dean P (1978) Visual acuity in hooded rats: effects of superior collicular or posterior neocortical lesions. Brain Res 156:17–31
13. Keller J, Strasburger H, Cerutti DT, Sabel BA (2000) Assessing spatial vision – automated measurement of the contrast-sensitivity function in the hooded rat. J Neurosci Methods 97:103–110
14. Jacobs GH, Fenwick JA, Williams GA (2001) Cone-based vision of rats for ultraviolet and visible lights. J Exp Biol 204:2439–2446
15. Morris R (1984) Developments of a watermaze procedure for studying spatial learning in the rat. J Neurosci Methods 11:47–60
16. Prusky GT, Silver BD, Tschetter WW, Alam NM, Douglas RM (2008) Experience-dependent plasticity from eye opening enables lasting, visual cortex-dependent enhancement of motion vision. J Neurosci 28:9817–9827
17. Douglas RM, Alam NM, Silver BD, McGill TJ, Tschetter WW, Prusky GT (2005) Independent visual threshold measurements in the two eyes of freely moving rats and mice using a virtual-reality optokinetic system. Vis Neurosci 22:677–684
18. Prusky GT, Harker KT, Douglas RM, Whishaw IQ (2002) Variation in visual acuity within pigmented, and between pigmented and albino rat strains. Behav Brain Res 136:339–348
19. McGill TJ, Douglas RM, Lund RD, Prusky GT (2004) Quantification of spatial vision in the Royal College of Surgeons Rat. Invest Ophthalmol Vis Sci 45:932–936
20. McGill TJ, Lund RD, Douglas RM, Wang S, Lu B, Prusky GT (2004) Preservation of vision following cell-based therapies in a model of retinal degenerative disease. Vision Res 44:2559–2566
21. Prusky GT, Alam NM, Beekman S, Douglas RM (2004) Rapid quantification of adult and developing mouse spatial vision using a virtual optomotor system. Invest Ophthalmol Vis Sci 45:4611–4616
22. Prusky GT, Alam NM, Douglas RM (2006) Enhancement of vision by monocular deprivation in adult mice. J Neurosci 26:11554–11561
23. McGill TJ, Lund RD, Douglas RM, Wang S, Lu B, Silver BD, Secretan MR, Arthur JN, Prusky GT (2007) Syngeneic Schwann cell transplantation preserves vision in RCS rat without immunosuppression. Invest Ophthalmol Vis Sci 48:1906–1912
24. McGill TJ, Prusky GT, Douglas RM, Yasumura D, Matthes MT, Nune G, Donohue-Rolfe K, Yang H, Niculescu D, Hauswirth WW, Girman SV, Lund RD, Duncan JL, LaVail MM (2007) Intraocular CNTF reduces vision in normal rats in a dose-dependent manner. Invest Ophthalmol Vis Sci 48:5755–5766
25. Schmucker C, Seeliger M, Humphries P, Biel M, Schaeffel F (2005) Grating acuity at different luminances in wild-type mice and in mice lacking rod or cone function. Invest Ophthalmol Vis Sci 46:398–407
26. Pinto LH, Vitaterna MH, Shimomura K, Siepka SM, McDearmon EL, Fenner D, Lumayag SL, Omura C, Andrews AW, Baker M, Invergo BM, Olvera MA, Heffron E, Mullins RF, Sheffield VC, Stone EM, Takahashi JS (2005) Generation, characterization, and molecular cloning of the Noerg-1 mutation of rhodopsin in the mouse. Vis Neurosci 22:619–629
27. Pinto LH, Vitaterna MH, Shimomura K, Siepka SM, Balannik V, McDearmon EL, Omura C, Lumayag S, Invergo BM, Glawe B, Cantrell DR, Inayat S, Olvera MA, Vessey KA, McCall MA, Maddox D, Morgans CW, Young B, Pletcher MT, Mullins RF, Troy JB, Takahashi JS (2007) Generation, identification and functional characterization of the

nob4 mutation of Grm6 in the mouse. Vis Neurosci 24:111–123
28. Morgans CW, Zhang J, Jeffrey BG, Nelson SM, Burke NS, Duvoisin RM, Brown RL (2009) TRPM1 is required for the depolarizing light response in retinal ON-bipolar cells. Proc Natl Acad Sci USA 106:19174–19178
29. Lagali P, Balya D, Awatramani G, Münch T, Kim D, Busskamp V, Cepko C, Roska B (2008) Light-activated channels targeted to ON bipolar cells restore visual function in retinal degeneration. Nat Neurosci 11:667–675

Chapter 3

Correlates and Analysis of Motor Function in Humans and Animal Models of Parkinson's Disease

Alexandra Y. Schang, Beth E. Fisher, Natalie R. Sashkin, Cindy Moore, Lisa B. Dirling, Giselle M. Petzinger, Michael W. Jakowec, and Charles K. Meshul

Abstract

The purpose of this chapter is to first describe common clinical and laboratory tests and measures used to capture alterations in motor control in individuals with Parkinson's disease (PD) and secondly, to detail both morphological and motor tests that are used in two rodent models of PD. For the description in humans, it is organized within the *body structure and function* and *activity* categories of the International Classification of Functioning, Disability, and Health (ICF). Specific tests discussed include the retropulsion test, turning test, Unified Parkinson's Disease Rating Scale, Timed Up-and-Go, Berg Balance Scale, electromyography, quantitative digitography, motion analysis, and force plate perturbation. Testing procedure, set-up, and interpretation are described and examples of application in the PD population are provided. We hope that clinicians and researchers develop a beginning understanding of the different methods available for examining alterations in motor control in individuals with PD. Using the rat model of PD, we first describe in detail a new ultrastructural processing method that is used not only to process tissue but also to localize specific proteins that can then be used to correlate synaptic changes with motor alterations that are observed following depletion of dopamine. Finally, using a mouse model of PD, we describe three locomotor tests that can be quantified and correlated with the loss of dopamine-labeled neurons in the substantia nigra.

1. Tests and Measures for Assessment of Motor Control in Individuals with Parkinson's Disease

Parkinson's disease (PD) is a progressive neurologic disorder that involves degeneration of the dopamine-producing neurons in the substantia nigra pars compacta, presenting clinically as a combination of motor and non-motor signs. The cardinal motor signs include resting tremor, rigidity, bradykinesia, and postural instability. With disease progression, individuals can be affected at multiple levels extending beyond motor symptoms or impairments (i.e., bradykinesia) to include deficits in functional abilities (i.e., ambulation). One classification system that has increasingly been utilized to gain a comprehensive understanding of how a disease process affects different dimensions of a patient's life is the International Classification of Functioning, Disability, and Health (ICF). It was developed by the World Health Organization as an international standard for the assessment of health and disability and takes into consideration personal as well as contextual factors (**Fig. 3.1**). The purpose of this chapter is to describe common tests and measures used to capture alterations in motor control in individuals with PD organized within the categories of the ICF framework (**Table 3.1**). An additional level of organization is the environment under which the test can be administered. Specifically, whether the test can be easily administered in a clinical setting or requires specialized equipment only available in a laboratory setting. Unique to PD is the common protocol of testing individuals both on medication (ON state) and off medication (OFF state). This enables clinicians and researchers to distinguish the effects of medication on changes in impairments and function from those resulting from study-specific interventions, such as exercise.

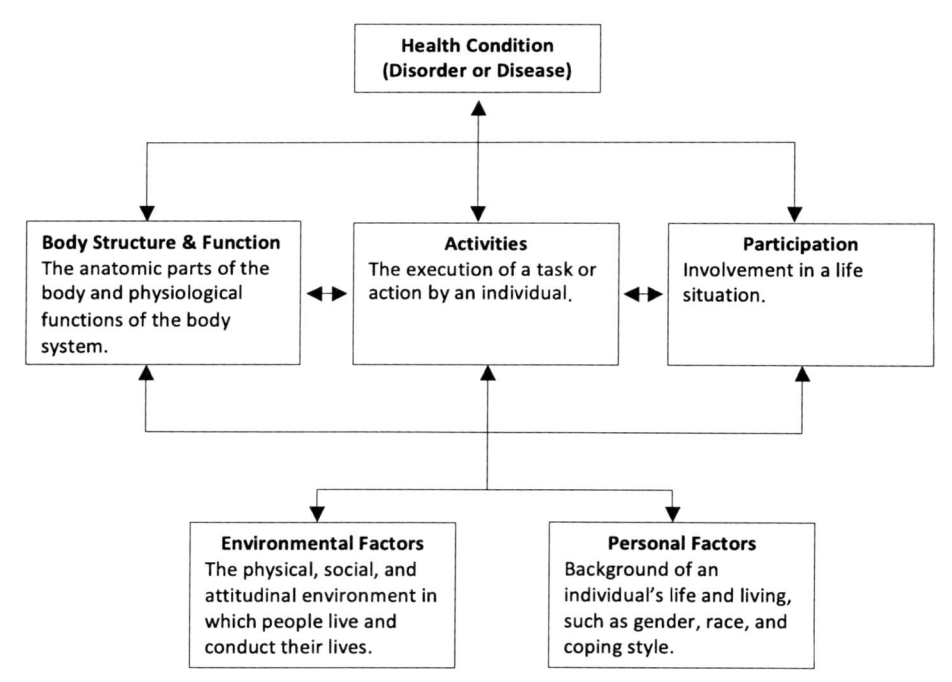

Fig. 3.1. International Classification of Functioning, Disability, and Health (modified from World Health Organization 2001).

Table 3.1
Classification of test and measures

Test and measure	Clinical	Laboratory	Body structure and function	Activity
Retropulsion test	x		Postural instability	
Turning	x		Freezing Bradykinesia	Turning while walking
Timed UpandGo	x			Functional mobility (sit-to-stand, ambulation, turning while walking)
Berg Balance Scale	x		Postural control Sensory integration	Functional mobility (sitting, standing, sit-to-stand, reaching, stepping, turning)
Unified Parkinson's Disease Rating Scale	x		Tremor Rigidity Bradykinesia Rapid alternating movement Speech Posture Postural control	Ambulation Sit-to-stand Activities of daily living
Electromyography		x	Tremor	
Quantitative digitography		x	Bradykinesia Freezing	
Motion analysis		x	Bradykinesia Tremor Gait and reach spatiotemporal characteristics Postural control	
Force platform perturbation		x	Postural control	

2. Clinical Assessment Tools

2.1. Body Structure and Function Level Measures

2.1.1. Retropulsion Test

The retropulsion test is a quick, simple test used by clinicians to assess postural instability. The test involves a posterior pull on the patient's shoulder and observation of the balance reaction, although there are many variations which differ in rating methods and the expectedness of the perturbation (1). A cross-sectional study that examined different techniques such as expected versus unexpected perturbation concluded that an unexpected pull rated on the 4-point ordinal scale proposed by Nutt and colleagues appears to have greatest predictive values for risk of falls (1). However, the test must be interpreted with caution as it has only been shown to have moderate correlation with performance on platform perturbation (*see* **Section 3** under laboratory tests and measures) during OFF state and no significant correlation during ON state (2) **Table 3.2**.

Table 3.2
Retropulsion rating proposed by Nutt et al. (114)

Rating	Description
0	Normal, may take two steps to recover
1	Takes two or more steps; recovers unaided
2	Would fall if not caught
3	Spontaneous tendency to fall or unable to stand unaided (test not executable)

2.1.2. Turning

Turning is a common, but complex postural control task because it requires that an individual initiate disequilibrium necessary to produce the turn during ongoing movement (3). While persons within 1 year of diagnosis of Parkinson's disease typically show minimal deficits of gait performance in a single direction, they often report difficulty with turning tasks (4, 5). Several studies that have examined turning in individuals with PD have utilized (1) counting the number of steps required for completing a 180° turn either on the spot or around an obstacle (making a u-turn); (2) quantifying the time it takes to complete a turn; (3) observing step length; and (4) observing the turning arc. Greater number of steps, increased time to complete a turn, as well as decreased step length and wider turning arc in individuals with PD have been observed (6, 7).

2.1.3. Unified Parkinson's Disease Rating Scale (UPDRS)

The UPDRS is the gold standard assessment most commonly used by both neurologists and researchers to determine disease severity and progression. It was developed in 1987 by a group of movement disorder specialists who derived components of existing PD rating scales to create a single tool that would allow for easier comparison of assessment (8). The test is partly based on patient/caregiver interview and partly on clinical testing by an examiner. It consists of six main sections: (1) mentation, behavior, and mood, (2) activities of daily living (ADLs), (3) motor, (4) motor complications, (5) modified Hoehn & Yahr staging, and (6) Schwab and England activities of daily living scale. A total score of 199 is possible on the first four subscales, with 0 representing no involvement and 199 representing severe disease. Within the context of the organization of this chapter, the UPDRS consists of elements that address both the body structure and function and activity levels of the ICF framework. Specifically, **Section 2** (ADLs) predominantly addresses the activity level while **Section 3** (motor) focuses on the body structure and function level, with a few items examining the level of assistance required to perform a task, therefore extending to the activity level (**Table 3.3**).

The UPDRS motor subscale includes 19 items examining speech, facial expression, resting and active tremor, rigidity, rapid

Table 3.3
Sample items from the Unified Parkinson's Disease Rating Scale motor subscale [Fahn et al. (115)]

Item	Scoring
Tremor at rest	0 = absent 1 = slight and infrequently present 2 = mild in amplitude and present most of the time 3 = moderate in amplitude and present most of the time 4 = marked in amplitude and present most of the time
Rigidity	0 = absent 1 = slight or detectable only when activated by mirror or other movements 2 = mild to moderate 3 = marked, but full range of motion easily achieved 4 = severe, range of motion achieved with difficulty
Gait	0 = normal 1 = walks slowly, may shuffle with short steps, but no festination (hastening steps) or propulsion 2 = walks with difficulty, but requires little or no assistance; may have some festination, short steps, or propulsion 3 = severe disturbance of gait, requiring assistance 4 = cannot walk at all, even with assistance
Postural stability	0 = normal 1 = retropulsion, but recovers unaided 2 = absence of postural response; would fall if not caught by examiner 3 = very unstable, tends to lose balance spontaneously 4 = unable to stand without assistance
Body bradykinesia and hypokinesia	0 = none 1 = minimal slowness, giving movement a deliberate character; could be normal for some persons. Possibly reduced amplitude 2 = mild degree of slowness and poverty of movement which is definitely abnormal. Alternatively, some reduced amplitude 3 = moderate slowness, poverty, or small amplitude of movement 4 = marked slowness, poverty, or small amplitude of movement

alternating movement, arising from a chair, posture, gait, postural stability, and bradykinesia. Each question is rated on an ordinal scale of 0 (normal)–4 (severe limitation) based on qualitative descriptions. Action and postural tremor of hands, arising from a chair, gait, and postural stability are four items within the motor scale that span the body structure and function and activity levels. They examine characteristics of the movement, such as speed and amplitude (body structure and function level), as well as interference with function and assistance required for safe performance (activity level). Studies that have examined the psychometric properties of the motor subscale suggest high internal consistency (Chronbach's alpha 0.88–0.91) (9, 10), interrater reliability (ICC = 0.82) (11), and test–retest reliability (ICC = 0.90) (12). The motor subscale has moderate-to-good correlation with Hoehn & Yahr classification (H&Y), a staging instrument based on progression of PD, and with timed functional performances (9, 10, 13).

Recently, the Movement Disorders Society (MDS) reviewed the clinimetric properties of the UPDRS and published a revised MDS-UPDRS. Several changes have been made to the motor subscale, including specific items added to address toe-tapping, freezing in gait, constancy of resting tremor, as well as more detailed instructions and descriptions in effort to better standardize test administration and grading (14). The internal consistency remains high (Chronbach's alpha = 0.93) and further studies need to be performed to determine the reliability of the MDS-UPDRS (14).

2.2. Activity Level Measures

2.2.1. Timed Up- and-Go

The Timed Up-and-Go (TUG) test was developed by Podsiadlo and Richardson to assess basic mobility skills in frail community-dwelling elderly (15). The patient is instructed to stand up from an armchair, walk a distance of 3 m, turn, walk back, and sit down. A stopwatch is used to determine the amount of time required to perform this task and interpretation is made based on the average of three trials (15). Although there exists some controversy in the literature (16) the vast majority of studies suggest excellent (ICC = 0.81–0.99) test–retest reliability and interrater reliability for the TUG in elderly adults (15, 17, 18). TUG times demonstrate correlation with the Berg Balance Scale ($r = -0.81$), gait speed ($r = -0.61$), and Barthel Index of activities of daily living ($r = -0.78$) (15). This functional mobility test has been shown to be a sensitive and specific tool to predict fall risk in elderly adults, with a score higher than 13.5 s indicating high risk of falls (19).

Several studies have examined the use of TUG in populations other than elderly adults, including individuals with strokes,

PD, arthritis, and amputations. The psychometric properties of the TUG when applied to patients with PD appear to closely mimic those observed in elderly adults. Interrater reliability has been shown to be good (ICC $\geq$ 0.73) regardless of whether the patient is ON or OFF medication and internal consistency is high ($r = 0.90$–0.97) (20). The TUG scores demonstrate significant difference during ON and OFF states, suggesting usefulness in detecting change in medication-related change in status in individuals with PD (20). The minimal detectable change (MDC) using a 95% confidence interval (MDC_{95}) has been reported to be 1.63 s (21).

In recent years, there has been a growing interest in learning about dual task performance in individuals with PD. A variation of the TUG that includes a high level, concurrent cognitive task (TUG_{cog}) (counting the days of the week backward) has been used to assess the interaction of cognitive demand on functional mobility. Individuals with PD demonstrate significant increase in their TUG_{cog} time and number of steps taken during the test – whereby healthy older adults do not demonstrate such changes (22).

2.2.2. Berg Balance Scale

The Berg Balance Scale (BBS) is a 14-item test that was initially developed to assess balance in the geriatric population. Activities examined include standing (with eyes open, eyes closed, feet close together, feet in tandem, single leg), sitting, transfers (sit-to-stand, chair-to-chair), turning (360° in place, weight shift without moving feet), alternate stepping, and reaching (level and to the floor). The test requires two standard height seating surfaces (one with arm rests, both approximately 18″ high), a yard stick, a stopwatch, and an average height foot stool or step (approximately 7″). Specific instructions are provided to increase standardization and administration time ranges from 15 to 20 min, depending on examiner's proficiency and the patient's level of function. Each item is rated on a 0–4 ordinal scale, with 0 representing poor ability to perform task. The BBS has been shown to be a valid and reliable test for predicting falls in community-dwelling adults, as well as length of stay, motor function, and disability level in individuals post-stroke (23–26).

Current research supports the use of the BBS as an ongoing assessment tool to determine fall risk and functional mobility in individuals with PD. It has good internal consistency (ICC = 0.86–0.88) and demonstrates moderate-to-high correlation (Spearman correlation = 0.50–0.78) with other functional measures (the forward and backward functional reach test, and Timed Up-and-Go, normal and fast gait speed), as well as with the UPDRS motor subscale (27–29). The MDC_{95} has been reported to be 5 points (29). Different cut-off scores have been proposed to maximally increase specificity and sensitivity of the BBS in this population to detect falls (43.5 and 54, respectively) (24, 30). Regardless of which cut-off score is used, it is important to consider the multifactorial nature of falls and examine personal (physical, cognitive, emotional, medication use, co-morbidities) and contextual factors (home environment, social support available) to most accurately predict fall risk.

3. Laboratory Assessment Tools

3.1. Body Structure and Function Level Measures

3.1.1. Electromyography (EMG)

EMG has been used to characterize upper extremity tremor presentation, specifically tremor occurrence, symmetry, frequency, and intensity, in individuals with PD. Long-term recording (up to 24 h) has been used for both differentiation of parkinsonian tremor from essential tremor, as well as for assessment of medication efficacy (31, 32). Forearm tremor can be recorded using surface electrodes and a commercial portable recorder, with electrode placement on the wrist or finger extensors (extensor carpi radialis, extensor digitorum) and wrist or finger flexors (flexor carpi ulnaris, flexor digitorum superficialis) (27, 33). The advantages of using solely EMG to quantify tremor include portability and the ability to obtain tremor recording without need to restrict limb movement, as would be necessary if using accelerometry (32). However, often times researchers seek to obtain both neuromuscular and kinematic information, therefore EMG is utilized in conjunction with motion analysis (31, 33, 34).

3.1.2. Quantitative Digitography (QD)

Researchers have used quantitative digitography as a technique to quantify bradykinesia in individuals with PD. The system, adopted from the electronic music industry, consists of a portable keyboard with optical sensors and a computer interface (called musical instrument digital interface, MIDI) (35, 36). The sensors are able to identify the key struck, time of strike and release, and velocity of strike, thereby allowing determination of the frequency and duration of digital movement (35).

Bronte-Stewart and colleagues utilized QD to examine digital control in individuals with idiopathic PD. They determined the technique to be useful in providing objective measures for bradykinesia, fatigue, and freezing during a 60-s alternating finger-tapping task (35). Performance was significantly different in subjects ON and OFF medication (35). QD has also been used to assess improvements in bradykinesia after microelectrode recording in patients undergoing deep brain stimulation surgery (37).

3.1.3. Motion Analysis (MA)

Motion analysis is commonly used to quantify movement kinematics and has been applied to the examination of bradykinesia, tremor, reaching, gait kinematics, and postural control in PD research (33, 38–41). Common methods employed include accelerometry, motion analysis system, and the GAITRite system. Accelerometry is a method used for motion analysis which involves application of acceleration sensors on a specific body part of interest, such as on the L3 spinous process for measuring trunk acceleration or on the forearm for measuring upper extremity bradykinesia. Depending on the type of sensor, signals produced are proportional to the linear acceleration or angular velocity of movement (37, 42). Selection of a sensor is based on relative frequency of the movement measured as well as the range of dynamic movement (42). The use of accelerometry for quantifying spatiotemporal gait kinematics, including gait speed, step length, and cadence in individuals with PD, has been shown to have good concurrent validity with measurements obtained using the GAITRite system (43). Reliability and validity of accelerometry for measuring gait speed, cadence, stride length, single/double limb support times, swing time, and stance time have been established across various gait speeds in healthy adults and individuals with hemiparesis (44). The root mean square velocity of angular movement is indicative of average speed and demonstrates significant inverse correlation with the bradykinesia subscore on the UPDRS (37). This technique has been used to study bradykinesia in forearm supination and pronation (41).

Motion analysis systems, such as Vicon®, involve placement of infrared-emitting markers on landmarks of interest and capturing movement with use of a motion capturing system, often consisting of multiple cameras and specific processing software. This particular technique allows for calculation of spatiotemporal parameters (magnitude, duration, onset, frequency of movement), movement trajectory, estimation of joint centers, and center of mass (COM) changes during movement. There is evidence that the accuracy of this technique may vary, depending on camera set-up, marker size, and lens filter application (45). Motion capturing systems have been used to examine reach velocity and accuracy during ON/OFF medication status (38), tremor after subthalamic deep brain stimulation surgery (34), head–trunk rotation during turning (46), and the relationship between alterations in gait kinematics and disease severity (13).

The GAITRite system is specific for measurement of gait performance. It consists of a 4.6-m electronic, portable walkway in which an array of 16,128 sensors is embedded and organized in a grid-like pattern to identify foot contact during gait (47, 48). It is connected to a computer and generates spatiotemporal recordings, including cadence, step time, step length, mean normalized velocity, step length ratio, heel-to-heel base of support, and single and double limb support time as a percentage of the gait cycle (48). The GAITRite system demonstrates the ability to differentiate gait performance in individuals with PD compared to healthy controls, as well as detect changes in gait parameters during ON and OFF medication states that correlate with changes in UPDRS motor section scores (47). It has been recommended as a tool to assist monitoring and reassessment of treatment efficacy (48).

In conjunction with a motion analysis system, use of force platforms embedded in the floor can provide ground reaction force and center of pressure (COP) data to measure postural control (i.e., COP displacement) during gait initiation, gait termination, and turning (49, 50). An emerging method for assessing dynamic postural control in PD is to measure the anterior–posterior and medial–lateral distance between the extrapolated center of mass (eCOM) and center of pressure (COP) [*see* (51) for specific method]. Simply stated, the greater the distance between eCOM *and* COP the more the movement is facilitated. However, these conditions require more active postural control to counteract the disequilibrium. Therefore an individual can be stable through smaller differences between eCOM and COP, but at the expense of movement (i.e., too stable); conversely an individual can produce movement with greater distance between eCOM and COP but at the risk of falling. The ideal situation then is to have a "controlled disequilibrium" in order to facilitate movement with control. Our data from a recent pilot study suggest that while measures of gait performance using motion analysis were not sensitive enough to differentiate between individuals with early PD (within 3 years of diagnosis) and age-matched healthy control subjects, significantly decreased distance between eCOM and COP during turning tasks was observed in our subjects with PD (Song et al., unpublished findings, 2008 (52)).

3.1.4. Force Platform Perturbation (FPP)

Individuals with PD frequently report difficulty with balance and frequent falls. An objective measure used to quantify dynamic postural control involves use of a platform that can provide perturbation via multidirectional linear translations and tilts. Subjects stand on a force plate that is embedded in the platform to provide information on center of pressure changes. FPP has been used alone and in conjunction with EMG to quantify a number of different aspects of postural control, including (1) muscle activation pattern; (2) magnitude of muscle response; (3) latency of muscle response; (4) direction of ground reaction force; and (5) magnitude of ground reaction force (2, 53–55). Although these studies utilize sophisticated equipment set-ups, a commercially available product – the SMsART Balance Master® – has become increasingly popular in the clinical setting and has enabled examination and monitoring of postural control changes in patients with PD.

4. Morphological and Neurochemical Correlates to Changes in Locomotion in the Rat 6-Hydroxydopamine Model of Parkinson's Disease

Although it is well established that in Parkinson's disease there is a significant loss of dopamine within nerve terminals in the striatum, the depletion of dopamine most likely influences other neurotransmitter systems. There is growing interest in the interactions between dopamine and glutamate and it may be the lack of dopamine in Parkinson's disease that results in dynamic changes in glutamate within at least the striatum (56, 57). In the rodent, the sensorimotor cortex provides the primary excitatory, glutamatergic input to the dorsolateral striatum (58, 59). This projection utilizes the vesicular glutamate transporter-1 (*VGLUT-1*). Recent data suggest that glutamate input from many nuclei within the thalamus may also be playing an important role (60–62). This thalamostriatal projection utilizes the vesicular glutamate transporter-2 (*VGLUT-2*). The dopamine terminals originating from the substantia nigra pars compacta make a symmetrical synaptic contact not only on the dendritic shaft of the medium spiny neuron but also on the neck of the dendritic spine (63, 64). The asymmetrical synaptic contact on the head of that same spine within the dorsolateral striatum originates from not only the motor cortex but also the thalamus (60, 62) and the nerve terminals contain the neurotransmitter, glutamate (63, 65, 66). Not only are dopamine and glutamate terminals anatomically located next to each other, these two neurotransmitters can control their own release and also the release from each other's nerve terminals (67–70). In addition, a small percentage of the glutamate nerve terminals originating from the cortex contain presynaptic dopamine D-2 receptors (71, 72). When these dopamine D2 receptors are activated or blocked, we and others have reported that glutamate release decreases or increases, respectively (67, 69, 73). Therefore, alterations in the level of striatal dopamine can have profound effects on nearby glutamate synapses.

In rodent models of nigrostriatal dopamine loss, either using the neurotoxins 6-OHDA or MPTP, exercise is neuroprotective and enhances behavioral recovery from injury (74). Immediately following the unilateral loss of striatal dopamine, rats were forced to use the impaired limb by the placement of the unimpaired limb in a cast (75). Improvement in the use of the impaired limb occurred, leading to increased levels of striatal dopamine compared to the non-casted animals. However, it was critical that the casting occur within a short-time frame after the nigrostriatal lesion (75). This same group reported, using the MPTP mouse model, that treadmill exercise started within 24 h of the MPTP lesion attenuated the loss of striatal dopamine and the motor behavioral deficits compared to the lesioned but non-exercised group (74). Following intracerebral injection of 6-OHDA into the striatum, if treadmill exercise was initiated up to 7 days post-lesion, there was some recovery of striatal dopamine levels and a decrease in the number of apomorphine-induced contralateral rotations (76). However, 7 days is a time point in which the loss of dopamine neurons within the substantia nigra is not complete (77). Other measures of behavioral deficit were not improved by exercise. It must be recognized that in all of the above studies, exercise was initiated prior to the complete loss of striatal dopamine as a model of neuroprotection. In a recent study in which exercise was initiated just after intrastriatal injection of 6-OHDA, there was significant behavioral/motor improvement but no change in the number of dopamine cells remaining in the substantia nigra or dopamine transporter levels in the striatum (78). This suggests that other neurotransmitters, such as glutamate, could be compensating for the loss of striatal dopamine in terms of locomotor improvement.

The effect of physical activity in both rats and mice on changes in striatal glutamate synapses following the loss of nigrostriatal dopamine is of particular interest in terms of how whether such therapy may be able to further compensate for the loss of dopamine cells in the substantia nigra. In a mouse model of partial bilateral nigrostriatal dopamine loss, treadmill exercise was initiated 4 days after the acute administration of the neurotoxin, MPTP, a time point at which dopamine cell body death has been completed (79). We find that exercise improved motor performance of the MPTP-treated group and resulted in a *reversal* in the changes in striatal nerve terminal glutamate immunolabeling to the level observed in the control group. In this same study, exercise resulted in a further decrease in striatal dopamine transporter immunolabeling compared to the lesioned but non-exercised group (79).

We used a combination of in vivo microdialysis and quantitative immunogold electron microscopy to determine changes in striatal glutamate following a lesion of the nigrostriatal pathway, followed by exercising the rats for 4 weeks. However, there is controversy as to the origin of the basal levels of glutamate that are measured in brain and whether this extracellular glutamate is derived from the (1) calcium-dependent vesicular pool, (2) calcium-independent, cytoplasmic pool associated with the glutamate/cystine antiporter, or (3) glial pool (80). We and others have reported that about 30% of basal extracellular glutamate is calcium dependent (80–82) and that over 60% of the K$^+$-depolarized extracellular level of glutamate is calcium dependent (83). This suggests a role for the synaptic vesicle pool within the nerve terminal contributing to the extracellular level of glutamate. Replacement of calcium with the divalent chelating agent, EGTA, and increasing the concentration of magnesium resulted in a decrease in the baseline level of glutamate (81), suggesting that a portion of the resting level of striatal glutamate is of neuronal and not glial origin. Therefore, not only is the basal level of glutamate measured, but also using ultrastructural immunocytochemistry, we quantify the relative density of glutamate immunogold labeling within nerve terminals making an identified asymmetrical (excitatory) synaptic contact. In addition, we have developed a double-labeling procedure in which the nerve terminal is first labeled with an antibody against either the vesicular glutamate transporter 1 or 2, followed by immunogold labeling for glutamate.

In this particular rodent model of Parkinson's disease, rats are administered the neurotoxin 6-hydroxydopmaine (6-OHDA), into the medial forebrain bundle on one side of the brain, in order to destroy the dopamine pathway from the substantia nigra to the striatum (83). This results in a greater than 90% loss of dopamine levels in the striatum and a nearly complete elimination of dopamine cells in the substantia nigra pars compacta (81, 83, 84). In order to determine the effects of exercise on motor behavior related to the loss of dopamine neurons, two tests were performed. The loss of dopamine on one side of the striatum will result in an increase in dopamine receptors on that side of the striatum (denervation supersensitivity). Administration of dopamine-receptor agonist drugs, such as apomorphine, will result in the animals turning away from the side of the lesion (contralateral rotations) (**Fig. 3.2**). This is a measure of the loss of dopamine/dopamine function and of turning behavior. We find that in the 6-OHDA-lesioned rats, exercise results in a significant decrease in the number of contralateral rotations compared to the lesioned but non-exercised group (**Fig. 3.2**). As another measure of motor function, exercise results in an increase in the time the rats can run on the treadmill at a maximum speed of 18 m/min (**Fig. 3.3**).

The data suggest that following the loss of dopamine, there is a significant decrease in the extracellular levels of striatal glutamate (**Fig. 3.4**). We also demonstrate that, using quantitative immunogold electron microscopy (**Fig. 3.5**) there is an accumulation of glutamate within the nerve terminals of the striatum (**Fig. 3.6**), inversely correlating with the decrease in extracellular glutamate as seen using in vivo microdialysis (**Fig. 3.4**) (83). We have hypothesized that this decrease in striatal glutamate may be related to the decrease in motor movement that

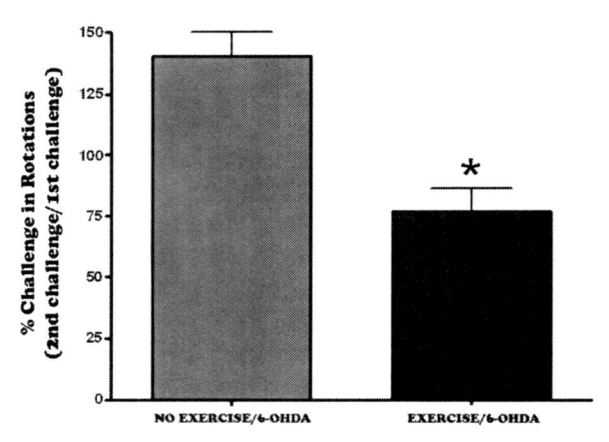

Fig. 3.2. Exercise decreases apomorphine-induced contralateral rotations. Two months following a 6-OHDA lesion, the animals were injected with apomorphine (0.05 mg/kg, s.c.) and the number of contralateral rotations counted (first challenge with apomorphine). One month following exercising of the rats (1 h/day, 5 days/week), all the animals were tested for apomorphine-induced contralateral rotations (second challenge with apomorphine). The percent change in contralateral rotations between the second and first challenge (i.e., number of rotations at second challenge/number of rotations at first challenge) was then determined. An overall group mean was calculated (mean percentage ± SEM) and the lesioned but no exercise group compared against the lesioned plus exercise group using the student's t-test. *p <0.05 compared to the other group.

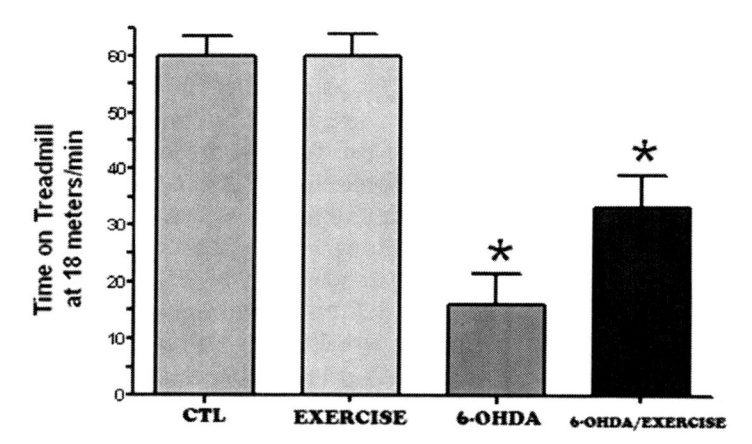

Fig. 3.3. Exercise increases the time running on the treadmill. Two months following a 6-OHDA lesion or sham lesion (control), the animals exercised for 1 h/day for 1 month (5 days/week). At the end of the 1-month time period, the rats were tested on the treadmill at the maximum speed of 18 m/min to determine how long the rats could stay on the treadmill. The time it took for them not to be able to maintain their running was determined. *p <0.05 compared to all other groups using an ANOVA, followed by Tukey–Kramer for comparison of multiple means.

is observed in patients with Parkinson's disease (83). Following exercise, there is a surprising decrease in striatal extracellular glutamate and that in the 6-OHDA-lesioned rats, there is an additional or additive decrease in extracellular glutamate (**Fig. 3.4**).

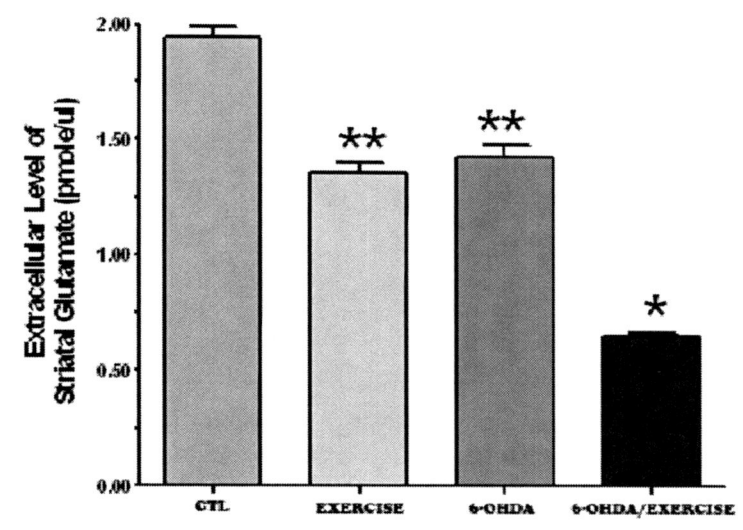

Fig. 3.4. Exercise decreases the extracellular levels of striatal glutamate. Two months following a 6-OHDA lesion or sham lesion (control), the animals exercised for 1 h/day for 1 month (5 days/week). At the end of the 1-month time period, the basal extracellular levels of glutamate were determined in the dorsolateral striatum. Exercise alone (exercise) resulted in a decrease in striatal glutamate levels, similar to that seen after a 6-OHDA lesion (6-OHDA). The combination of a 6-OHDA lesion plus exercise resulted in a further decrease in striatal extracellular glutamate. *p <0.05 compared to all other groups using an ANOVA, followed by Tukey–Kramer for comparison of multiple means. **p < 0.05 compared to the control and 6-OHDA/exercise group using an ANOVA, followed by Tukey–Kramer for comparison of multiple means.

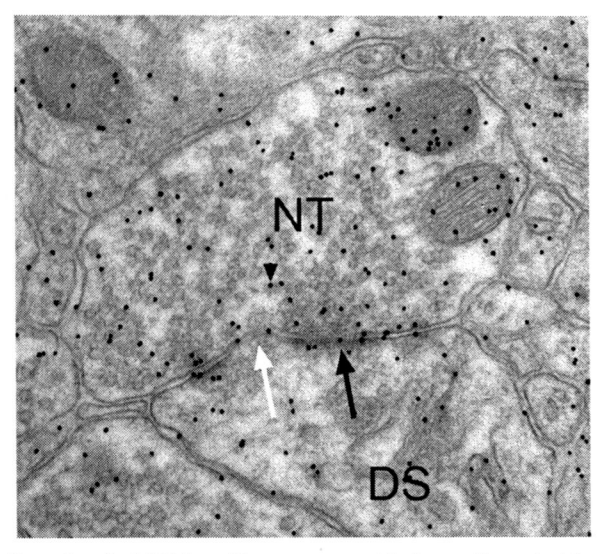

Fig. 3.5. Nerve terminal (NT) is making an asymmetrical synaptic contact (*arrow*, pointing to the post-synaptic density) with an underlying dendritic spine (DS). The post-synaptic density is discontinuous or perforated (*white arrow*). Within the NT are numerous 10-nm gold particles (*arrowhead*) indicating the location of the neurotransmitter, glutamate.

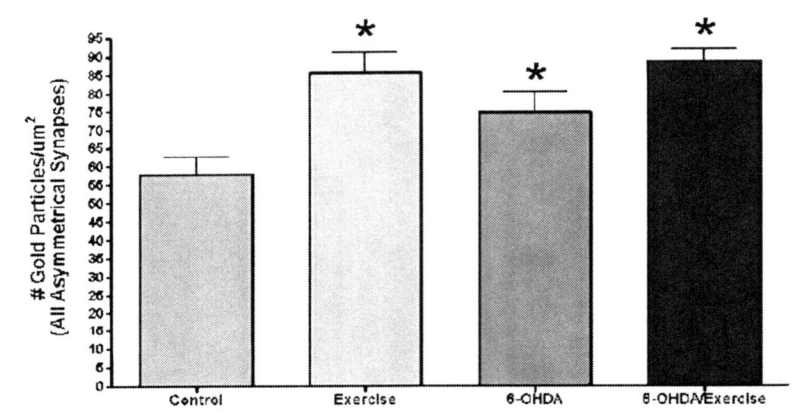

Fig. 3.6. Exercise the density of glutamate immunogold labeling within nerve terminals making an asymmetrical (excitatory) synaptic contact within the dorsolateral striatum. Two months following a 6-OHDA lesion or sham lesion (control), the animals exercised for 1 h/day for 1 month (5 days/week). At the end of the 1-month time period, the animals were perfused with fixative and the density of glutamate immunogold labeling within identified nerve terminals of the dorsolateral striatum calculated. In all three experimental groups, there was an increase in the density of immunogold labeling compared to the control group. This is inversely correlated to the decrease in extracellular glutamate as determined by in vivo microdialysis (see **Fig. 3.4**). *$p < 0.05$ compared to the control group using an ANOVA, followed by Tukey–Kramer for comparison of multiple means.

However, this is associated with a decrease in the apomorphine-induced contralateral rotations (**Fig. 3.2**), a finding consistent with a similar decrease in rotations following a lesion of the motor cortex (85). The finding of a further decrease in extracellular glutamate in the 6-OHDA/exercise group is also consistent with a recent report that exercise results in a decrease in the regional cerebral blood flow in the motor cortex and striatum (86). This suggests that as the animal continues to exercise, the corticostriatal pathway requires less activation in order to perform either the same task or it becomes more efficient at carrying out that same locomotor activity. This has also been shown to occur in humans (87, 116). This could result in changes in either the number of corticostriatal synapses, increased efficiency in neurotransmitter release (88, 89), and the induction of long-term potentiation (90), or an increase in the number of post-synaptic glutamate receptors (91).

The procedure used to carry out the nerve terminal glutamate immunogold labeling (**Fig. 3.5**) is detailed below and is a modification of the procedure previously reported by Phend et al. (92).

(1) Animals are first anesthetized and perfused transcardially with the following fixative at room temperature: 2.5% glutaraldehyde/0.5% paraformaldehyde/0.1% picric acid, in 0.1 M HEPES buffer, pH 7.3. The brain is removed and placed in fixative overnight. It is critical that the brain be fixed overnight or the post-embed immunogold procedure

will not work. However, do not fix the tissue for more than just overnight.

(2) The brain is then washed in HEPES buffer several times and the brain can continue to be washed in buffer for up to 2 weeks before being processed for electron microscopy.

(3) The tissue is cut either with a vibratome (50–300 μm slices) or by hand.

(4) The tissue is then processed using the new Biowave/microwave oven technology (**Fig. 3.7**) as described in **Table 3.4**.

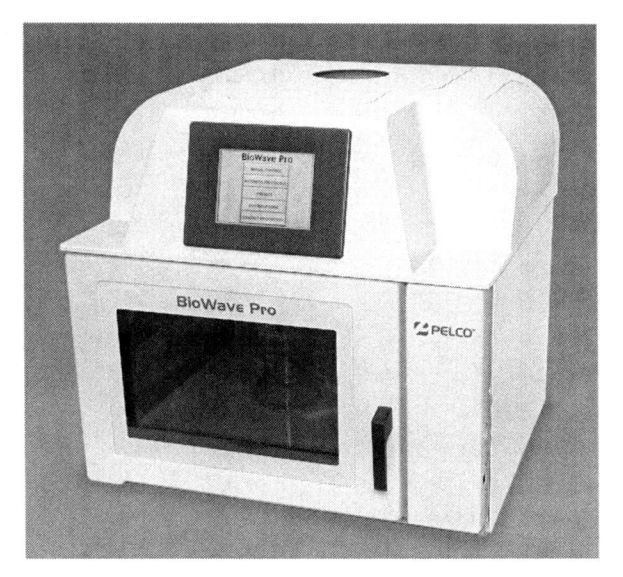

Fig. 3.7. PELCO BioWave Pro model from Ted Pella, Inc., used for ultrastructural immunolabeling (Cat #36500). Photograph used by permission from Ted Pella, Inc.

The post-embed immunogold procedure is detailed below (**Fig. 3.5**):

Immunogold protocol

Solutions

| TBST 7.6 | 0.05 M Tris (pH 7.6) 0.9% NaCl 0.1% Triton X-100 | TBST 8.2 | 0.05 M Tris (pH 8.2) 0.9% NaCl 0.1% Triton X-100 |

1. Thin (gold) sections are collected on nickel-coated, 75 mesh grids (formvar coated).

2. Grids are allowed to dry at room temperature for a minimum of 1.5 h. Grids should be cut the same day the

Table 3.4
Microwave procedure for tissue processing

PROTOCOL: NORMAL EM TISSUE PROCESSING
PROGRAM #6
NAME: EM PROCESSING

Steps no.	Description	User prompt	Time (h:min:s)	Watts	Temp	Load cooler	Vacuum
1	OSMIUM ON	ON	0:03:00	100	60	AUTO	CYCLE
2	OSMIUM OFF	OFF	0:02:00	0	60	AUTO	CYCLE
3	OSMIUM ON	OFF	0:03:00	100	60	AUTO	CYCLE
4	OSMIUM OFF	OFF	0:02:00	0	60	AUTO	CYCLE
5	OSMIUM ON	OFF	0:03:00	100	60	AUTO	CYCLE
6	RINSE	ON	0:00:40	150	60	AUTO	OFF
7	RINSE	ON	0:00:40	150	60	AUTO	OFF
8	UA	ON	0:02:00	100	60	AUTO	CYCLE
9	UA	OFF	0:02:00	0	60	AUTO	CYCLE
10	UA	OFF	0:02:00	100	60	AUTO	CYCLE
11	50% ETOH	ON	0:00:40	150	60	AUTO	OFF
12	75% ETOH	ON	0:00:40	150	60	AUTO	OFF
13	95% ETOH	ON	0:00:40	150	60	AUTO	OFF
14	100% ETOH	ON	0:00:40	150	60	AUTO	OFF
15	100% ETOH	ON	0:00:40	150	60	AUTO	OFF
16	PROPYLENE OXIDE (PO)	ON	0:00:40	150	60	AUTO	OFF
17	1:1 PO/RESIN	ON	0:03:00	150	60	AUTO	CONT
18	100% RESIN	ON	0:03:00	150	60	AUTO	CONT
19	100% RESIN	ON	0:03:00	150	60	AUTO	CONT
20	100% RESIN	ON	0:03:00	150	60	AUTO	CONT
	*IF TISSUE IS HAND CUT OR THICKER THAN 100 μM CONTINUE WITH THE FOLLOWING STEPS						
21	100% RESIN	ON	0:03:00	150	60	AUTO	CONT
22	100% RESIN	ON	0:03:00	150	60	AUTO	CONT

Place tissue in molds or flat embed using ACLAR film; place in 60°C oven overnight

Solutions/resins used for EM tissue processing:

Osmium
 1:1
2% OSMIUM TETROXIDE in H_2O
3% POTASSIUM FERRICYANIDE IN H_2O

UA
 0.5% URANYL ACETATE IN DEIONIZED H_2O

Mix Spurr:Epon resin:

Add all ingredients on a balance into a plastic disposable beaker in order given below. Stir with disposable tongue depressor

Cover resin with parafilm until ready to use

Table 3.4 (continued)

RESIN SPURR	Amt. (g)	Amt. (g)	Amt. (g)	Amt. (g)	Amt. (g)	Amt. (g)	Amt. (g)	Amt. (g)
VCD (ERL 4206)	1	2	4	5	7	10	15	20
DER 736	0.4	0.8	1.6	2	2.8	4	6	8
NSA	2.6	5.2	10.4	13	18.2	26	39	52
DMAE	0.04	0.08	0.16	0.2	0.28	0.4	0.6	0.8
TOTAL	4.04	8.08	16.16	20.2	28.28	40.4	60.6	80.8
EPON								
EPON 812 (Embed 812)	140	2.8	5.6	7	9.8	14	21	28
DDSA	0.65	1.3	2.6	3.25	4.55	6.5	9.75	13
NMA	0.66	1.32	2.64	3.3	4.62	6.6	9.9	13.2
DMP-30	0.11	0.21	0.43	0.54	0.75	1.07	1.61	2.14
TOTAL	2.82	5.63	11.27	14.09	19.72	28.17	42.26	56.34
Total SPURR + EPON	6.86	13.71	27.43	34.29	48	68.57	102.86	137.14

primary incubation in the antibody is started since labeling is greatly reduced if sections are cut the day before.

3. All incubations and rinses are carried out at room temperature, with grids submerged, tissue side up, in *drops of solution* (approximately 50 µl) on silicon grid pads or parafilm. All rinse and diluted solutions are filtered (0.22 µm) before use.

4. Drain excess solution from grids by touching the edge of the grid and tweezers to a kimwipe between each step.
   (a) 2% Aqueous periodic acid: 7 min
   (b) Water rinse grids
   (c) 2% Aqueous sodium metaperiodate: 7 min
   (d) Water rinse grids

5. Create a moist chamber (place wet kimwipes along the sides of the Petri dish) for the primary antibody incubation.

6. Wash with TBST 7.6, 5 min

7. Primary antibody [1:10,000 for anti-glutamate from Sigma (cat #: G-6624); for double-labeling we use a 1:250 dilution; for GABA (Sigma: cat #2052) we use a 1:250 dilution either alone or with the double-labeling procedure]: overnight in a covered/moist Petri dish.

8. Next day, wash grids in TBST 7.6, 5 min.

9. TBST 7.6, 5 min.
10. TBST 7.6, 30 min.
11. TBST 8.2, 5 min.
12. Secondary antibody: incubate for 1.5 h. Secondary antibody (Jackson ImmunoResearch Lab, Inc., Goat anti-rabbit IgG,) is conjugated to 12 nm gold and diluted 1:50 in TBST 8.2.
13. TBST 7.6, wash grids for 5 min.
14. TBST 7.6, wash grids for 5 min.
15. Dip grids gently 3 times in each of three 10-ml beakers filled with millipore filtered deionized water.
16. Float grids tissue side down on top of separate drops of millipore filtered deionized water for 5 min, then dry at room temperature. The sections are then photographed at a final magnification of 40,000×. The number of gold particles and the area of the nerve terminal are calculated using Image-Pro Plus imaging software (Media Cybernetics).

Since the alterations in glutamate immunolabeling following a 6-OHDA lesion, with or without exercise, have been associated with terminals making an asymmetrical synaptic contact onto dendritic spines, we speculated that these glutamate synapses primarily originated from the motor cortex. However, because striatal nerve terminals making an axospinous contact can also originate from the thalamus (60–62, 90), it is possible that the thalamic input may also be contributing to the changes in striatal glutamate following the loss of dopamine. Examples of VGLUT-1-labeled striatal terminals that have been double labeled for glutamate immunogold are shown in **Fig. 3.8**. It has been recently reported that after MPTP administration in the non-human primate, there is an increase in the number of nerve terminals containing VGLUT-1 protein (90), suggesting a change in the glutamate input from the cortex (61, 91, 93). For VGLUT-2, there was a shift in the number of terminals making an axospinous versus axodendritic contact, suggesting some sort of glutamate synaptic reorganization originating from the thalamus. Therefore, it will be essential to distinguish the input from the cortex and thalamus in terms of determining the effects of physical activity on glutamate synapses within the striatum.

In order to process tissue for both pre-embed immunolabeling for VGLUT-1 or VGLUT-2 localization using diaminobenzidine histochemistry and post-embed immunogold to localize glutamate, the following procedure is carried out using the new microwave procedure (**Table 3.5**). For double labeling, we have found that the animal needs to be perfused with the following fixative: 1% glutaraldehyde/0.5% paraformaldehyde/0.1% picric acid in 0.1 M phosphate buffer. The brain is left in the fixative

Analysis of Motor Function in Humans 75

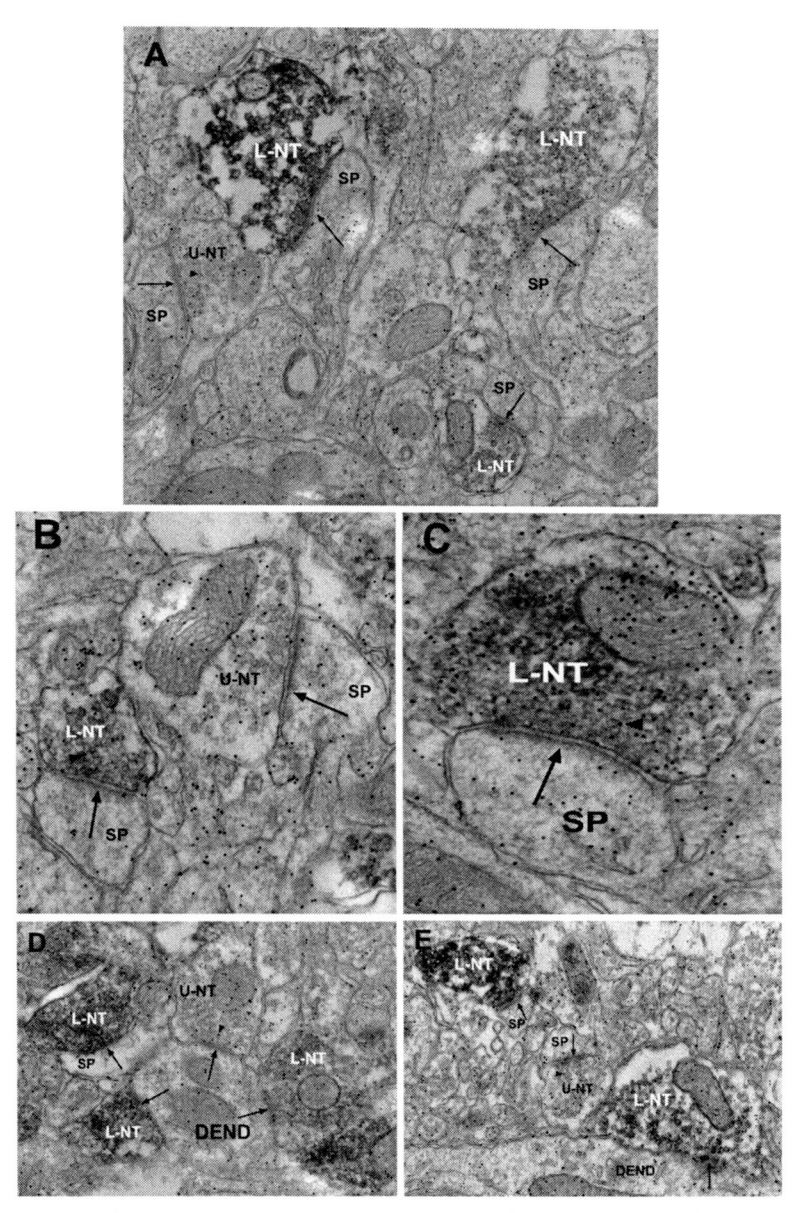

Fig. 3.8. Double labeling for the vesicular glutamate transporter-1 (VGLUT-1) (**a–d**) and VGLUT-2 (**e**) and glutamate. The darkened reaction product (DAB) inside labeled terminals (L-NT) shows the localization for the VGLUT-1 or VGLUT-2 protein. There are also unlabeled nerve terminals (U-NT) within the field. Both labeled and unlabeled terminals are shown making an asymmetrical synaptic contact (*arrow*) onto an underlying dendritic spine (SP) or dendrite (DEND). Within the nerve terminals are numerous 10-nm gold particles (*arrowhead*) indicating the location of the neurotransmitter, glutamate. (**a**) Low power view where three labeled terminals (L-NT) containing the darkened reaction product for VGLUT-1 localization and one unlabeled terminal (U-NT) can be seen making contact with an underlying spine. Immunogold labeling can be seen concentrated within the nerve terminals. (**b**) Slightly higher power view versus **a** above showing one labeled and one unlabeled terminal making an asymmetrical contact onto dendritic spines. (**c**) Higher power view showing a single-labeled nerve terminal, containing numerous 10-nm gold particles, making an asymmetrical contact onto a dendritic spine. (**d**) Two labeled and one unlabeled terminals are seen making an asymmetrical synaptic contact onto a dendrite, while a nearby labeled terminal is making an asymmetrical synaptic contact onto a dendritic spine. (**e**) VGLUT-2 labeling of nerve terminals making a synaptic contact onto either a dendrite (DEND) or spine (sp), while there is one unlabeled terminal contacting a dendritic spine.

Table 3.5
Microwave procedure for EM immunohistochemistry

PBS: phosphate buffer saline (0.1 M phosphate, 0.9% sodium chloride, pH 7.3)

ABC: avidin–biotin complex (Vector Labs, diluted according to the manufacturer's instructions)

DAB: diaminobenzidine [22 mg DAB in 100 ml Tris-buffered saline (0.1 M Tris, 0.9% sodium chloride) plus 10 ml 30% H_2O_2]

Description	User prompt	Time (h:min:s)	Watts	Temp	Load Cooler	Vacuum
ANTIGEN RETRIEVAL	ON	0:05:00	550	60	AUTO	OFF
PBS RINSE	ON	0:01:00	150	60	AUTO	OFF
30% HYDROGEN PEROXIDE	ON	0:01:00	150	60	AUTO	OFF
PBS RINSE	ON	0:01:00	150	60	AUTO	OFF
PBS RINSE	ON	0:01:00	150	60	AUTO	OFF
BLOCKING STEP	ON	0:01:00	150	60	AUTO	OFF
PRIMARY PRE-VAC	ON	0:00:10	200	60	AUTO	CYCLE
Primary Antibody (Ab) ON	OFF	0:02:00	200	60	AUTO	CONT
Primary Ab OFF	OFF	0:03:00	0	60	AUTO	CONT
Primary Ab ON	OFF	0:02:00	200	60	AUTO	CONT
PBS RINSE	ON	0:01:00	150	60	AUTO	OFF
PBS RINSE	ON	0:01:00	150	60	AUTO	OFF
Secondary Ab ON	ON	0:04:00	200	60	AUTO	CYCLE
Secondary Ab OFF	OFF	0:03:00	0	60	AUTO	CYCLE
Secondary Ab ON	OFF	0:04:00	200	60	AUTO	CYCLE
PBS RINSE	ON	0:01:00	150	60	AUTO	OFF
PBS RINSE	ON	0:01:00	150	60	AUTO	OFF
ABC ON	ON	0:04:00	150	60	AUTO	CYCLE
ABC OFF	OFF	0:03:00	0	60	AUTO	CYCLE
ABC ON	OFF	0:04:00	150	60	AUTO	CYCLE
PBS RINSE	ON	0:01:00	150	60	AUTO	OFF
PBS RINSE	ON	0:01:00	150	60	AUTO	OFF

DAB on bench ~ 10 min.

Stop reaction by placing tissue in PBS. Put tissue in change of PBS until ready to process for (electron microscopy).

Antigen retrieval: 10 mM sodium citrate, pH 6.0 or 10 mM TRIS, 1 mM EDTA, pH 9.0.

overnight in the cold and then cut the next day. We have tried using a slightly less harsh fixative consisting of 1% acrolein/0.5% glutaraldehyde/2% paraformaldehyde in phosphate buffer and the results have been less than satisfying in terms of both the pre-embed DAB labeling and the post-embed immunogold labeling. It appears that in order to obtain good immunogold labeling, it is essential that the fixative contain 1% glutaraldehyde. For the antigen retrieval, a pH of 6.0 generally works, but for a few antibodies, a pH of 9.0 works even better. We find that after a nearly complete loss of dopamine (~90%), there is a significant decrease in the density of nerve terminal glutamate immunogold labeling in VGLUT-1-labeled and in unlabeled nerve terminals making an asymmetrical synaptic contact in the striatum compared to the control group. This decrease was 35% for the VGLUT-1-labeled terminals and 58% for the unlabeled terminals (control vs. lesioned group). This decrease in nerve terminal glutamate labeling was associated with an increase in the apomorphine-induced contralateral turning (83).

For the antibodies we have tested using the pre-embed/DAB method above, the following dilutions were used: VGLUT-1 (Synaptic Systems, polyclonal: cat #135-303: 1:2,000); VGLUT-2 (Synaptic Systems, polyclonal: cat #135-403, 1:100); Activity Related Complex (ARC: Synaptic Systems, polyclonal: cat #156-003, 1:200); Tyrosine Hydroxylase (ImmunoStar, monoclonal: cat #22941, 1:1,000); and Glial Fibrillary Acidic Protein (Sigma, monoclonal: cat #G3893, 1:500).

5. Histological Correlates to Changes in Locomotion in the 1-Methyl-4-Phenyl-1,2,3,6-Tetrahydropyridine (MPTP) Mouse Model of Parkinson's Disease

The MPTP mouse model of Parkinson's disease is one of the most widely used. Several administration protocols ranging from acute to chronic mimic various stages in the progression of dopamine (DA) cell neurodegeneration in the substantia nigra pars compacta (SNpc). Because each MPTP model results in a unique degree of DA cell degeneration (anywhere from 20 to 80% cell loss) and deprivation of DA to the striatum, it is necessary to develop testable behavioral tasks which are differently sensitive to DA loss. The Free-Standing Rear and Parallel Rod Activity Chamber (PRAC) tests described below are two such assessments of locomotor function. We demonstrate their application in both young adult and aged C57Bl/6 J male mice using both subacute (30 mg/kg/d i.p. in young adult; 15 mg/kg/d i.p. in aged animals × 7 d) and chronic administration models (7 mg/kg/d i.p. in young adult; 4 mg/kg/d i.p. in aged animals × 28 d). Both of these tests are particularly sensitive to DA loss; reaching, forepaw placement, and foot-faulting are motor responses that most consistently reflect the damage to the basal ganglia circuits central to movement control (94).

As examples of how these behavioral tests have been used in the MPTP model, we present their correlation with depleted tyrosine hydroxylase (TH)-labeled DA cells of the SNpc in a non-invasive therapeutic intervention study and in a novel progressive model. Few studies have investigated the effects of an enriched environment (EE) on restoration of brain chemistry or motor behavior due to the loss of DA in the nigrostriatal pathway (95–98). The EE consists of housing conditions in which social and novel physical stimulation are promoted by an excess of toys (e.g., tubes, wheels, ladders) and peers in a large cage. It was first reported that exposure to an EE for 2 months, followed by acute administration of MPTP (20 mg/kg × 4/d, fifth injection on day 7), resulted in a decrease in the loss of tyrosine hydroxylase-immunoreactive (TH-ir) DA cells in the SNpc (99). Acute MPTP resulted in a nearly 70% loss of TH-ir cells in the SNpc, but in the EE-exposed group, this loss was only about 40%. However, there was *no behavioral correlate* to this reduced loss of TH-ir cells. We present evidence that 1 week *after* the partial loss of TH-ir cells within the SNpc, exposure to an EE results in a significant restoration in the number of TH-ir cells and in partial recovery of motor behavior. We also find that the extent of recovery in both TH-ir cells and behavior is dependent on the MPTP model used and the age of the animal.

6. Free-Standing Rears as a Measurement of Locomotor Deficit

Several behavioral tests are commonly used to assess locomotor impairments in the MPTP mouse model of Parkinson's disease; among them, the Rotarod, the Grid test, and the Beam Traversal Challenge (94, 100). Although vertical rearing behavior has been measured in the MPTP-lesioned mouse, a distinction between wall-assisted and free-standing rears has never been reported. We have observed that healthy mice, both young adult and aged,

Fig. 3.9. Rearing test cylinder.

tend to rear freely and wall-assisted 50% of the time (**Fig. 3.9**). When lesioned with a subacute dose of MPTP (30 mg/kg/d in young adults; 15 mg/kg/d in aged animals × 7 d), mice rear freely only 30–50% of their baseline (**Fig. 3.10a**). In the chronic MPTP model (7 mg/kg/d in young adults; 5 mg/kg/d in aged animals × 28 d), this frequency increases to 30% indicating a lesser locomotor deficit (**Fig. 3.10a**). Additionally, total number of rears decreases by 50% with subacute MPTP (controls: 101.69 ± 1.81% of baseline; subacute MPTP: 71.35 ± 5.38% of baseline) and by 45% in the chronic model (controls: 22.5 ± 1.4 total rears; chronic MPTP: 12.37 ± 1.13 total rears). This observation supports findings in previous studies using similar MPTP models (101–103) and suggests that vertical exploration is significantly impaired in both subacute and chronic MPTP-lesioned mice.

1. *The cylinder*: A cylinder of 3 cm radius and at least 9 cm height with no ceiling is used. The walls should be translucent; this behavior does not depend on directed stimuli and is not skewed by visualization of surroundings. In a solid

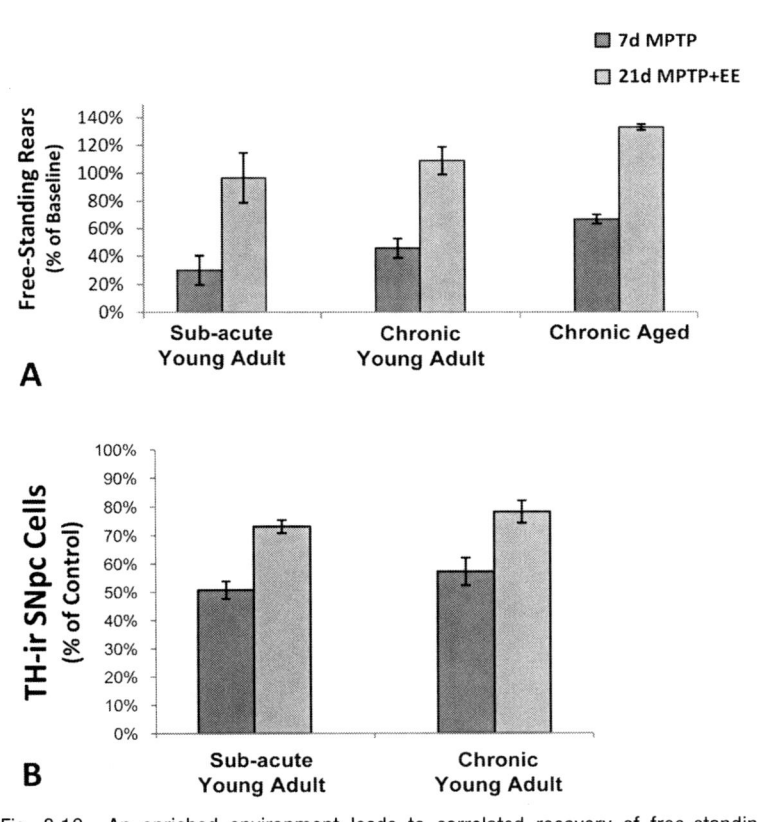

Fig. 3.10. An enriched environment leads to correlated recovery of free-standing rears and TH-immunoreactive SNpc cells. The decrease in free-standing rears (**a**) and 40–50% decrease in TH-ir cells (**b**) caused by MPTP is reversed by EE in both subacute and chronic models.

cylinder, animals tend to show behavioral phenotypes consistent with anxiety: attempting to jump over the wall of the cylinder, sitting still at the base of the cylinder wall.

2. *Collecting baseline data*: Mice are moved into the testing room 1 h prior to testing to acclimate to the conditions of the room. Prior to any experimental treatment, animals should be tested inside the cylinder to collect baselines. The number of times over 2 min each animal rears using the cylinder wall for support (wall-assisted) and without touching forepaws to the wall (free-standing) should be recorded. It is not necessary to acclimate the animal to the cylinder prior to baseline collection because rearing is an exploratory

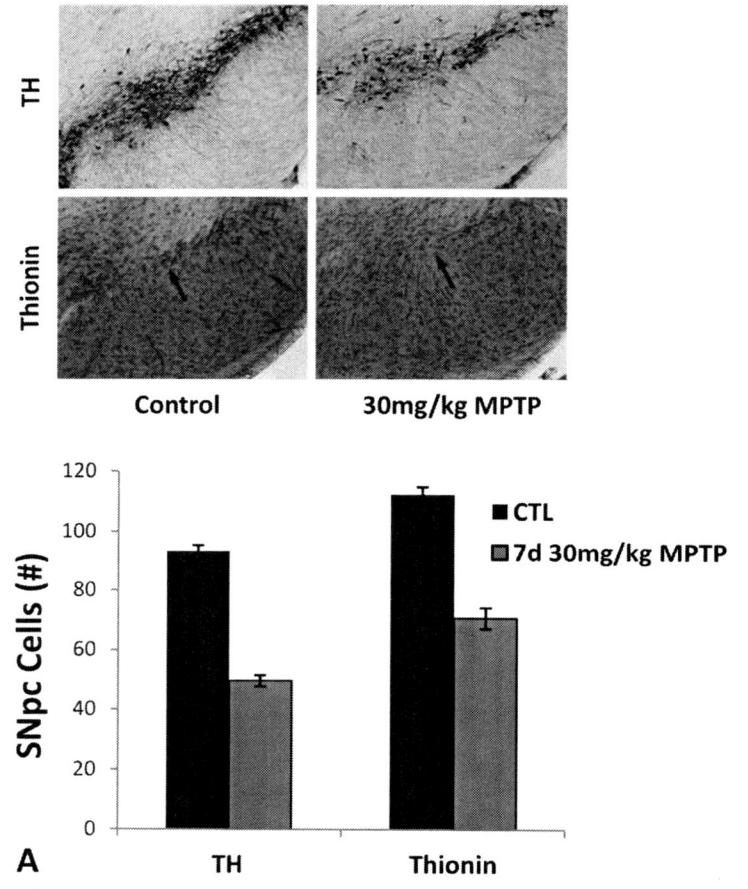

Fig. 3.11. Subacute and chronic MPTP doses have a differential effect on TH-ir cells in the SNpc of young adult and aged mice. In the subacute young adult model (30 mg/kg/d × 7 d), there is a 50% TH-ir cell depletion and a comparable 40% decrease in Thionin-stained cells (**a**). Alternatively, the chronic MPTP dose (4 mg/kg/d × 21 d) causes a 40% TH-ir loss and only a 15% decrease in Thionin-stained cells by day 7 in aged mice (**b**).

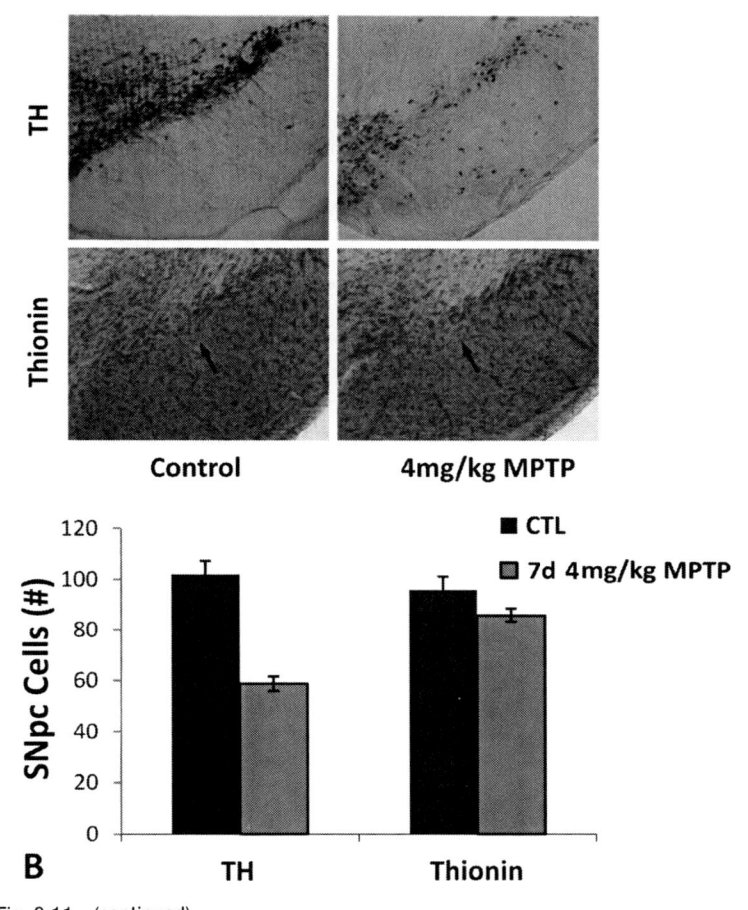

Fig. 3.11. (continued)

behavior and does not depend on familiarization with the testing environment.

3. *Testing*: Mice are moved into testing room 1 h prior to testing to acclimate to the conditions of the room. Following treatment, the number of wall-assisted and free-standing rears over 2 min should be counted. These numbers are compared to baseline levels to assess (1) the change in percentage of free-standing rears and (2) the change in total rears due to treatment.

Decreases in both total rears and percent free-standing rears have shown to correlate with loss of expression of the rate-limiting enzyme, TH, within DA cells located in the SNpc (**Fig. 3.11**). This suggests that free-standing rearing behavior is a sufficient indicator of DA deficiency in subacute MPTP mouse model of Parkinson's disease. Although immunohistochemical TH-labeling is standard for detecting DA cells, it has been suggested that in MPTP models of Parkinson's disease, a decrease in immunoreactive neurons in the SNpc can be due to either dysfunction or degeneration (104, 105). For this reason, it is necessary to compare cell number between TH-labeled and thionin-stained cells to confirm that non-immunoreactive cells are degenerated as opposed to unable to express TH. With the subacute MPTP dose, comparison of TH-ir and thionin-stained neurons in serial slices of the SNpc shows comparable loss (**Fig. 3.11a**). This suggests that the subacute dose of MPTP is sufficient to cause DA cell degeneration as opposed to rendering them dysfunctional. In the chronic model, however, the thionin-stained cell loss is only 15% compared to the 40–50% TH-ir loss, indicative of dysfunctional but not degenerated DA cells (**Fig. 3.11b**).

We have also found that exposure to an EE can restore the percentage of free-standing rears back to nearly the control level (**Fig.3.11**). In both the subacute and chronic models, mice injected for 7 days with MPTP were subsequently put into an EE 24 h/d (MPTP + EE). After 21 days, the MPTP + EE group shows an increase in both free-standing rears and TH-ir cells (**Fig. 3.11**). This suggests that an EE is not only resulting in a recovery of DA cell expression but that this increase in TH-ir cells has an influence on motor behavior.

7. Foot-Faults and Activity as a Measurement of Locomotor Deficit

Open-field activity is a traditional measurement of spontaneous exploratory behavior. In the Parallel Rod and Activity Chamber (PRAC) test, it has been combined with the measurement of foot-faulting behavior to determine frequency of foot-faults in healthy and MPTP-lesioned mice. The task measures the number of times a mouse paw slips through a chamber floor consisting of a series of parallel metal rods, as previously described (106). Simultaneously, photo-beams affixed to the walls of the chamber record activity inside the chamber by the number of times a mouse passes in front of a beam, breaking the signal.

It is important to measure these two locomotor variables together, as an increase or decrease in one does not imply the same change in the other. For instance, we have found that when comparing foot-faults alone, one group may fault significantly less than another. This can either be due to more faulting per unit of movement, or because the animals in one group are less active giving them less opportunity to fault. For this reason, it is necessary to also express foot-faults relative to total activity.

The parallel rod chamber described by Kamens and Crabbe (106) was originally designed to test ataxia in ethanol-treated mice. For the purposes of their studies, animals were acclimated to the parallel rod chambers for 5 min prior to testing. However, in the subacute and chronic MPTP models of Parkinson's disease we have found that acclimation for 10 min/day over 2 days prior to measuring baseline behavior yields more consistent data. This method is in agreement with previous reports of cocaine-treated rats (107).

1. *The Parallel Rod and Activity Chamber (PRAC)*: Equipment is set-up as previously described (106) with the modification of eight photo-beams affixed in pairs across each chamber face to record total activity (**Fig. 3.12**). When the mouse's paw slips through the parallel rods and contacts the metal plate below, a circuit is closed and an error recorded by the computer. Slips of the tail or feces through the rods are not recorded. Similarly, an error is recorded each time the mouse breaks a photo-beam in each quadrant.

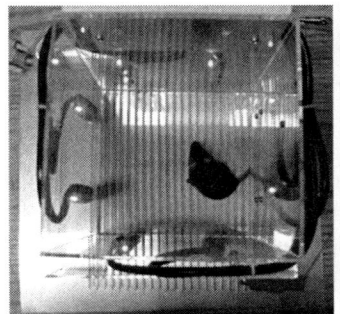

Fig. 3.12. Parallel Rod and Activity Chamber (PRAC). Similar to that described by Kamens and Crabbe (106) but modified with eight photo-beams.

2. *Collecting baseline data and testing*: Over 2 days prior to collecting baseline behavior, acclimate animals to chambers for 10 min a day. Mice should be moved into the testing room 1 h prior to testing to acclimate to the conditions of the room. On the third day, baseline data are collected: one mouse should be placed in each PRAC and behavior recorded for a minimum of 5 min. Between each recording session, metal plates and parallel rods should be cleaned with 10% isopropyl ethanol to remove feces and urea; if allowed to accumulate in large quantities these will become conductive and break the circuit such that the computer will not record mouse paw slips.

It has been reported that chronic, daily administration of MPTP in mice for up to 20 days results in a decline in the percentage of remaining TH-ir cells in the SNpc (108). The maximum loss of TH-ir cells was about 65%, which occurred by day 15 and was maintained through day 20. The advantage of such a chronic model for Parkinson's disease is that intervention therapy such as an EE can be started at any time point after neurotoxin administration. Unlike the acute and subacute models, there is no concern with chronic administration that there will be compensatory sprouting of new DA nerve terminals (109). However, the TH-ir loss in the chronic model plateaus after 15 days such that studies of long-term therapeutic effects cannot intervene at the more acute stages of cell loss. Therefore, we have developed a progressive MPTP model for the purposes of studying EE intervention at varying stages of TH-ir cell loss.

There are several MPTP models characterized as progressive (110–112). We present evidence for a workable chronic, progressive MPTP model using intraperitoneal injection (**Fig. 3.13**). For a period of 4 weeks, young adult mice are injected daily with a dose of MPTP increasing each week (4 mg/kg/d for 1 week; 8 mg/kg/d for 1 week; 12 mg/kg/d for 1 week; 16 mg/kg/d for 1 week). Foot-faults per total activity, as measured by the PRAC test, increased in a dose-dependent manner from weeks 1–4 (**Fig. 3.13a**). One week of 4 mg/kg/d MPTP did not affect the foot-fault per beam break ratio (FF/BB), but was sufficient to decrease TH-ir SNpc cells by 22% (**Fig. 3.13b**). The maximum cell loss that occurred by day 28 was 60%.

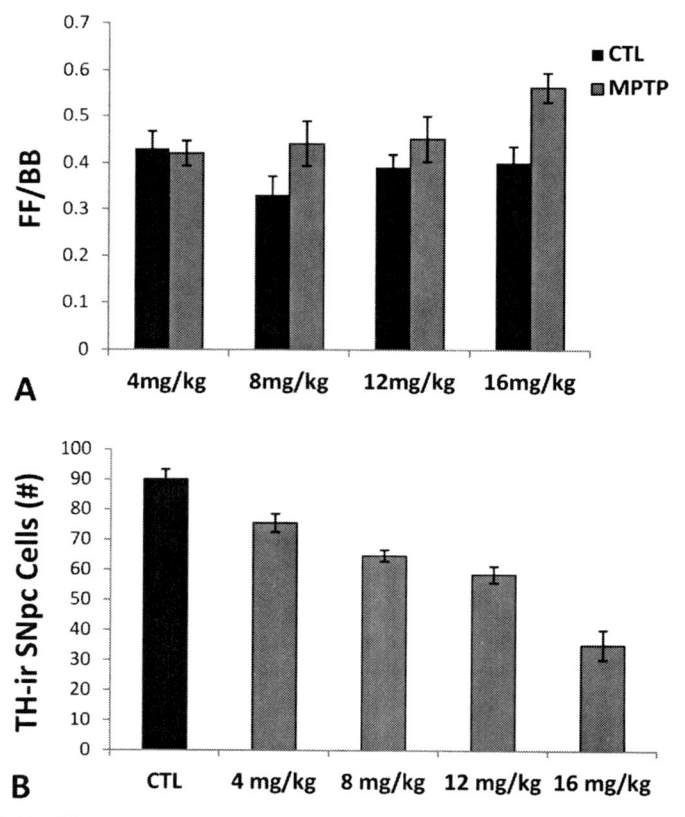

Fig. 3.13. Effects of increasing MPTP dose on PRAC test and TH-ir cells in the SNpc. In young adult mice, foot-fault per beam break ratio (FF/BB) increases in a dose-dependent manner (**a**). Similarly, TH-ir neurons of the SNpc decrease as MPTP dose increases over 4 weeks (**b**).

8. Comparison of Motor Function Tests in Humans with Parkinson's Disease Versus the Rodent Model

In the rodent model of Parkinson's disease that is used in the author's lab, the motor deficiencies are not as apparent as they are in humans with this disease. By the time humans show the obvious behavioral signs of Parkinson's disease, there is a severe loss (~70–80%) of striatal DA and about a 50% or greater loss of DA neurons in the SNpc. Rodents, in general, do not develop obvious tremor or freezing with the loss of nigrostriatal DA. Motor deficiencies in the MPTP non-human primate model are far closer to those seen in humans with Parkinson's disease, but this chapter has focused exclusively on the rodent model. With rodents, the possible rigidity and bradykinesia/hypokinesia can be measured using the Parallel Rod and Activity Chamber as described above. The use of the cylinder test, which probably measures a number of factors, such as balance, coordination, and vestibular/sensory function, may be related to the postural control and sensory integration that can be measured in humans (*see* **Table 3.1**).

Gait is an important measure of motor function that is tested in humans (**Tables 3.1** and **3.3**). Recently, such a motor function test has been developed for use in various rodent models of disorders within the central nervous system (113). This gait analysis apparatus can accurately measure a number of functions, such as stride length, stance width, and foot placement angle while the rodent is running on treadmill. Whether this apparatus can measure changes in gait dynamics when there is a more subtle loss of striatal DA over an extended period of time versus the more acute loss of DA is yet to be determined. However, this gait apparatus is a significant advancement in terms of being able to translate some motor function tests in humans to rodents.

Acknowledgments

This work was supported by the Department of Veterans Affairs Merit Review program to CKM, Team Parkinson/Parkinson Alliance, NIH (RO1 NS44327), and the US Army NETRP (#W81XWH-04-1-0444) to GMP and MWJ.

References

1. Visser M, Marinus J, Bloem BR, Kisjes H, van den Berg BM, van Hilten JJ (2003) Clinical tests for the evaluation of postural instability in patients with Parkinson's disease. Arch Phys Med Rehabil 84: 1669–1674
2. Bloem BR, Beckley DJ, van Hilten BJ, Roos RAC (1998) Clinimetrics of postural

instability in Parkinson's disease. J Neurol 245:669–673
3. Xu D, Carlton LG, Rosengren KS (2004) Anticipatory postural adjustments for altering direction during walking. J Mot Behav 36:316–326
4. Stack E, Jupp K, Ashburn A (2004) Developing methods to evaluate how people with Parkinson's disease turn 180 degrees: an activity frequently associated with falls. Disabil Rehabil 26:478–484
5. Stack EL, Ashburn AM, Jupp KE (2006) Strategies used by people with Parkinson's disease who report difficulty turning. Parkinsonism Relat Disord 12:87–92
6. Crenna P, Carpinella I, Rabufetti M, Calabrese E, Mazzoleni P, Nemni R, Ferrarin M (2007) The association between impaired turning and normal straight walking in Parkinson's disease. Gait Posture 26:172–178
7. Willems A, Nieuwboer A, Chavret F, Desloovere K, Dom R, Rochester L, Kwakkel G, van Ween E, Jones D (2007) Turning in Parkinson's disease patients and controls: the effect of auditory cues. Mov Disord 22:1871–1878
8. Movement Disorders Task Force (2003) The Unified Parkinson's Disease Rating Scale (UPDRS): status and recommendations. Mov Disord 18:738–750
9. Stebbins GT, Goetz CG (1998) Factor structure of the Unified Parkinson's Disease Rating Scale: motor examination section. Mov Disord 13:633–636
10. van Hilten JJ, van der Zwan AD, Zwinderman AH, Roos RAC (1994) Rating impairment and disability in Parkinson's disease: evaluation of the Unified Parkinson's Disease Rating Scale. Mov Disord 9:84–88
11. Richards M, Marder K, Cote L, Richard M (1994) Interrater reliability of the Unified Parkinson's Disease Rating Scale motor examination. Mov Disord 9:89–91
12. Siderworft A, McDermott M, Kieburtz K, Blindauer K (2002) Test–retest reliability of the Unified Parkinson's Disease Rating Scale in patients with early Parkinson's disease: results from a multicenter clinical trial. Mov Disord 17:758–763
13. Song J, Fisher BE, Petzinger G, Wu A, Gordon J, Salem GJ (2009) The relationship between the Unified Parkinson's Disease Rating Scale and lower extremity functional performance in persons with early-stage Parkinson's disease. Neurorehabil Neural Repair 23:657–661
14. Goetz CG, Tilley BC, Shaftman SR et al (2008) Movement disorder society-sponsored revision of the Unified Parkinson's Disease Rating Scale (MDS-UPDRS): scale presentation and clinimetric testing results. Mov Disord 23:2129–2170
15. Podsiadlo D, Richardson S (1991) The timed "up & go": a test of basic functional mobility for frail elderly persons. J Am Geriatr Soc 38:142–148
16. Rockwood K, Awalt E, Carver D, MacKight C (2000) Feasibility and measurement properties of the functional reach and the timed up and go tests in the Canadian Study of Health and Aging. J Gerontol A Biol Sci Med Sci 55A:M70–M73
17. Steffen TM, Hacker TA, Mollinger L (2002) Age- and gender-related test performance in community-dwelling elderly people: six-minute walk test, Berg Balance Scale, timed up & go test, and gait speeds. Phys Ther 82:128–137
18. Thompson M, Medley A (1995) Performance of community dwelling elderly on the timed up and go test. Phys Occ Ther Ger 13:17–30
19. Shumway-Cook A, Brauer S, Woolacott M (2000) Predicting the probability for falls in community-dwelling older adults using the timed up & go test. Phys Ther 80:896–903
20. Morris S, Morris ME, Iansek R (2001) Reliability of measurements obtained with the timed "up & go" test in people with Parkinson disease. Phys Ther 81:810–818
21. Lim LIIK, van Wegen EEH, de Goede CJT, Jones D, Rochester L, Hetherington V, Nieuboer A, Willems AM, Kwakkel G (2005) Measuring gait and gait-related activities in Parkinson's patients own home environment: a reliability, responsiveness and feasibility study. Parkinsonism Relat Disord 11:19–24
22. Campbell CM, Rowse JL, Ciol MA, Shumway-Cook A (2003) The effect of cognitive demand on timed up and go performance in older adults with and without Parkinson disease. Neurol Rep 27:2–6
23. Blum L, Korner-Bitensky N (2008) Usefulness of the Berg Balance Scale in stroke rehabilitation: a systematic review. Phys Ther 88:559–566
24. Landers MR, Backlund A, Davenport J, Fortune J, Schuerman S, Alktenburger P (2008) Postural instability in idiopathic Parkinson's disease: discriminating fallers from nonfallers based on standardized clinical measures. J Neurol Phys Ther 30:60–67
25. Muir SW, Berg K, Chesworth B, Speechley M (2008) Predicting multiple falls in community-dwelling elderly people: a prospective study. Phys Ther 88:449–459

26. Wee JY, Wong H, Palepu A (2003) Validation of the Berg Balance Scale as a predictor of length of stay and discharge destination in stroke rehabilitation. Arch Phys Med Rehabil 84:731–735
27. Bacher M, Scholz E, Diener HC (1989) 24 hour continuous tremor quantification based on EMG recording. Electroencephalogr Clin Neurophysiol 72:176–183
28. Qutubuddin AA, Pegg PO, Cifu DX, Brown R, McNamee S, Carne W (2005) Validating the Berg Balance Scale for patients with Parkinson's disease: a key to rehabilitation evaluation. Arch Phys Med Rehabil 86:789–792
29. Steffen T, Seney M (2008) Test–retest reliability and minimal detectable change on balance and ambulation tests, the 36-item short-form health survey, and the Unified Parkinson Disease Rating Scale in people with parkinsonism. Phys Ther 88:733–746
30. Dibble LE, Lange M (2006) Predicting falls in individuals with Parkinson disease: a reconsideration of clinical balance measures. J Neurol Phys Ther 30:60–67
31. Breit S, Spieker S, Schulz JB, Gasser T (2008) Long-term EMG recordings differentiate between parkinsonian and essential tremor. J Neurol 255:103–111
32. Scholz E, Bacher M, Diener HC, Dichgans J (1988) Twenty-four-hour tremor recordings in the evaluation of the treatment of Parkinson's disease. J Neurol 235:475–484
33. Sturman MM, Vaillancourt DE, Metman LV, Bakay RAE, Corcos DM (2004) Effects of subthalamic nucleus stimulation and medication on resting and postural tremor in Parkinson's disease. Brain 127:2131–2143
34. Blahak C, Bazner H, Capelle H, Wohrle JC, Weigel R, Henneriei MG, Krauss JK (2009) Rapid response of parkinsonian tremor to STN-DBS changes: direct modulation of oscillatory basal ganglia activity? Mov Disord 24:1221–1225
35. Bronte-Stewart HM, Ding L, Alexander C, Zhou Y, Moor GP (2000) Quantitative digitography (QDG): a sensitive measure of digital motor control in idiopathic Parkinson's disease. Mov Disord 15:36–47
36. Brusse KJ, Zimdars S, Zalewski KR, Steffen TM (2005) Testing functional performance in people with Parkinson disease. Phys Ther 85:134–141
37. Kelly VE, Hyngstrom AS, Rundle MM, Bastian AJ (2001) Interaction of levodopa and cues on voluntary reaching in Parkinson's disease. Mov Disord 17:38–44
38. Koop MM, Andrezejewski A, Hill BC, Heit G, Bronte-Stewart HM (2006) Improvement in a quantitative measure of bradykinesia after microelectrode recording in patients with Parkinson's disease during deep brain stimulation surgery. Mov Disord 21:673–678
39. Latt MD, Menz HB, Fung VS, Lord SR (2009) Acceleration patterns of the head and pelvis during gait in older people with Parkinson's disease: a comparison of fallers and nonfallers. J Gerontol A Biol Sci Med Sci 64A:700–706
40. Koop MM, Shjivitz N, Bronte-Stewart H (2008) Quantitative measures of fine motor, limb, and postural bradykinesia in very early stage, untreated Parkinson's disease. Mov Disord 23:1262–1268
41. Louie SL, Koop MM, Frenklach A, Bronte-Stewart H (2009) Quantitative lateralized measures of bradykinesia at different stages of Parkinson's disease: the role of the less affected side. Mov Disord 24:1991–1997
42. Kavanagh JJ, Menz HB (2008) Accelerometry: a technique for quantifying movement patterns during walking. Gait Posture 28:1–15
43. Lord S, Rochester L, Baker K, Nieuwboer A (2008) Concurrent validity of accelerometry to measure gait in Parkinsons disease. Gait Posture 27:357–359
44. Saremi K, Marehbian J, Yan X, Regnaux J, Elashoff R, Bussel B, Dobkin BH (2006) Reliability and validity of bilateral thigh and foot accelerometry measures of walking in healthy and hemiparetic subjects. Neurorehabil Neural Repair 20:297–305
45. Windolf M, Gotzen N, Morlock M (2008) Systematic accuracy and precision analysis of video motion capturing systems – exemplified on the Vicon-460 system. J Biomech 41:2776–2780
46. Huxham F, Baker R, Morris ME, Iansek R (2008) Head and trunk rotation during walking turns in Parkinson's disease. Mov Disord 23:1391–1397
47. Chien S, Lin S, Liang C, Soong Y, Lin S, Hsin Y, Lee C, Chen S (2006) The efficacy of quantitative gait analysis by the GAITRite system in evaluation of parkinsonian bradykinesia. Parkinsonims Relat Disord 12:438–442
48. Nelson AJ, Zwick D, Brody S, Doran C, Pulver L, Rooz G, Sadownick M, Nelson R, Rothman J (2002) The validity of the GAITRite and the functional ambulation performance scoring system in the analysis of Parkinson gait. NeuroReabilitation 17:255–262
49. Ferrarin M (2006) Locomotor disorders in patient at early stages of Parkinson's disease:

a quantitative analysis. Conf Proc IEEE Eng Med Biol Soc 1:1224–1227
50. Oates AR, Frank JS, Patla AE, VanOoteghem K, Horak FB (2008) Control of dynamic stability during gait termination on a slippery surface in Parkinson's disease. Mov Disord 23:1977–1983
51. Hof AL, Gazendam MG, Sinke WE (2005) The condition for dynamic stability. J Biomech 38:1–8
52. Song J, Fisher B, Sigward S, Petzinger G, Salem GJ (2008) The effect of early stage Parkinson's disease on dynamic postural stability during turning activities. Paper presented at Society for Neuroscience Conference</pub>, Washington, DC
53. Dimitrova D, Horak FB, Nutt JG (2004) Postural muscle responses to multidirectional translations in patients with Parkinson's disease. J Neurophysiol 91:489–501
54. Dimitrova D, Nutt J, Horak FB (2004) Abnormal force patterns for multidirectional postural responses in patients with Parkinson's disease. Exp Brain Res 156:183–195
55. Horak FB, Dimitrova D, Nutt JG (2005) Direction-specific postural instability in subjects with Parkinson's disease. Exp Neurol 193:504–521
56. Starr MS (1995a) Glutamate/dopamine D1/D2 balance in the basal ganglia and its relevance to Parkinson's disease. Synapse 19:264–293
57. Starr MS (1995b) Antiparkinsonian actions of glutamate antagonists-alone and with L-DOPA: a review of evidence and suggestions for possible mechanisms. J Neural Transm [P-D Sect] 10:141–185
58. Fonnum F (1984) Glutamate: a neurotransmitter in mammalian brain. J Neurochem 42:1–11
59. McGeorge AJ, Faull RLM (1989) The organization of the projection from the cerebral cortex to the striatum in the rat. Neurosci. 29:503–537
60. Lacey CJ, Boyes J, Gerlach O, Chen L, Magill PJ, Bolam JP (2005) GABA-B receptors at glutamatergic synapses in the rat striatum. Neurosci 136:1083–1095
61. Raju DV, Shah DJ, Wright TM, Hall RA, Smith Y (2006) Differential synaptology of vGluT2-containing thalamostriatal afferents between the patch and matrix compartments. J Comp Neurol 499:231–243
62. Smith Y, Raju DV, Pare JF, Sidibe M (2004) The thalamostriatal system: a highly specific network of the basal ganglia circuitry. Trends Neurosci 27:520–527
63. Bouyer JJ, Park DH, Joh TH, Pickel VM (1984) Chemical and structural analysis of the relation between cortical inputs and tyrosine hydroxylase-containing terminals in rat neostriatum. Brain Res 302:267–275
64. Freund TF, Powell JF, Smith AD (1984) Tyrosine hydroxylase-immunoreactive boutons in synaptic contact with identified striatonigral neurons, with particular reference to dendritic spines. Neurosci 13:1189–1215
65. Gundersen V, Ottersen OP, Storm-Mathisen J (1996) Selective excitatory amino acid uptake in glutamatergic nerve terminals and in glia in the rat striatum: quantitative electron microscopic immunocytochemistry of exogenous (D)-aspartate and endogenous glutamate and GABA. Eur J Neurosci 8:758–765
66. Meshul CK, Stallbaumer RK, Taylor B, Janowsky A (1994) Haloperidol-induced synaptic changes in striatum are associated with glutamate synapses. Brain Res 648:181–195
67. Bamford NA, Robinson S, Palmiter R, Joyce JA, Moore C, Meshul CK (2004) Presynaptic modulation of corticostriatal terminals in dopamine-deficiency. J Neurosci 24:9541–9552
68. Morari M, O'Connor WT, Ungerstedt U, Fuxe J (1994) Dopamine D_1 and D_2 receptor antagonism differentially modulates stimulation of striatal neurotransmitter levels by N-methyl-D-aspartic acid. Eur J Pharm 256:23–30
69. Nithianantharajah J, Hannan AJ (2006) Enriched environments, experience-dependent plasticity and disorders of the nervous system. Nature Rev 7:687–709
70. Yamamoto BK, Davy S (1992) Dopaminergic modulation of glutamate release in striatum as measured by microdialysis. J Neurochem 58:1736–1742
71. Zhang H, Sulzer D (2003) Glutamate spillover in the striatum depresses dopaminergic transmission by activating group I metabotropic glutamate receptors. J Neurosci 19:10585–10592
72. Sesack SR, Aoki C, Pickel VM (1994) Ultrastructural localization of D_2 receptor-like immunoreactivity in midbrain dopamine neurons and their striatal targets. J Neurosci 14:88–106
73. Wang H, Pickel VM (2002) Dopamine D_2 receptors are present in prefrontal cortical afferents and their targets in patches of the rat caudate-putamen nucleus. J Comp Neurol 442:392–404
74. Calabresi P, De Murtas M, Mercuri NB, Bernardi G (1992) Chronic neuroleptic treatment: D_2 dopamine receptor

supersensitivity and striatal glutamatergic transmission. Ann Neurol 31:366–373
75. Tillerson JL, Caudle WM, Reveron ME, Miller GW (2003) Exercise induces behavioral recovery and attenuates neurochemical deficits in rodent models of Parkinson's disease. Neurosci 119:899–911
76. Tillerson JL, Cohen AD, Philhower J, Miller GW, Zigmond MJ, Schallert T (2001) Forced limb-use effects on the behavioral and neurochemical effects of 6-hydroxydopamine. J Neurosci 21: 4427–4435
77. Poulton NP, Muir GD (2005) Treadmill training ameliorates dopamine loss but not behavioral deficits in hemi-Parkinsonian rats. Exp Neurol 193:181–197
78. Metz GA, Whishaw IQ (2002) Drug-induced rotation intensity in unilateral dopamine-depleted rats is not correlated with end point or qualitative measures of forelimb or hindlimb motor performance. Neuroscience 111:325–336
79. O'Dell SJ, Gross NB, Fricks AN, Casiano BD, Nguyen TB, Marshall JF (2007) Running wheel exercise enhances recovery from nigrostriatal dopamine injury without inducing neuroprotection. Neuroscience 144:1141–1151
80. Fisher BE, Petzinger GM, Nixon K, Hogg E, Bremmer S, Meshul CK, Jakowec MW (2004) Exercise-induced behavioral recovery and neuroplasticity in the 1-methyl-4-phenyl-1,2,3,6-tetrahydropyridine-lesioned mouse basal ganglia. J Neurosci Res 77:378–390
81. Baker DA, Xi ZX, Shen H, Swanson CJ, Kalivas PW (2002) The origin and neuronal function of in vivo nonsynaptic glutamate. J Neurosci 22:9134–9141
82. Meshul CK, Kamel D, Moore C, Kay TS, Krentz L (2002) Nicotine alters striatal glutamate function and decreases the apomorphine-induced contralateral rotations in 6-OHDA lesioned rats. Exp Neurol 175:257–274
83. Bland ST, Gonzales RA, Schallert T (1999) Movement-related glutamate levels in rat hippocampus, striatum, and sensorimotor cortex. Neurosci Lett 277:119–122
84. Wolf ME, Xue CJ, Li Y, Wavak D (2000) Amphetamine increases glutamate efflux in the rat ventral tegmental area by a mechanism involving glutamate transporters and reactive oxygen species. J Neurochem 75: 1634–1644
85. Meshul CK, Emre N, Nakamura CM, Allen C, Donohue MK, Buckman JF (1999) Time-dependent changes in striatal glutamate synapses following a 6-hydroxydopamine lesion. Neuroscience 88:1–16
86. Meshul CK, Allen C (2000) Haloperidol reverses the changes in striatal glutamatergic immunolabeling following a 6-OHDA lesion. Synapse 36:129–142
87. Cenci MA, Bjorklund. A (1993) Transection of corticostriatal afferents reduces amphetamine- and apomorphine-induced striatal Fos expression and turning behaviour in unilaterally 6-hydroxydopamine-lesioned rats. Eur J Neurosci 5:1062–1072
88. Holschneider DP, Yang J, Guo Y, Maarek JM (2007) Reorganization of functional brain maps after exercise training: importance of cerebellar–thalamic–cortical pathway. Brain Res 1184:96–107
89. Meeusen R, Smolders I, Sarre S, De Meirleir K, Keizer H, Serneels M, Ebinger G, Michotte Y (1997) Endurance training effects on neurotransmitter release in rat striatum: an in vivo microdialysis study. Acta Physiol Scand 159:335–341
90. Farmer J, Zhao X, van Praag H, Wodtke K, Gage FH, Christie BR (2004) Effects of voluntary exercise on synaptic plasticity and gene expression in the dentate gyrus of adult male Sprague-Dawley rats in vivo. Neuroscience 124:71–79
91. Raju DV, Ahern TH, Shah DJ, Wright TM, Standaert DG, Hall RA, Smith Y (2008) Differential synaptic plasticity of the corticostriatal and thalamostriatal systems in an MPTP-treated monkey model of parkinsonism. Eur J Neurosci 27:1647–1658
92. Fremeau RT Jr, Troyer MD, Pahner I, Nygaard GO, Tran CH, Reimer RJ, Belloccho EE, Fortin D, Storm-Mathisen J, Edwards RH (2001) The expression of vesicular glutamate transporters defines two classes of excitatory synapse. Neuron 31:247–260
93. Dietrich MO, Mantese CE, Porciuncula LO, Ghisleni G, Vinade L, Souza DO, Portela LV (2005) Exercise affects glutamate receptors in postsynaptic densities from cortical mice brain. Brain Res 1065:20–25
94. Phend KD, Weinberg RJ, Rustioni A (1992) Techniques to optimize post-embedding single and double staining for amino acid neurotransmitters. J Histochem Cytochem 40:1011–1020
95. Kaneko T, Fujiyama F, Hioki H (2002) Immunohistochemical localization of candidates for vesicular glutamate transporters in the rat brain. J Comp Neurol 444: 39–62
96. Meredith GE, Kang UJ (2006) Behavioral models of Parkinson's disease in rodents: a

97. Laviola G, Hannan AJ, Macri S, Solinas M, Jaber M (2008) Effects of enriched environment on animal models of neurodegenerative diseases and psychiatric disorders. Neurobiol Dis 31:159–168
98. Mora F, Segovia G, del Arco A (2007) Aging, plasticity, and environmental enrichment: structural changes and neurotransmitter dynamics in several areas of the brain. Br Res Rev 55:78–88
99. Mora F, Segovia G, del Arco A (2008) Glutamate–dopamine–GABA interactions in the aging basal ganglia. Br Res Rev 58: 340–353
100. Bezard E, Dovero S, Belin D, Duconger S, Jackson-Lewis V, Przedborski S, Piazza PV, Gross CE, Jaber M (2003) Enriched environment confers resistance to 1-methyl-4-phenyl 1,2,3,6-tetrahydropyridine and cocaine: involvement of dopamine transporter and trophic factors. J Neurosci 23:10999–11007
101. Fleming SM, Salcedo J, Fernagut PO, Rockenstein E, Maslia E, Levine MS, Chesselet MF (2004) Early and progressive sensorimotor anomalies in mice overexpressing wild-type human α-synuclein. Neurobio Dis 24:9434–9440
102. Fredriksson A, Danysz W, Quack G, Archer T (2001) Co-administration of memantine with sub/suprathreshold doses of L-Dopa restores motor behavior of MPTP-treated mice. J Neural Transmission 108:1435–1463
103. Sundström E, Fredriksson A, Archer T (1990) Chronic neurochemical and behavioral changes in MPTP-lesioned C57Bl/6 mice: a model for Parkinson's disease. Brain Res 528:181–188
104. Jackson-Lewis V, Jakowec M, Burke RE, Przedborski S (1995) Time course and morphology of dopaminergic neuronal death caused by the neurotoxin 1-methyl-4-phenyl-1,2,3,6-tetrahydropyridine. Neurodegen 4:257–269
105. Unal-Çevik I, Kilinç M, Gürsoy-Özdemir Y, Gurer G,, Dalkara T (2004) Loss of NeuN immunoreactivity after cerebral ischemia does not indicate neuronal cell loss: a cautionary note. Brain Res 1015:169–174
106. Kamens HM, Crabbe JC (2007) The parallel rod floor test: a measure of ataxia in mice. Nat Protocols 2:277–281
107. Kozell LB, Meshul CK (2001) The effects of acute or repeated cocaine administration on nerve terminal glutamate within the rat mesolimbic system. Neuroscience 106: 15–25
108. Bezard E, Dovero S, Bioulac B, Gross CE (1997) Kinetics of nigral degeneration in a chronic model of MPTP-treated mice. Neurosci Lett 234:47–50
109. Ricaurte GA, Irwin I, Forno LS, DeLanney LE, Langston E, Langston JW (1987) Aging and 1-methyl-4-phenyl-1,2,3,6-tetrahydropyridine-induced degeneration of dopaminergic neurons in the substantia nigra. Br Res 403:43–51
110. Fornai F, Schluter OM, Lenzi P, Gesi M, Ruffoli R, Ferrucci M, Lazzeri G, Busceti CL, Pontarelli F, Battaglia G, Pellegrini A, Nicoletti F, Ruggieri S, Paparelli A, Sudhof TC (2004) Parkinson-like syndrome induced by continuous MPTP infusion: convergent roles of the ubiquitin–proteasome system and α-synuclein. Proc Natl Acad Sci (USA) 102:3413–3418
111. Novikova L, Garris BL, Garris DR, Lau YS (2006) Early signs of neuronal apoptosis in the substantia nigra pars compacta of the progressive neurodegenerative mouse 1-methyl-4-phenyl-1,2,3,6-tetrahydropyridine/probenecid model of Parkinson's disease. Cell Neurosci 140: 67–76
112. Petroske E, Meredith GE, Callen S, Totterdell S, Lau YS (2001) Mouse model of Parkinsonism: a comparison between subacute MPTP and chronic MPTP/probenecid treatment. Neuroscience 106:589–601
113. Amende I, Kale A, McCue S, Glazier S, Morgan JP, Hampton TG (2005) Gait dynamics in mouse models of Parkinson's disease and Huntington's disease. J Neuroeng Rehabil 2:20–33
114. Nutt JG, Hammerstad JP, Gancher ST (1992) Parkinson's disease: 100 maxims. Edward Arnold, London
115. Fahn S, Elton RL and UPDRS Development Committee (1987) Recent developments in Parkinson's disease. Macmillan, Florham Park, NJ, pp 153–163
116. Pascual-Leone A, Amedi A, Fregni F, Merabet LB (2005) The plastic human brain cortex. Ann Rev Neurosci 28: 377–401

new look at an old problem. Movement Dis 21:1595–1606

Chapter 4

Spatial Learning and Memory in Animal Models and Humans

Gwendolen E. Haley and Jacob Raber

Abstract

Spatial learning and memory requiring navigation has been widely assessed as a part of traditional rodent cognitive testing. Significantly fewer studies have examined spatial learning and memory requiring navigation in nonhuman primates and humans. While rodent spatial tasks utilize navigation and an allocentric frame of reference, nonhuman primate and human spatial tasks often utilize an egocentric frame of reference, lacking a navigational component. Due to this difference, cross species comparisons cannot be easily made. In rodent models, both spatial learning and memory and object recognition tasks requiring navigation are used to assess hippocampus-dependent learning and memory. Furthermore, addition of a spatial component to the traditional object recognition task in the mouse and human model has increased the sensitivity of the task to detect cognitive changes. Based on hippocampus-dependent cognitive tests used in our mouse studies, we developed spatial learning and memory tests requiring navigation for nonhuman primates (Spatial Foodport Maze) and humans (Memory Island) as well as the object recognition test Novel Image Novel Location (NINL) for humans. Here, we discuss these translational cognitive tests that are being used to bridge the gap between object recognition and spatial learning and memory tasks across species.

1. Introduction

Cognitive tests in animals have been developed to examine performance between different experimental conditions and to explore the neurobiological processes of memory. Ideally, findings from animal tests would translate to understanding of cognitive problems and pathologies that arise in the human population. A variety of sensitive cognitive tests, such as hippocampus-dependent spatial learning and memory tests requiring navigation which are a hallmark of sensitive rodent cognitive test batteries and routinely used in animal behavioral laboratories, have mostly not been

J. Raber (ed.), *Animal Models of Behavioral Analysis*, Neuromethods 50,
DOI 10.1007/978-1-60761-883-6_4, © Springer Science+Business Media, LLC 2011

translated for developing sensitive cognitive tests in humans. Such translational tests might be sensitive to detect cognitive changes in nonhuman primates and humans as well. They are also required for cross-species comparison of this type of learning memory and the underlying neurobiological mechanisms under physiological and pathological conditions.

1.1. Spatial Memory in Rodents

The most widely used animals for assessing spatial learning and memory requiring navigation are rodents. Rodents are relatively small and their environment is easily manipulated and controlled. There is one important difference between spatial learning and memory tests generally used for rodents, nonhuman primates, and humans. The rodent spatial learning and memory navigational tests are administered with an allocentric, or world-centered, frame of reference in which animals are required to use navigation to orient within a large space. In contrast, most nonhuman primate and human spatial learning and memory tests do not assess the subject's ability to navigate through space. Instead, these tests tend to have all the information within one field of view, giving the subject an egocentric, or a self-centered, frame of reference. Two well-known spatial learning and memory tests requiring navigation in rodents are the water maze and the Barnes maze.

Perhaps the most well known and used rodent spatial learning and memory test requiring navigation is the water maze, developed by Richard Morris (1). In the water maze, rodents are placed in a large tub containing an escape platform beneath opaque water. Rodents are natural swimmers and will actively search for the platform to escape swimming and the maze. There are various versions of water maze designs. In some designs, the animals are first trained to locate a visible platform and are subsequently trained to locate one or more hidden platform locations. In some water maze designs, "probe" trials, trials in which the platform is removed, are used to assess spatial memory retention. The water maze has been extensively used for cognitive testing of rodents (for examples *see* (2–6)).

While not as popular as the water maze, the Barnes maze (7) was developed chronologically before the water maze to assess spatial memory. The Barnes maze consists of a circular disk with holes around the perimeter, and one of the holes leads to an escape box. Rodents are placed on top of the disk to start the trial. Rodents do not like to be in an open bright-lit field and are trained to search for the escape box. During hidden escape training, like during hidden platform training in the water maze, fixed visual cues are present in the room to allow the animal to generate a spatial map to locate the escape box. In the Barnes maze analysis, different search strategies can be distinguished. There are spatial, serial, and random search strategies (8). When an animal has learned the task and is able to make a spatial map, a "spatial search" pattern might be used; the mouse searches within one hole of the target location. In contrast, if the animal has not learned the target location (yet) but has learned the task it might use a "serial search;" the mouse searches each location in a consecutive order (either direction) until the target location is found. A third search strategy an animal can use is a random search; neither a spatial nor a serial search is being used and there are no clear search patterns detectable. While in principle the Barnes maze is useful to assess spatial learning and memory (7, 9–11), a potential concern is motivation to perform the task. Unlike the water maze in which the animal is motivated to locate the platform and escape the water, motivational issues can arise, even when bright light, loud noise, and forced air are used to motivate the animal to search for the escape tunnel.

Video tracking software can be used to record and analyze performance in both the water maze and Barnes maze. In the water maze, distance traveled, latency, cumulative distance to the target, and swim speed are generally used as outcome measures. In the Barnes maze, search strategies, as described above, and numbers of errors (searching a hole not containing the escape tunnel) are usually used as outcome measures.

In addition to the water maze and Barnes maze, spatial learning and memory can also be assessed using object recognition tests with a spatial component, which increases sensitivity compared to more traditional tasks. A number of variations of object recognition tests have been used in laboratories for over 20 years. The first one was reported in 1988 (12), although it was known that rodents explore novel objects more than familiar objects from as early as 1950. In general, the design of the object recognition test involves habituation to the testing arena without objects, habituation to the testing arena with the objects, and testing in the arena containing either a novel object replacing a familiar one or containing a familiar object moved to a novel location. The addition of a spatial component challenges the animal to recall not only the object, but also its location within a testing arena. Rodents have an innate preference to explore novel objects and exploring objects moved to a novel location (5). Adding a spatial component, it increases the sensitivity of the object recognition test (3, 13–15).

Example of a mouse object recognition test design: First, mice are habituated to an open field (40.6 cm × 40.6 cm) for 5 min for 3 consecutive days. On the fourth consecutive day, three different plastic toys are placed in the open field, consistently in the same arrangement during three consecutive trials, and the animals are allowed to explore the toys. Each trial lasts 10 min with a 5-min inter-trial interval. After these three training trials, mice are tested in two consecutive trials of similar time and containing the same inter-trial interval. Testing involves moving one of the familiar objects to a different location (trial 4) or replacing one of the objects with a new object (trial 5). Objects are used only once and replicas of the objects are used in subsequent trials to eliminate potential remaining odors or other potential markings on the objects serving as additional cues to the animals. Each trial can be recorded and analyzed with multiple body point or whole body video-tracking software (16). Outcome measures are the total time spent exploring all objects and the percent of time spent exploring the individual objects in the different trials.

Lesion studies in rodents have demonstrated that spatial learning and memory as assessed in the water maze, Barnes maze, and in some versions of the object recognition tasks are hippocampus dependent (17), yet reliant on different areas of the structure (18). Notably, hippocampus-dependent tasks are sensitive to the effect of environmental toxins (3, 4) and in utero exposure to drugs of abuse such as methamphetamine (19). The hippocampus is also one of the first brain areas to decline in normal aging (20, 21) and one of the first brain areas affected by neurological diseases such as Alzheimer's disease (AD) (22, 23).

1.2. Nonhuman Primates

While rodent tests such as the water maze utilize navigation, nonhuman primate tests traditionally have not. For example, nonhuman primate object recognition tasks have been developed including the Wisconsin General Testing Apparatus (WGTA), first introduced into the literature by Harlow in 1959 (24). More recently, the WGTA has been advanced to computerized versions. Although the WGTA was developed with the safety of the experimenter as well as the monkey involved in the testing, it does not allow for much movement within space. The WGTA continues to provide valuable data on object recognition tests, particularly information on how well nonhuman primates can discriminate between two objects.

Until recently, spatial learning and memory was tested in the nonhuman primate in the WGTA or via similar, computerized tests. An example of a computerized nonhuman primate test is the Delayed Response task. In this task, a red block is presented in a specific location (i.e., the right side of the tray/screen). Following a delay period, two blocks are presented and the animal's task is to choose the correct location, based on the location of the first block. While this has traditionally been the task to assess spatial memory in nonhuman primates, recent evidence indicates that it is reliant on different neural pathways compared to the rodent spatial mazes requiring navigation. Human studies suggest that tasks without navigation recruit different neuronal networks than those with navigation. Therefore, direct inferences about the two types of tests cannot be made (25, 26). As a result, while nonhuman primate studies have yielded valuable information about learning and memory not requiring navigation, more specialized tests are required to allow for comparison across different species.

The most popular object recognition test for nonhuman primates is the delayed matching to sample (DMS) or delayed nonmatching to sample (DNMS), although DNMS might be reliant on alternate neural pathways than traditional object recognition tasks (27–29). Before testing, an object or computerized image is presented to the animal. Following a delay, the animal is then presented with two objects or computerized images and is allowed to choose the matching (DMS) or the nonmatching (DNMS) sample. Outcome measures include the number of correct answers. The WGTA has had an integral role in understanding object recognition in the nonhuman primate and different variations are being used to assess object recognition.

Recently, a few nonhuman primate tests have been developed to assess spatial learning and memory requiring navigation. In 1997, Rapp et al. developed a nonhuman primate test resembling the rodent radial arm maze (30). In this test, the animals, tethered to the floor, search food ports on the floor in an 8-point star formation guided by visual cues on the wall. Nonhuman primates were able to learn the task, evident in the recollection of the baited port. Another spatial maze was developed by Hampton et al. (31). In this test, the animals are tethered in an open room and allowed to search several "forage" sites. As a decrease in test performance is reported following hippocampal damage, this test seems hippocampus dependent. While these two tests are a vast improvement compared to previous spatial tests for nonhuman primates, the animals are tethered to sedentary objects and are thus limiting animals from freely moving.

Two spatial learning and memory tests requiring navigation for nonhuman primates have recently been developed. One foraging maze was developed for squirrel monkeys (32). In this particular task, food ports are mounted to a large cage. The ports are far enough away from each other that the animals have to move from port to port. Multiple food ports are baited with food rewards, and the animals consistently can find the reward in the baited ports. The squirrel monkeys decreased their number of errors (searching ports not baited) during repeated trials, supporting that they learned the task.

Another spatial learning and memory navigational test, the Spatial Foodport Maze, was developed for freely behaving rhesus macaques and is somewhat similar to the rodent Barnes maze. In this test, the animals are placed in an open room containing food ports mounted to the wall. Nonhuman primates are trained to retrieve a food reward from a single baited port (33). Recently, we reported that during the first half of the trials, old female rhesus macaques mostly use a serial search pattern to find the baited port. However, during the second half of the trials, the animals mostly

Table 4.1
Percentage of each search strategy used for the nonhuman primate Spatial Foodport Maze

	Percentage of spatial search	Percentage of serial search	Percentage of random search
First 1/2 of initial trials	36.6 ± 6.0	54.4 ± 5.6	9.1 ± 2.0
Last 1/2 of initial trials	62.0 ± 6.2[a,b]	34.3 ± 6.1	3.6 ± 1.2
First 1/2 of shift trials	28.9 ± 7.1	60.5 ± 6.9[b]	10.3 ± 2.9
Last 1/2 of shift trials	51.4 ± 5.9[a]	44.7 ± 6.1	5.5 ± 1.5

[a]Indicates a significant increase between the first half and the second half in a search strategy category.
[b]Indicates a significantly higher percentage between spatial and serial search strategy ($P < 0.05$).

use a spatial search strategy, indicating they learned the location of the baited port (see **Table 4.1**). The use of a spatial strategy to find the target port is similar to what has been reported for the use of a spatial search strategy in the rodent Barnes maze (8). This maze requires navigation in a larger testing arena, resembling the rodent spatial memory tasks more than traditional WGTA spatial memory tasks and does not require the animal to be tethered. Therefore, this kind of test might allow better comparisons between the two test paradigms than previous spatial tasks used in nonhuman primates.

1.3. Human Testing

Traditional tests of spatial memory in humans are the Spatial Span Forward and Spatial Span Backward tests (34). These tasks use a series of ten blocks glued to a plastic tray on a tabletop. The experimenter "taps" the blocks in a specific order, increasing the number of blocks tapped in each trial. The subject is then asked to tap the blocks in the same order. The complexity increases until the person is unable to recall the order. Although these tests are challenging, they have a different design compared to traditionally spatial learning and memory tasks used in rodents.

Similar to nonhuman primate studies, most human neuropsychological testing does not involve navigation. While this issue is partly due to a lack of space, recent technological advances allow tasks with navigation. Computerized virtual reality training modules have been developed for this purpose. Such training modules are utilized for teaching skills such as driving or flying an airplane. In the clinical setting, computerized modules have been created to desensitize people, especially with post-traumatic stress syndrome and phobias (35–38). In addition, virtual reality mazes can be useful to test spatial learning and memory requiring navigation in humans. For instance, a virtual water maze for humans based on the rodent water maze has been developed (39). Results generated using the virtual water maze for humans does parallel results generated using the rodent water maze (39, 40). It should be kept in mind that in the virtual water maze there is a different motivation than the one in the water maze used in rodents.

Neurobiological networks of learning and memory in the human brain are often studied through case studies. Perhaps the most famous neuroscience patient is H.M., notable from the removal of his medial temporal lobe as a result of treatment for epilepsy (41). Although the removal of the brain tissue did alleviate the seizures, H.M. had major cognitive deficits. Interestingly, not all of his cognitive capabilities were affected; rather, it appeared that it was isolated to his declarative memory. Declarative memory consists of episodic and semantic memory or knowledge of "facts." These types of memory include personal facts, history (cultural and personal), objects, and words. Episodic memory is the recollection of events that occur in a given time period. On the other hand, semantic memory is general knowledge accrued over time without being related to specific experiences. It is important to realize that these memories overlap and share common information. Moreover, each person has a unique episodic and semantic memory. Therefore, object recognition tasks were developed to standardize the cognitive paradigm.

Based on results from human studies including H.M. observations, declarative memory is hypothesized to be dependent on the hippocampus, but some debate remains on the way to effectively test episodic and semantic memory. Supporting the results from studying H.M., animal models of lesions of the hippocampus have demonstrated that disruption of the hippocampus can alter performance on object recognition tests. Furthermore, performance on object recognition tests is dependent on the extent and location of the lesion (42, 43).

Object recognition tests for human testing have undergone an evolution from 3D object recognition to simply visual object recognition, and participants are asked to generate a "snapshot" of an image. Object recognition tests are different from another popular test for human neuropsychological paradigms, the facial recognition test. Unlike the hippocampus-dependent object recognition, facial recognition tests are targeting a different area of the brain, the fusiform gyrus, which is in close proximity to but not encompassed within the hippocampus structure (44). Therefore, facial recognition tests and object recognition are separated into different tests in order to identify function of distinct areas of the brain.

To address which neural pathways are involved in performing different human cognitive tasks, functional magnetic resonance imaging (fMRI) can be used. Thus far, imaging studies in humans have corroborated the findings from the H.M. case study and indicate that object recognition tests differentially affect the hippocampus (45). Moreover, human imaging studies support that the neural pathways involved in facial recognition are different from those involved in object recognition (46). fMRI is not only a valuable to study cognition in humans but also to study cognition in nonhuman primates (47).

1.4. Sex Differences

One important consideration in studying spatial learning and memory is the sensitivity to gonadal steroids; the hippocampus is particularly sensitive to such effects. Not only have studies demonstrated that there is a morphological difference between the male and female hippocampus (48–50), but there is also a difference at the cellular electrophysiological level; males show a greater excitability than females (51). Furthermore, synaptic plasticity, including LTP, is different between the sexes, with males having greater synaptic plasticity than females (52). As such, it is not surprising that effects of sex are observed in object recognition and spatial learning and memory requiring navigation in animals as well as humans (53–57), further supported by imaging studies (58, 59). For instance, castration decreases spatial memory in wild-type mice (60), but androgen treatment recovers performance in males and increases performance in old female mice (61, 62). Furthermore, in mice expressing human apoE4, performance on spatial memory tests is improved with androgen treatment (testosterone or dihydrotestosterone) (63) and with selective androgen receptor modulators (SARMs) treatment (64). Increased efforts are warranted to understand the influence of sex hormones on spatial learning and memory requiring navigation and object recognition across species.

1.5. Implications for Translational Research

Traditionally, translational efforts of spatial learning and memory requiring navigation have been relatively unsuccessful, mostly due to the nonhomologous neural systems tested. Because of the discrepancies in the design of spatial learning and memory tests between rodents, nonhuman primates, and humans, few parallels have been made across species. Not surprisingly, most data on spatial learning and memory requiring navigation have been generated in rodent models. However, the addition of new spatial learning and memory tests for nonhuman primates and humans in recent years will allow more comparisons with rodent models. This in turn helps our understanding of spatial learning and memory requiring navigation across species in health and disease.

In **Section 2**, we describe two new test designs that enable cross-species comparison of object recognition and spatial learning and memory requiring navigation between rodents, nonhuman primate, and humans. Protocols for Memory Island, a human spatial learning and memory test, and Novel Image Novel location, a human object recognition test, will be highlighted, described, and discussed in detail.

2. Methods

In any cognitive paradigm, three phases of testing should be considered; habituation, training, and testing. Habituation is necessary to familiarize the subject with the testing arena, since the testing arena is usually outside the home environment. A training phase allows the subject to learn the task. Finally testing determines how well the subject learned the task and the efficacy of the cognitive function of interest. These phases ensure that the outcome of the test is appropriate and allows for interpretable results.

2.1. Habituation

Habituation, a process of learning in which a subject stops responding when repeatedly presented with a stimulus (65), is an important part of cognitive testing paradigms. For example, testing paradigms place subjects in unfamiliar situations and if as a result of this novelty they have increased measures of anxiety that might increase performance on the test. Multiple studies suggest stress-inducing anxiety or anxiety-like behaviors can decrease cognitive function. Therefore, it is important for the subjects to not feel overly anxious about the task and to be relatively relaxed when tested. The necessary time for habituation ranges depending on the paradigm and subject, and the range should be predetermined experimentally for each experiment.

2.2. Training

The training phase of a testing paradigm should allow sufficient time for subjects to learn the task. There can be a standard number of trials or subjects can be trained to perform to a certain criterion level. Once the subject has demonstrated learning of the task, testing performance can be assessed.

2.3. Testing

The testing phase involves the knowledge previously acquired during the training phase to determine how well the test was learned. This is the key phase that is often reported in literature, but the foundation of the habituation and training phases must be included and described for proper interpretation of cognitive test performance.

2.4. New Human Tasks for Cross-species Comparison

2.4.1. Spatial Learning and Memory (Memory Island)

Spatial memory tests for humans similar to those used in rodent models have been few in number due to the differences in the size requirements of the testing environments. Manipulation of an environment for 30 g mouse is obviously much easier than manipulation of an environment for a much larger and heavier 60 kg human. For example, if a test were developed with similar dimensions as the mouse water maze (tub diameter: 141 cm) for a human, the tub of water would need to be 3.2 km in diameter, which is not practical! An alternative is using virtual reality computer technology, which has revolutionized the ability to develop spatial learning and memory tests for humans using navigation without requiring a large arena. The virtual water maze for humans holds translational capabilities, as does the more entertaining and immersive spatial learning and memory test in humans involving navigation of a virtual island environment, Memory Island (13, 66). A key feature of Memory Island, seen in **Fig. 4.1**, is the various cues and distinct landmarks that offer the subject orientation within the island. A Memory Island testing paradigm is described below:

1. Participants are placed in front of a 19-inch computer monitor with a stereo speaker and subwoofer.

Fig. 4.1. View of Memory Island through screen shots of the virtual reality maze. Image **a** is from the start point, directed toward a flag marking the location of one of the target items during a visible trial, observed next to a lighthouse. *Panel* **b** is an image of the glass structure, a unique structure in quadrant 1. Image **c** shows the *arrow* that will appear if subjects are having difficulty finding the visible or hidden target. The *arrow* appears 2 min after the start of the trial and points in the direction of the target. Image **d** is a view of quadrant 3 from the starting location.

2. Using a joystick, participants are trained to navigate through the virtual world simulating an island environment of 347 × 287 m² and for analysis composed of four quadrants containing a unique target item (**Fig. 4.2**). A computer-generated coordinate file is used to calculate the before-described outcome measures.

3. In the first four trials, subjects are trained to navigate to clearly visible target items, one in each quadrant. In these "visible" trials, each target item is marked with a flag that is clearly visible from the distance. Starting orientation is varied for the four trials but kept the same across all participants. If the target is not located within 2 min, a directional arrow appears on the screen to guide the participant to the target (**Fig. 4.1c**).

4. Once all four visible target items have been located, the subject is trained to locate a hidden target (no flag).

Fig. 4.2. Screen shots of the target items in Memory Island: image **a**, black moving statue, quadrant 1; image **b**, seal, quadrant 2; image **c**, seagull, quadrant 3; image **d**, water fountain, quadrant 4.

5. The starting orientation is varied for each trial but kept the same for all participants. If the target has not been located within 2 min, a directional arrow appears on the screen to guide the participant to the target.

6. After a delay (5–90 min) the subject is asked to locate the item in a "probe" trial (no target or flag).

2.4.1.1. Data Analysis

For each trial, the computer software program records a time-stamped coordinate file. The parameters that are recorded during acquisition allow for calculation of total distance moved (virtual feet), velocity (virtual feet per second), latency (seconds), cumulative distance to the target (virtual feet), and percentage time spent in the target quadrant. If the subject locates the target item within 2 min the trial is deemed a success.

For statistical analyses of outcome measures of Memory Island, ANOVAs, ANCOVAs, and Mann–Whitney U tests can be used (13). Performance on Memory Island can also be used to assess potential correlations with performance measures on other cognitive tests using Pearson or Spearman correlations.

2.4.2. Object Recognition (Novel Image Novel Location)

Based on the Novel Object Novel Location test developed in the mice, a similar Novel Image Novel Location (NINL) test has been developed for human testing, which is described in detail here:

1. Three sets of 12 panels are shown, each containing 3 images. The two sets of three-image panels are similar in complexity, but different in content and arrangement within the four quadrants of the panel.

2. Before beginning the training phase, the study participants are given an example of the test (habituation).

3. For the training phase of the study, the first set of 12 panels is presented. Each panel is presented for 8 s. The study participant is asked to memorize the panels. Examples of the panel layout for no change, novel image, and novel location are shown in **Fig. 4.3**.

4. In the testing phase of the study, the second set is presented. The testing set contains panels either identical or slightly changed compared to the reference set by containing one novel image or one image in a novel location. The subject is asked to identify if the panel is identical to or changed compared to the original panel. Then, if a change is indicated, the subject is asked to identify the type of change and where the change occurred. In this set, four panels are identical to the reference set, four panels have one image replaced with a novel image, and four panels have identical images with one image moved to a novel location.

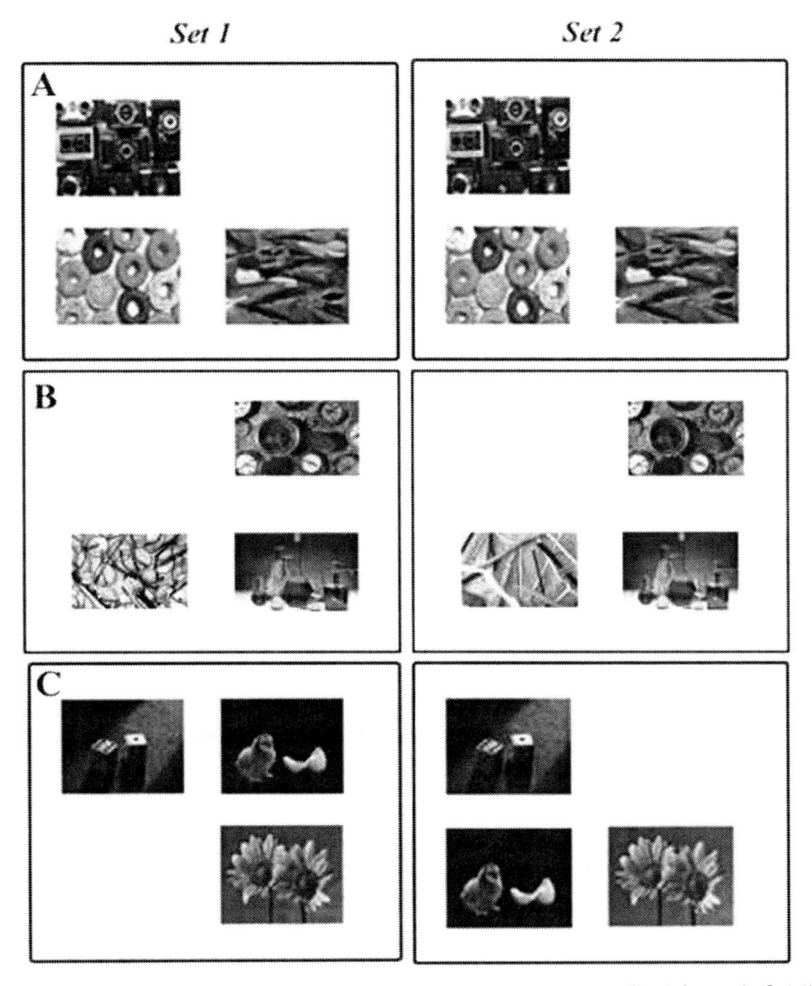

Fig. 4.3. Examples of the panels used in the Novel Image Novel Location (NINL) test. The *left panels* (Set 1) are examples from the reference set, the set that the subjects are presented first and asked to memorize. The *right panels* (Set 2) are examples from the test set. Panels in row **a** are an example of a set of "no change" panels. *Panels* in row **b** are an example of "novel image"' panels. *Panels* in row **c** are an example of "novel location."

5. Following a delay period (5–90 min in length) and without seeing the training set again, the study participants are presented with a third set of 12 panels. Once again, the subjects are asked to identify the same criteria as in the first testing set, making reference back to the original reference set. The third set is identical to the second set but the order of the panels is rearranged.

2.4.2.1. Data Analysis

To score NINL test performance, a point system is used. Score sheets record responses of the subjects, i.e., no change, novel image, or novel location. In order to score a point, the subject has to correctly identify if there was a change, the type of change (novel image or novel location), and where the change occurred (quadrant of panel). A maximum total of 12 points is possible for both the immediate (first testing set) and delayed (second testing set). In addition to the total score, subscores are analyzed with a total of 4 points each for the no change, novel image, and novel location conditions.

NINL scores can be analyzed with one-way ANOVAs and correlations between NINL scores and subscores and other behavioral measures.

3. Notes

3.1. Memory Island

1. Memory Island is a computerized task. Those individuals who have not been previously exposed to computerized graphics may be wary of the graphics or intimidated by the computer-based task. Some people with vertigo type symptoms or those easy to motion sickness might not be willing to perform the task. If the researcher suggests to the subject to watch the distant scenery instead of the immediate path, it can decrease potential unpleasant emotions. In our experience, these potential issues hardly occur in adult, elderly, or children.

2. Although children appear to be very efficient at using the joystick, older generations, or those without previous exposure to using joysticks, might not be as predisposed to use the joystick effectively. Thorough explanation of the use of the joystick helps if this is a concern. Also, aiding in joystick direction during the first visible trial can increase the confidence and performance of the subject. Therefore, in some studies it is important to include data on pertinent computer use.

3. As time to complete Memory Island is an essential performance measure, it is important to make sure there are no distractions for the subjects. Removing all external stimuli will enable the subject to focus on the test and provide the best opportunity for successful completion of the test.

4. Learning curves can be made using any of the parameters measured. For instance, if learning occurs over the four hidden trials, total distance to target should be the highest in the first hidden trial and decrease with each consecutive hidden trial (**Fig. 4.4**). The learning curve can then be compared across experimental groups. Percentage time spent in each quadrant is a useful tool to establish a subject's spatial memory retention.

5. This test is sensitive to the effects of sex (66).

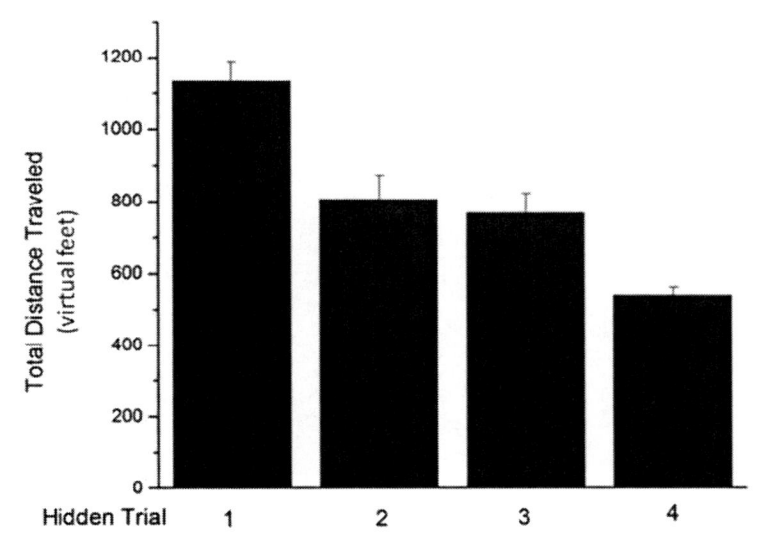

Fig. 4.4. An example of a learning curve on Memory Island using the Total Distance Traveled as the outcome measure. In the hidden trials, the distance traveled (measured in virtual feet) is the greatest in the first hidden trial and decreases with consecutive trials, indicative of learning. Data represented in the graph is from nondemented elderly individuals that successfully located the target item in all four hidden trials.

6. Memory Island has been used to test spatial memory among young children to elderly individuals (13, 66, 67) and can be used to assess spatial memory across ages.

7. Although some people might want to "explore" the island, time to complete the trial is an outcome measure. As such, it is important to emphasize to the subjects that the goal is to locate the target items as quickly as possible.

8. A trial is defined as a "successful trial" if the target is located within 2 min. However, with some individuals, especially on some of the hidden trials, the target item might not be located within the 2-min time frame. Therefore, the percentage of successful trial can be used as an outcome measure.

9. With a computer-based task, it is feasible to use it in a home computer. This will decrease the time in a doctor office and might allow the subjects to be assessed at their convenience and more frequently.

3.2. NINL

1. When administering the NINL task, it is best to use nonemotional images. Images that are familiar but do not elicit a strong emotional response are the best.

2. Like Memory Island, NINL task has been successfully used for testing of humans 7–95 years (13, 66).

3. The NINL task can predict dropouts in a longitudinal study of healthy nondemented elderly, suggesting the NINL task

Table 4.2
Novel Image, Novel Location (NINL) scores in elderly participating in a longitudinal study

	Baseline NINL	6 month NINL	18 month NINL
Non-ε4 finishers	18.40 ± 0.52	16.82 ± 0.56	16.65 ± 0.58
ε4 finishers	18.93 ± 0.90	19.47 ± 0.98	19.13 ± 1.17
Non-ε4 dropouts	18.28 ± 1.25	16.60 ± 1.20	
ε4 dropouts	14.27 ± 1.28[a]	13.38 ± 1.42[a]	

[a] Indicates ε4 dropouts scored significantly lower on the NINL test ($P < 0.05$). Data represented in the table comes from the same study of nondemented elderly seen in **Fig. 4.4**.

could be a measure to identify cognitive decline prior to any clinical signs (*see* **Table 4.2**) (68).

3.3. Summary and Implications for Cross-species Comparison

Virtual reality tasks have proven to be a valuable tool in human neuropsychological treatment. Now, virtual reality tasks have been developed to assess spatial memory in human and could become an added dimension in human neuropsychological testing. Cross-species comparison between the rodent water maze, the rodent Barnes maze, the nonhuman primate Spatial Foodport Maze, and the human Memory Island can be made as they all use navigation. As the Spatial Foodport Maze and Memory Island are implemented into cognitive battery testing for nonhuman primate and human, respectively, the ability to have cross-species comparison for spatial learning and memory tasks will increase. Potentially, these tests could provide insight into changes in brain function without aberrant brain lesions.

Traditional object recognition tests in human have allowed for cross-species comparison. However, the advance of the object recognition test described here, with the added novel location dimension, will allow a more specific comparison across species. Although a comparable study design has yet to be developed in the nonhuman primate, rodent and human studies have utilized the comparison across test paradigms. Indeed, findings from the human version of the task parallel findings from the mouse model (5, 66). Moreover, reproducibility of the human NINL test was determined to be as strong as established facial recognition tests. With these new tasks, cross-species comparisons will be more widely available and the translational capabilities of spatial learning and memory tasks will greatly increase.

Acknowledgments

This work was funded by EMF AG-NS-0201, the Medical Research Foundation of Oregon, a pilot project of the Layton Center for Aging and Alzheimer's disease, PHS grants 5 M01 RR000334 and NIA T32-AG023477.

References

1. Morris R (1984) Developments of a water-maze procedure for studying spatial learning in the rat. J Neurosci Methods 11(1): 47–60
2. Lukoyanov NV et al (1999) Effects of age and sex on the water maze performance and hippocampal cholinergic fibers in rats. Neurosci Lett 269(3):141–144
3. Villasana L et al (2006) Sex- and APOE isoform-dependent effects of radiation on cognitive function. Radiat Res 166(6): 883–891
4. Villasana L, Rosenberg J, Raber J (2009) Sex-dependent effects of (56)Fe irradiation on contextual fear conditioning in C57BL/6 J mice. Hippocampus 20(1): 19–23
5. Duvoisin RM et al (2005) Increased measures of anxiety and weight gain in mice lacking the group III metabotropic glutamate receptor mGluR8. Eur J Neurosci 22(2):425–436
6. Mendez IA et al (2008) Long-term effects of prior cocaine exposure on Morris water maze performance. Neurobiol Learn Mem 89(2):185–191
7. Barnes CA (1979) Memory deficits associated with senescence: a neurophysiological and behavioral study in the rat. J Comp Physiol Psychol 93(1):74–104
8. Raber J et al (2004) Radiation-induced cognitive impairments are associated with changes in indicators of hippocampal neurogenesis. Radiat Res 162:39–47
9. Inman-Wood SL et al (2000) Effects of prenatal cocaine on Morris and Barnes maze tests of spatial learning and memory in the offspring of C57BL/6 J mice. Neurotoxicol Teratol 22(4):547–557
10. Barnes CA (1988) Spatial learning and memory processes: the search for their neurobiological mechanisms in the rat. Trends Neurosci 11(4):163–169
11. Bach ME et al (1995) Impairment of spatial but not contextual memory in CaMKII mutant mice with a selective loss of hippocampal LTP in the range of the theta frequency. Cell 81(6):905–915
12. Ennaceur A, Delacour J (1988) A new one-trial test for neurobiological studies of memory in rats. 1: behavioral data. Behav Brain Res 31(1):47–59
13. Berteau-Pavy F, Park B, Raber J (2007) Effects of sex and APOE ε4 on object recognition and spatial navigation in the elderly. Neuroscience 147:6–17
14. Besheer J, Jensen HC, Bevins RA (1999) Dopamine antagonism in a novel-object recognition and a novel-object place conditioning preparation with rats. Behav Brain Res 103(1):35–44
15. de Lima MN et al (2005) Reversal of age-related deficits in object recognition memory in rats with l-deprenyl. Exp Gerontol 40(6):506–511
16. Benice TS, Raber J (2008) Object recognition analysis in mice using nose-point digital video tracking. J Neurosci Methods 168(2):422–430
17. Eichenbaum H (2004) Hippocampus: cognitive processes and neural representations that underlie declarative memory. Neuron 44(1):109–120
18. Murray EA, Bussey TJ, Saksida LM (2007) Visual perception and memory: a new view of medial temporal lobe function in primates and rodents. Annu Rev Neurosci 30: 99–122
19. Smith AM, Chen WJ (2009) Neonatal amphetamine exposure and hippocampus-mediated behaviors. Neurobiol Learn Mem 91(3):207–217
20. Geinisman Y et al (1995) Hippocampal markers of age-related memory dysfunction: behavioral, electrophysiological and morphological perspectives. Prog Neurobiol 45: 223–252
21. Winocur G, Gagnon S (1998) Glucose treatment attenuates spatial learning and memory deficits of aged rats on tests of hippocampal function. Neurobiol Aging 19: 233–241

22. Gallagher M, Nicolle MM (1993) Animal models of normal aging: relationship between cognitive decline and markers in hippocampal circuitry. Behav Brain Res 57(2):155–162
23. Hardy JA et al (1986) An integrative hypothesis concerning the pathogenesis and progression of Alzheimer's disease. Neurobiol Aging 7(6):489–502
24. Harlow H (1959) The development of learning in the rhesus monkey. Am Sci 45:459–479
25. McCarthy RA, Evans JJ, Hodges JR (1996) Topographical amnesia: spatial memory disorder, perceptual dysfunction, or categoric specific semantic memory impairment?. J Neurol Neurosurg Psychiatry 60:318–325
26. Habib M, Sirigu A (1987) Pure topographical disorientation: a definition and anatomical basis. Cortex 23:73–85
27. Zola-Morgan S, Squire LR, Amaral DG (1989) Lesions of the hippocampal formation but not lesions of the fornix or the mammillary nuclei produce long-lasting memory impairment in monkeys. J Neurosci 9(3):898–913
28. Meunier M et al (1993) Effects on visual recognition of combined and separate ablations of the entorhinal and perirhinal cortex in rhesus monkeys. J Neurosci 13(12):5418–5432
29. Murray EA, Mishkin M (1998) Object recognition and location memory in monkeys with excitotoxic lesions of the amygdala and hippocampus. J Neurosci 18(16):6568–6582
30. Rapp PR, Kansky MT, Roberts JA (1997) Impaired spatial information processing in aged monkeys with preserved recognition memory. NeuroReport 8:1923–1928
31. Hampton RR, Hampstead BM, Murray EA (2004) Selective hippocampal damage in rhesus monkeys impairs spatial memory in an open-field test. Hippocampus 14:808–818
32. Ludvig N et al (2003) Spatial memory performance of freely-moving squirrel monkeys. Behav Brain Res 140(1–2):175–183
33. Haley GE et al (2009) Circadian activity associated with spatial learning and memory in aging rhesus monkeys. Exp Neurol 217(1):55–62
34. Wechsler D (1997) WMS-III Administration and Scoring Manual. Psychological Corporation, San Antonio, TX
35. Difede J, Hoffman H, Jaysinghe N (2002) Innovative use of virtual reality technology in the treatment of PTSD in the aftermath of September 11. Psychiatr Serv 53(9):1083–1085
36. Rothbaum BO et al (1995) Effectiveness of computer-generated (virtual reality) graded exposure in the treatment of acrophobia. Am J Psychiatry 152(4):626–628
37. Wiederhold BK, Wiederhold MD (2004) The future of cybertherapy: improved options with advanced technologies. Stud Health Technol Inform 99:263–270
38. Wiederhold BK, Wiederhold MD (2003) Three-year follow-up for virtual reality exposure for fear of flying. Cyberpsychol Behav 6(4):441–445
39. Astur RS et al (2002) Humans with hippocampus damage display severe spatial memory impairments in a virtual Morris water task. Behav Brain Res 132(1):77–84
40. Astur RS et al (2004) Sex differences and correlations in a virtual Morris water task, a virtual radial arm maze, and mental rotation. Behav Brain Res 151(1–2):103–115
41. Scoville WB, Milner B (1957) Loss of recent memory after bilateral hippocampal lesions. J Neurol Neurosurg Psychiatr 20(1):11–21
42. Ainge JA et al (2006) The role of the hippocampus in object recognition in rats: examination of the influence of task parameters and lesion size. Behav Brain Res 167(1):183–195
43. Zhang WN et al (2004) Dissociation of function within the hippocampus: effects of dorsal, ventral and complete excitotoxic hippocampal lesions on spatial navigation. Neuroscience 127(2):289–300
44. Allison T et al (1994) Human extrastriate visual cortex and the perception of faces, words, numbers, and colors. Cereb Cortex 4(5):544–554
45. Price CJ et al (1996) The neural regions sustaining object recognition and naming. Proc Biol Sci 263(1376):1501–1507
46. Marotta JJ, Genovese CR, Behrmann M (2001) A functional MRI study of face recognition in patients with prosopagnosia. Neuroreport 12(8):1581–1587
47. Zhang Z et al (2000) Functional MRI of apomorphine activation of the basal ganglia in awake rhesus monkeys. Brain Res 852:290–296
48. Juraska JM (1984) Sex differences in developmental plasticity in the visual cortex and hippocampal dentate gyrus. Prog Brain Res 61:205–214
49. Diamond MC et al (1983) Age-related morphologic differences in the rat cerebral cortex and hippocampus: male–female; right–left. Exp Neurol 81(1):1–13

50. Diamond MC et al (1982) Morphologic hippocampal asymmetry in male and female rats. Exp Neurol 76(3):553–565
51. Smith MD, Jones LS, Wilson MA (2002) Sex differences in hippocampal slice excitability: role of testosterone. Neuroscience 109(3):517–530
52. Romeo RD, Waters EM, McEwen BS (2004) Steroid-induced hippocampal synaptic plasticity: sex differences and similarities. Neuron Glia Biol 1(3):219–229
53. Petersen K, Sherry DF (1996) No sex difference occurs in hippocampus, food-storing, or memory for food caches in black-capped chickadees. Behav Brain Res 79(1–2):15–22
54. Bucci DJ, Chiba AA, Gallagher M (1995) Spatial learning in male and female Long-Evans rats. Behav Neurosci 109(1):180–183
55. Einon D (1980) Spatial memory and response strategies in rats: age, sex and rearing differences in performance. Q J Exp Psychol 32(3):473–489
56. Joseph R, Gallagher RE (1980) Gender and early environmental influences on activity, overresponsiveness, and exploration. Dev Psychobiol 13(5):527–544
57. Basso MR et al (2000) Sex differences on the WMS-III: findings concerning verbal paired associates and faces. Clin Neuropsychol 14(2):231–235
58. Hugdahl K, Thomsen T, Ersland L (2006) Sex differences in visuo-spatial processing: an fMRI study of mental rotation. Neuropsychologia 44(9):1575–1583
59. Coffey CE et al (1998) Sex differences in brain aging: a quantitative magnetic resonance imaging study. Arch Neurol 55(2):169–179
60. Pfankuch T et al (2005) Role of circulating androgen levels in effects of apoE4 on cognitive function. Brain Res 1053:88–96
61. Benice TS, Raber J (2009) Testosterone and dihydrotestosterone differentially improve cognition in aged female mice. Learn Mem 16(8):479–485
62. Benice TS, Raber J (2009) Dihydrotestosterone modulates spatial working-memory performance in male mice. J Neurochem 110(3):902–911
63. Raber J et al (2002) Androgens protect against apolipoprotein E4-induced cognitive deficits. J Neurosci 22:5204–5209
64. Acevedo S et al (2008) Selective androgen receptor modulators antagonize apolipoprotein E4-induced cognitive impairments. Lett Drug Des Discov 5:271–276
65. Rose JK, Rankin CH (2001) Analyses of habituation in *Caenorhabditis elegans*. Learn Mem 8(2):63–69
66. Rizk-Jackson A et al (2006) Effects of sex on object recognition and spatial navigation in humans. Behavl Brain Res 173:181–190
67. Acevedo S et al (2010) Apolipoprotein e4 and sex affect neurobehavioral performance in primary school children. Pediatr Res 67(3):293–299
68. Haley GE et al (2010) Effects of ε4 on object recognition in the non-demented elderly. Curr Aging Sci 3(2):127–137

Chapter 5

Fear Conditioning in Rodents and Humans

Mohammed R. Milad, Sarah Igoe, and Scott P. Orr

Abstract

Fear conditioning is an experimental tool that has been, and continues to be, widely used in the field of neuroscience. It is used to understand the neural and psychological bases for fear learning and more recently for fear extinction, along with several other phenomena such as reinstatement and spontaneous recovery. Like any other experimental paradigm, there are several variants of fear conditioning that are employed by investigators. The parameters utilized, such as the type of conditioned stimuli and the unconditioned stimuli, vary from one study to another depending on the scientific question being tested. In this chapter, we will provide an overall summary of the most commonly used parameters and discuss the reasons for changing and/or modifying such parameters. We discuss technical problems that may arise when using the fear-conditioning paradigm in both rodents and humans and how best to resolve them.

1. Introduction

Fear may well be the most important of all human emotions, both for its role in human survival as well as its role in psychopathology. To date, there is a very substantial amount of published animal and human research that has focused on fear and used fear-conditioning procedures to advance our understanding of its nature. Fear-conditioning procedures have been, and continue to be, commonly used across a variety of different fields, including neuroscience, psychology, and psychiatry. The usefulness of these procedures is enhanced by the fact that it is relatively easy to train rodents in fear-conditioning procedures; fear conditioning can be easily implemented in human studies and it provides an animal-based model that has helped guide neuroscientists over the past several decades to begin exploring the neural circuits that mediate conditioned fear and its extinction. What are the historical roots of experimental fear conditioning?

Fear conditioning is based on the same principles of Classical (also known as Pavlovian) Conditioning first described in 1927 (1). In the early 1900s, Ivan Pavlov, a Russian physiologist, was studying the interaction between salivation and the action of the stomach. He predicted that reflexes of the autonomic nervous system would link them both. Pavlov wanted to know if external stimuli would affect the process of salivation. So he rang a bell at the same time he gave his dog a bit of food. After a number of pairings of the bell with food, the dog would salivate when the bell rang, even though no food was present. Pavlov named this type of learning *conditioning*, in which the dog came to associate the sound of the bell with food. He referred to the bell as the "conditioned stimulus (CS)," the food as the "unconditioned stimulus (US)," and salivation induced by the CS as the "conditioned response (CR)." This type of classical conditioning came to be known as "Appetitive Conditioning." The primary difference between appetitive conditioning and fear conditioning is the type of US that is used to support the conditioning. In the case of fear conditioning, a cue such as a light or tone is paired with the presentation of an aversive US, such as an electric footshock. Results of this pairing can produce a range of conditioned responses such as freezing, change in blood pressure or heart rate, and analgesia in animals, and increased sweat activity (skin conductance response) and facial muscle tension in humans. Once a conditioned response has been established, it can be diminished ("extinguished") by repeatedly presenting the CS without the US.

Fear conditioning has been used as an experimental tool for exploring the neurobiological basis of fear learning in many species. It has also been used to understand the additional phenomena associated with fear learning such as extinction, spontaneous recovery, renewal, reinstatement, and extinction of conditioned fear. Pavlov was the first to demonstrate spontaneous recovery. Following extinction of a CR, he allowed some time to pass and then rang the bell again without food. Pavlov noted that the dog would again salivate at the sound of the bell. Based on this observation, Pavlov argued that extinction *cannot be regarded as an irreparable destruction of the conditioned reflex, due to disruption of the respective nervous connections, as evidenced by the fact that the extinguished reflexes spontaneously regenerate in course of time*. Pavlov also observed that following extinction of a CR, exposure to the US alone could restore (or reinstate) the conditioned responses to the CS (1). Reinstatement has been subsequently observed and studied by a number of investigators (e.g., 2, 3). The process of renewal was demonstrated by Bouton and colleagues (4) whereby following extinction of a CR, placing the animal in a context different from that in which extinction took place renewed the CR to the CS.

In this chapter, we provide an overview of how fear-conditioning procedures have been implemented in rodent studies. We will describe the different indices of fear and how such are measured. We will then describe how some of these tools have been translated to fear-conditioning experiments in humans. More recent modifications to fear-conditioning procedures have been implemented so that it can be used in conjunction with neuroimaging in order to examine the neural circuits of fear learning and extinction in the human brain. These modifications will also be summarized in this chapter.

2. Materials and Methods

2.1. Fear Conditioning in Rodents

2.1.1. Conditioned and Unconditioned Stimuli

The duration of the conditioned stimulus presentation varies from one study to the other, ranging from 2 to 30 s (5–8). Commonly used CSs are auditory tones in the 2–10 kHz range (9, 10) with an intensity of about 80 dB (10–12). For auditory-conditioning paradigms, LED indicator lights that are invisible to the animals are typically mounted within the chambers. These serve to indicate tone presentations on video recordings without audio tracks. The most commonly used US is a mild electric shock that may vary in intensity from 0.4 to 2 mA (13–15) with durations that typically range from 0.5 to 2 s (15–19).

2.1.2. Measuring Conditioned Responding

Conditioned fear responding in rodents can be measured by several physiological indicators. Aside from fear-potentiated startle, fear can manifest itself in terms of an increased release of stress hormones, increased heart rate, increased arterial blood pressure, and hypoalgesia (decreased sensitivity to painful stimuli) (9).

There are two customary methods used to measure conditioned responding that do not require physiological monitoring, both of which are widely used in cued and contextual fear conditioning. Both *freezing* and *bar-press suppression* have become standard practice and are heavily relied upon in rodent studies (20, 21). Freezing is defined as the absence of all movements except respiration and is a well-documented behavior associated with fear expression. Methods for examining and recording freezing behavior will be discussed in detail later in the chapter. Bar-press suppression is the decrease in learned food-seeking behavior. Further discussion of the rationale behind including operant conditioning in fear behavior paradigms will also appear later in the chapter.

There are three general types of fear-conditioning models: cued fear conditioning, contextual fear conditioning, and fear-potentiated startle. There are several variations with respect to the implementation of fear conditioning in rodents. We will describe the most commonly used methods for each of these types of fear conditioning.

1. Cued Fear Conditioning

As noted above, the majority of cued fear-conditioning studies use an auditory cue for the CS, such as a 10-kHz tone, although visual cues such as lights have been used (22). Most studies use an electric shock for the US, the intensity of which varies among studies. In general, fear-conditioning experiments are conducted in Plexiglas chambers inside sound-attenuated boxes. These chambers typically contain a single overhead house light, video camera, and a speaker mounted on the wall through which tone presentations are delivered. The video camera is used to record the animal's behavior throughout the experiment, and the behavioral outcome (such as freezing) is later scored offline. The floors of these chambers consist of stainless steel bars capable of delivering a mild electric shock (**Fig. 5.1**).

Fear-conditioning studies typically take place over the course of 2 or 3 consecutive days and many involve

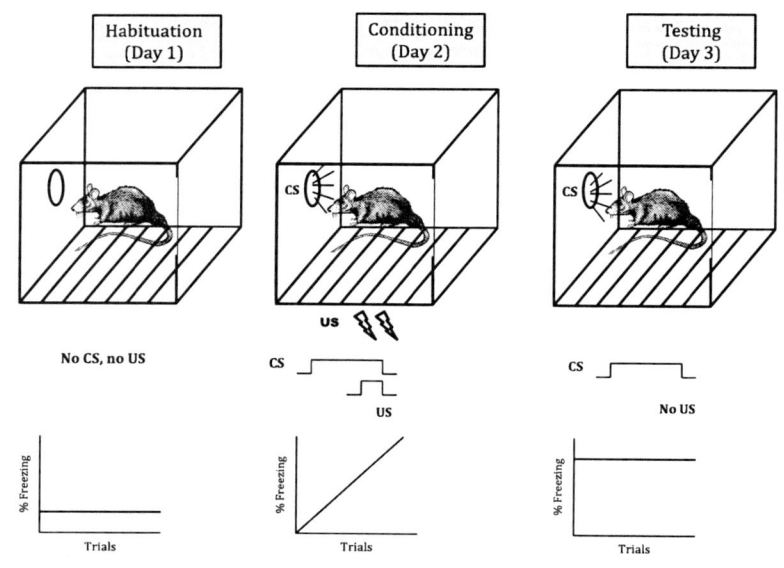

Fig. 5.1. Schematic illustration of the fear-conditioning chamber and the different experimental phases during delayed auditory fear conditioning. Note that pairing the conditioned stimulus (CS) with the unconditioned stimulus (US) results in gradual increase of freezing during the conditioning phase which can be observed 24 h later during a test phase.

follow-up tests after a predetermined amount of time has elapsed. They often begin with a *habituation* phase during which the CS is presented without the US in order to exclude the novelty element of the cue and to alleviate any potential fear-inducing qualities intrinsic to the tone itself. The *conditioning* or *acquisition* phase, in which the CS–US pairings are presented, follows. The number of paired presentations varies depending on the desired strength of the CS–US association – some studies will present the pairing only once or twice, others will present it as many as 75 times (9). Various tactics can be used to establish an especially reliable or enduring CR, the most common of which is simply increasing the number of paired presentations. Other methods include increasing the intensity of the US or implementing the conditioning phase on 2 consecutive days instead of 1 (10). The amount of time between CS presentations is referred to as the *inter-trial interval* or ITI and is either *fixed* at a certain number of seconds, or set to be *variable* throughout the session according to a predetermined minimum, maximum, and average time. **Figure 5.2** summarizes the three experimental phases of this type of conditioning.

Trace versus delay fear conditioning: Two variations of cued conditioning are commonly used. The previously mentioned technique, in which the US co-terminates with

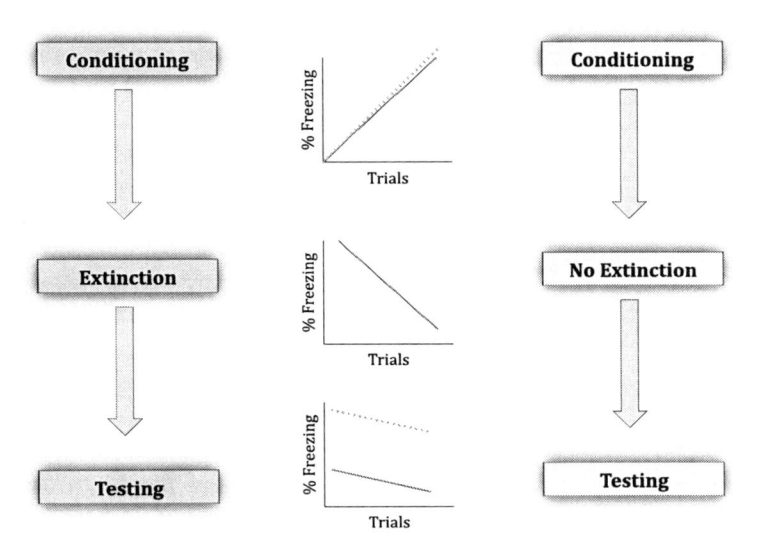

Fig. 5.2. Schematic illustration of expected behavior of animals undergoing the different phases of fear conditioning, including extinction phase (in which the CS is repeatedly presented in the absence of the US). If extinction training is not to take place, then fear to that CS will remain high during test. The difference in fear responses to the extinguished CS and the unextinguished CS could be referred to as the extinction memory.

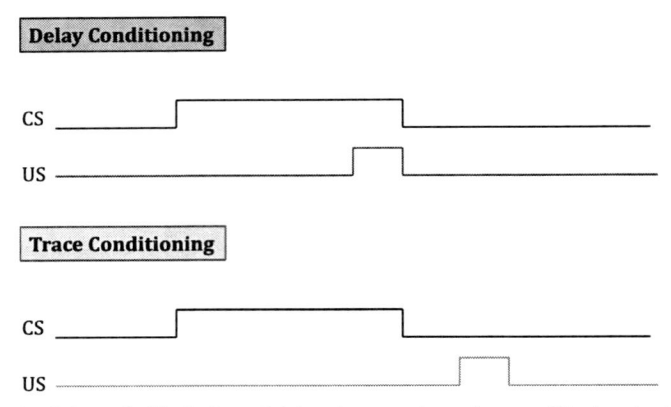

Fig. 5.3. Schematic illustration of delayed versus trace fear conditioning. In delayed fear conditioning, the US presentation co-terminates with that of the CS. In trace conditioning, the US presentation occurs seconds after the offset of the CS presentation.

the CS (i.e., the shock is delivered precisely as the CS ends) is referred to as *delay* conditioning (*see* **Fig. 5.3**). *Trace* conditioning occurs when both a CS and a US are presented repeatedly throughout a session, but not at the same time (**Fig. 5.3**). Exact timing varies depending on the ITI, but in trace conditioning the US usually occurs about 10–20 s after the end of the CS presentation (23). Under normal circumstances, CR to the cue is stronger in delay conditioning than in trace conditioning, as the CS–US pairing is more closely linked in time. Trace conditioning requires the activation of the hippocampus whereas delay conditioning predominantly activates the amygdala. Thus the choice of using one paradigm over the other will depend on the scientific question being addressed.

To summarize the above-mentioned parameters, in a typical fear-conditioning experiment, seven presentations of 30-s tones (10 kHz) are presented with an intensity of 80 dB. Each of these tones is followed by 0.6 mA electric shock delivered to the grid floor. This phase is typically referred to as the "conditioning" phase or the "fear acquisition" phase.

2. Contextual Fear Conditioning

A noteworthy confound of cued conditioning is the fact that the fear observed during the testing phase is not entirely attributable to the presence of the CS. During cue conditioning, the animal forms an association between the CS and the US. However, it has also been repeatedly demonstrated that an association is formed between *contextual cues* and the US, and that this association serves as a basis for some of the fear behavior (23–25). The concept of "context" incorporates a range of sensory stimuli related to the circumstances under which conditioning occurs, including visual appearance, tactile cues, background noise, scent, and time of day. Experimenters have relied on a variety of creative methods to establish discernible contexts that animals will identify as different from one another. Contextual variations may include switching the metal grid floor for a solid one, spraying scented oils like peppermint or vanilla, playing tapes of static or other white noise, and using flashing or colored lights.

A commonly used method for studying contextual fear conditioning is quite similar to that used for cued conditioning. A rat is placed in a novel environment and allowed to explore for a short time, usually about 3 min and no less than 1 min. It has been documented that successful context conditioning requires that the rodent is placed in the conditioning context and allowed to explore the context for at least 60 s between placement in the chamber and administration of shock (26). After this time, the aversive stimulus is presented. It is typically brief (0.5–2 s), mild (0.3–1.5 mA), and infrequent (one to four shocks) (27). As with cued conditioning, freezing is monitored following the administration of the US for 30–90 s (28). After the passage of some amount of time, extinction training and/or administration of pharmacological agents, animals undergo a *test* phase which, similar to cued fear conditioning, commonly uses freezing behavior as the measure of fear. However, unlike cued conditioning the behavioral monitoring begins immediately after the animal is placed in the cage and not after the presentation of a previously conditioned stimulus. In short, in contextual conditioning, the *entire environment* functions as the CS.

3. Fear-Potentiated Startle

As an alternative to directly measuring conditioned responding to specific cues or contexts, fear-potentiated startle has been established as an equally reliable method of monitoring conditioned fear in animals. It is based on the notion that the natural startle reflex is augmented by a fearful stimulus. The paradigm begins with a *matching* procedure, during which rats are presented with several (10–30) loud tones of varying intensities inside a cage specifically designed to measure movement using an accelerometer (29). The mean startle response amplitude is used to divide the animals into groups with similar baseline levels of sensitivity. One to 3 days later, animals undergo a *training* session inside a chamber capable of delivering a CS–US pairing, normally a tone (but sometimes a light) and shock (29, 30). After about 5 min of acclimation to this chamber, the CS and US will be repeatedly paired to establish an association. As in cued conditioning, the CS may co-terminate with the US precisely or the US may occur after several seconds. Similarly, the typical range of shock intensity, ITI, noise level, and number of trials fall in the same range as for cued conditioning.

Following experimental manipulations (extinction trials, time lapse, surgery, drug administration, etc.), the animals are returned to the startle testing apparatus for the *testing* phase. This can occur 1 day after training, or even up to a month later, depending on the conditions of the experiment. In this phase, animals are presented with about 60 "startle stimuli," typically loud bursts of white noise. When a light is used as the CS during training, a pure tone can be used as the startle stimulus (31, 32). Half of the startle stimuli are presented alone and half are presented in concert with the CS. The theory behind fear-potentiated startle is that when a strong association has formed between the CS and US during training, the presence of the CS will produce fear and the animal will exhibit a heightened startle response when the startle stimulus is paired with this CS, compared to when the startle stimulus is presented alone. A larger discrepancy between CS-noise versus noise-alone reflects greater conditioned fear (33, 34). Like cued conditioning, fear-potentiated startle can be used to measure spontaneous recovery and reinstatement. However, renewal tends to be more difficult to establish because the shocks are administered in a different setting than the startle testing apparatus.

2.2. Fear Conditioning in Humans

2.2.1. Early Conditioning Studies

As with cued conditioning in animal studies, human fear conditioning seeks to develop an association between a CS and an aversive US. For example, in an early study of conditioning to social cues investigators used slides of photographs of human facial expressions as CSs and a mild shock to the fingers as the US. Not surprisingly, it was found that stronger conditioning occurred when the reinforced conditioned stimulus was a picture of a fearful face than when it was a happy face (35). (Further discussion of reinforced and non-reinforced conditioned stimuli follows in the next section.) The focus of a considerable amount of the early human conditioning research was directed toward understanding the influence of biological preparedness (35, 36) in the acquisition and extinction of conditioned fear responses. Consequently, many early studies used fearful or angry faces as CSs because they conditioned quickly and without subjecting participants to a large number of, or high intensity, US presentations to achieve a measurable level of fear acquisition (37–39).

2.2.2. The CS, US, and Reinforcement Schedules

Recent studies of human conditioning have trended toward using experimental protocols that more closely follow those established

Fear Conditioning in Rodents and Humans 119

in the animal literature. Commonly, studies use a short (about half a second), mild shock delivered by electrodes to two fingers on the subject's dominant hand (the non-dominant hand is used for recording physiological measures) (40–42). Other studies deliver the aversive shock to the wrist or ankle of the participant (43). Because of the ethical considerations surrounding shock delivery in human subjects, participants typically predetermine the level of shock to be used as "highly annoying but not painful" (44, 45). Therefore, human fear conditioning in its simplest form involves the association of a CS (normally a colored picture of a geometric shape such as a square or circle) and a mild shock. For example, participants will be seated in front of a computer monitor programmed to display a red circle (the image is typically displayed for about 8–12 s) (46, 47). If the shock is always delivered after presentation of this cue, the subject will quickly learn that it predicts the shock and will demonstrate a conditioned response (e.g., a change in sweat activity) when it appears.

2.2.3. Differential Conditioning

In addition to the CS that is paired with the US, many studies of human conditioning use a second stimulus, such as a blue circle, which is *never* paired with the US. This type of conditioning is known as differential conditioning and rarely has been used in animal studies. In differential conditioning, the CS that is not paired with the US (e.g., a yellow square) is referred to as the *CS–* and the CS that is paired with the US (e.g., a blue circle) is referred to as the *CS+* (*see* **Fig. 5.4**). A typical differential conditioning study might involve several, initial exposures to both the CS+

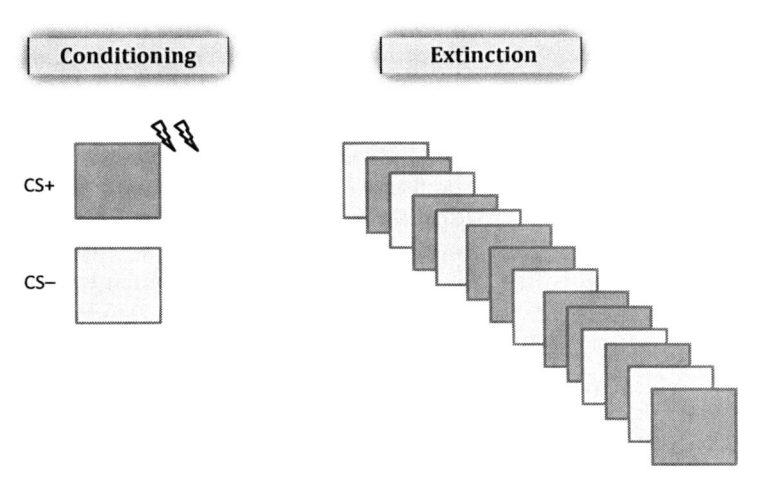

Fig. 5.4. Schematic illustration of stimuli typically used in human fear conditioning. This type of paradigm is known as differential conditioning, in which two CSs are used: one followed by the US (CS+, *dark square* in this example), and the other is not followed by the US (CS–, *light square*). Extinction training in this paradigm involves repeated presentation of both the CS+ and CS– in the absence of the US presentation.

and the CS− (perhaps 4–5 times each) in random order and with no shocks administered. This "habituation" phase reduces orienting responding and establishes physiological baselines. During the "conditioning" phase, the CS+ and CS− might be presented 5–10 times each with the US being administered immediately following CS+ offset. As with animal conditioning, the duration of time between cues is set to vary around a predetermined average, with an overall range of about 12–25s for studies using skin conductance change as the measure of fear (48, 49).

Other types of unconditioned stimuli have also been employed in human conditioning studies. For example, some studies have used aversive noises rather than mild shocks (50, 51). The study designs are similar to those described, commonly using simple visual images as the conditioned stimuli. However, during the fear acquisition phase, a burst of 90–100 dB of white noise will be delivered through headphones for 0.5–2s as the US (52, 53). Studies that include a CS− may pair the CS− with a non-aversive tone (50–70 dB) with the same duration as the aversive US. Other studies have used small blasts of air directed at the participant's throat as the US (54–56). This is accomplished by connecting a compressed air tank to plastic tubing with a solenoid valve set up so as to deliver puffs of air for about 100–250 ms, typically at an intensity of about 140 psi (55).

2.2.4. The Index of CR in Humans

Since the 1960s electrodermal measures, e.g., skin resistance, skin potential, and skin conductance, have been the most popular indices of aversive conditioning in humans. Over the years, various general terms have been used to refer to electrodermal activity, such as galvanic skin response (GSR), psychogalvanic reflex (PGR), and electrodermal response (EDR). Currently, *skin conductance* is the preferred measure of electrodermal activity. Ohm's Law serves as the underlying principle for measuring skin conductance. Skin conductance level primarily reflects the amount of sweat present in the sweat glands. It is commonly measured from the fingers or palm of the hand; glands on the palms of the hands (and the soles of the feet) are known to be particularly responsive to emotional arousal. An increase in skin conductance is caused by sympathetic nervous system activation, which increases secretion from the eccrine sweat glands. When a human being becomes emotionally aroused, he or she produces small amounts of sweat that are more electrically conductive than dry skin. Skin conductance is measured by attaching two electrodes, separated by about 14 mm, on the subject's non-dominant hand. A skin conductance coupler then passes a small electrical current at a constant voltage (typically, 0.5 V) through electrodes filled with isotonic paste. An analog-to-digital converter "digitizes" the voltage output of the skin conductance coupler, which is then recorded and translated into a numeric measure of skin conductance level by a computer program. A change in skin conductance level can then be calculated so as to provide a measure of the magnitude of the skin conductance response to a given stimulus. For example, one might calculate the average skin conductance level during the 1-s interval prior to onset of a CS and subtract this value from the peak skin conductance level during the CS interval. This change score would provide a measure of skin conductance *response* to the CS presentation. For an excellent overview and discussion of the electrodermal system and assessment of electrodermal activity, the reader is referred to Dawson et al. (57). In addition to skin conductance, some investigators record other physiological indices such as heart rate and facial electromyogram (EMG) of the corrugator muscle, a small muscle at the medial end of the eyebrow that is associated with furrowing of the eyebrows or frowning.

The human startle response is commonly measured from eyeblink-related activity using electromyography (EMG) of the right orbicularis oculi (e.g., 55, 58), the muscle that controls closing of the eyelids. To do so, Ag/AgCl electrodes filled with electrolyte gel are positioned about 1 cm under the pupil and 1 cm below the lateral canthus (the corner of the eye where the upper and lower lids meet). A reference electrode is then placed on the upper forehead, on the participant's arm, or behind one ear over the jawbone. EMG activity is amplified and recorded using hardware and computer software specifically designed to sample EMG activity at 1,000 Hz beginning at startle stimulus onset (or just prior to) and ending a few hundred milliseconds after stimulus onset.

Several studies have used heart rate (HR) as a measure of conditioned responding (e.g., 44). This can be done by electrocardiogram (ECG or EKG), which measures electrical impulses associated with contraction of the myocardial muscles of the heart. The ECG is easily measured using surface electrodes placed on particular areas of the body (e.g., chest, arms, legs). ECG output appears in the form of a voltage between electrode pairs. The amount of time between heart beats, recorded in milliseconds between successive R wave components of the ECG, can be converted to a measure of heart rate. Some studies have monitored blood pressure during conditioning paradigms (59–61). Blood pressure is monitored using a sphygmomanometer, or blood pressure monitor, which displays pressure in terms of mmHg, or millimeters of mercury, based on the original devices that used the height of a column of mercury to deduce circulating pressure. Whereas skin conductance, EMG, and heart rate measurements can be continuously and unobtrusively obtained, the need to inflate and deflate a blood pressure cuff makes blood pressure a more intrusive and difficult measure.

3. Notes

This chapter is intended to introduce the reader to methodologies used in rodent and human experiments that target various aspects of fear conditioning. Though the principles are similar, choices regarding the type, duration, and number of CSs and USs vary substantially across studies. In addition, the number of days and the implementation of additional test phases (e.g., extinction, retention, or renewal) vary depending on the scientific question(s) being addressed. In this section of the chapter, we discuss some additional and important considerations regarding the many variants of the classical fear-conditioning paradigm.

3.1. Freezing Measurements

In the early years of fear-conditioning research, freezing behavior was always scored manually. Manual scoring is a procedure whereby the entire experimental session is videotaped; the investigator later reviews the tape and meticulously scores freezing behavior with a stopwatch. The number of seconds the animal spends motionless will be recorded, usually as a percentage score. For context conditioning, the percentage would be referenced to the total time spent in the chamber (~3 min); for cued conditioning, it would be referenced to the duration of the cue. Often, if the cue is a tone, a light will be set up that is invisible to the animal, but serves to indicate the presence of the CS to the scorer. As previously mentioned, investigators using cued conditioning may also record freezing before the onset of the CS to confirm that the behavior is, at least in part, associative in nature.

Manual scoring requires a considerable amount of time and relies on subjective judgments of the experimenter. To address these issues, some labs will randomly select a subset of sessions and have a different researcher re-score them in order to corroborate the values. Some labs will require that two different researchers score every session and use the average of the raters' scores. Although the methods of managing human error may vary among labs, comparisons within each lab will not be tainted as long as the scoring methods remain consistent over time. An additional limitation of manual scoring arises from *experimenter bias*, a tendency (usually subconscious or unintentional) to score or make judgments that favor an expected or hoped for result. Researchers attempt to control this bias by conducting *blind experiments*, in which the individual doing the scoring is not informed as to the experimental condition of the animals. The scorer is thereby unable to form a specific prediction regarding the expected behavior for a particular animal.

In recent years, computer software has been developed to score freezing automatically. As with manual scoring, the session is recorded and incorporates an indicator light, invisible to the animal, for signaling CS onset during cued conditioning. Motion-sensing software is then able to calculate how much time the animal spends completely still during the cue or session. An important drawback to this approach is that not all absence of motion is attributable to fear behavior, especially when operant conditioning is not used in conjunction with fear conditioning. It can be argued that human input is necessary to discriminate true freezing from incidental pauses in movement. However, introducing these subjective determinations is done at the expense of increasing variability between scorers. In general, automated scoring programs are becoming more popular in labs that study fear conditioning because they remove the elements of human error and experimenter bias and are much more time-efficient.

3.2. Floor and Ceiling Effects

As explained above, freezing scores during cued conditioning are often recorded as a percentage of the overall presentation time for the CS. For example, if an animal freezes for 24s during a 30-s long tone presentation, it is said to exhibit 80% freezing for that trial. However, if the animals are consistently freezing for the entire duration of the cue (100% freezing for each trial), there is the risk of *ceiling effects*. Ceiling effects occur when variations in freezing time predominantly occur at time points beyond the CS presentation, thereby rendering such trials as unscorable by traditional methods. For example, if after conditioning an animal freezes for 45s following the onset of a 30-s CS, this will be scored as 100% freezing because it was motionless for the entire duration of the cue (and longer). If a subsequent extinction procedure or drug manipulation reduces the animal's freeze time to 30 s, a 33% reduction, it will again be scored as 100% freezing and it will appear as though the intervention had no effect whatsoever.

Floor effects, on the other hand, tend to occur at the end of extinction trials or after a pharmaceutical manipulation that reduces fear behavior nearly to zero. In such cases, it is difficult or impossible to decipher the relative effects of two different experimental conditions, because both groups repeatedly display little or no freezing and are therefore said to have reached the "floor" level of fear behavior. Ceiling and floor effects can typically be avoided by selecting the proper conditioning paradigm and conditioning parameters, as outlined in this chapter and as is well documented in the conditioning literature. For example, ceiling effects might be the result of "over conditioning," which could be corrected by reducing the frequency or intensity of the US. Alternatively, floor effects might be alleviated by reducing the number of extinction trials.

3.3. Context Conditioning

In some studies it is important to control for or measure the portion of the fear response that may be attributable to conditioning to contextual cues. This can be done by scoring fear behavior as a difference, rather than an absolute value. In other words, fear can be measured before the CS, i.e., when only the context is present, and after the CS presentation during the testing phase; a difference is then calculated that essentially removes the influence of fear associated with the context. Another way to control or measure the effects of context is to perform conditioning in one context and then test animals in either the same context or a novel one. If those tested in the novel environment express less-conditioned responding than those tested in the original environment, contextual conditioning is thought to be represented by difference between those groups.

The role of contextual stimuli on conditioned fear remains unclear. Several studies have found that testing in a context different from the one in which conditioning occurred causes very little, if any, disruption of conditioned responding (4, 62–64). In fact, one study found that the switch to a new context actually enhanced, rather than inhibited, conditioned responding to the CS (65). For this reason, there is little agreement as to how context should be regarded. On one hand, the Rescorla–Wagner model treats it merely as a second CS that is presented in concert with the intended CS (25). On the other hand, it has been suggested that contexts carry information about the relationships between events (i.e., CS–US pairings) that occur within them, but do not hold specific associations themselves.

3.3.1. Post-conditioning Tests

The conditioned responses that are triggered by a CS–US association can be diminished after repeated presentations of the CS in the absence of the US, a process known as *extinction*. Following conditioning, either the same day or the following day, fear extinction is produced by means of systematic presentations of the CS without the US, which gradually extinguishes the CR. While the amount of time between conditioning and extinction has not been standardized, long-term extinction memory is thought to develop only when extinction training occurs at least 6 h after conditioning (66). The number of extinction trials can vary greatly between studies. Sometimes there are fewer extinction trials than conditioning trials (9) and other times (6) the number of extinction trials exceeds that for conditioning. Contextual fear can be extinguished simply by repeatedly returning the animal to the chamber in the absence of shock. Typically, an extinction session will occur over about the same amount of time as the conditioning session (~3 min) and must be repeated over a number of days in order for there to be an enduring reduction in fear behavior. As noted previously, there are several different fear memory-related phenomena that can be tested. The simplest *test* phase would involve a single CS or startle presentation following extinction, surgery, drug administration, or other experimental manipulations specific to a given study. Depending on the goal and limitations of the study, testing could occur as soon as an hour or as long as a month after conditioning/extinction.

Spontaneous recovery is the return of a previously extinguished CR that requires no experimental procedure aside from the passage of time. *Reinstatement* typically involves one to three US presentations and is tested the following day by presenting the previously extinguished CS (67). *Renewal* is tested when the CS–US pairings occur in one context and the CS is subsequently extinguished, i.e., presented alone, in another context. Returning the animal to the original context during testing will produce a renewed fear response. Renewal can also be observed when the animal is tested in a third, neutral context (3).

3.4. Combining Fear-Conditioning Paradigms with Operant Conditioning Procedures

In some fear-conditioning experiments, researchers will also include an operant conditioning procedure, whereby animals are trained to press a metal bar for a food reward repeatedly during the conditioning procedure. In order to establish a consistent motivation for a food reward, animals are fed systematically by investigators such that they are kept at about 80% of their original body weight. The classic approach is based on the method of *successive approximations* first demonstrated with pigeons by B.F. Skinner in 1937. Also referred to as "auto-shaping," this method is executed by first manually delivering a food reward whenever the animal walks in the vicinity of the metal bar. As Skinner describes in *The Behavior of Organisms*, the investigator will subsequently deliver reward for "each of the following steps in succession: approaching the site of the lever, lifting the nose in the air toward the lever, lifting the fore-part of the body into the air, touching the lever with the feet, and pressing the lever downward." (68). Throughout this process the rat gradually learns which actions will produce a reward, thereby increasing the frequency of these actions and demonstrating operant conditioning. After about 10 days of training, most rats will reach a level at which they press the bar very consistently even when pellets are not delivered after each press. Rather, pellets are delivered at a rate predetermined by a computer program to vary at an interval of about 60 s. Most investigators will exclude rats from their experiments that do not meet this standard. Alternative training methods have been proposed. One such method involves shining a photocell beam onto the Plexiglass bar and delivering the reward when the animal interrupts the light beam with its body (69). Though some methods have claimed to expedite the learning process, most studies continue to rely on Skinner's classic method.

3.5. Why Use Operant Conditioning in Conjunction with Fear Conditioning?

There are two important reasons as to why an investigator might decide to train rats to press a bar for food while undergoing fear conditioning. First, this approach allows bar-press suppression to be used as a measure of fear, in addition to the classical measure of fear, freezing. Press-rate suppression has the benefit of providing a relative measure of fear. In other words, because the ratio is calculated by comparing conditions before and after the onset of the conditioned stimulus, a stronger argument can be made that changes in behavior result from the CS onset. Secondly, bar-pressing behavior serves to provide a consistent level of locomotion during the experiment, against which freezing can be measured. If the rodent does not have incentive to move about the chamber, it is more difficult to attribute a lack of motion as representing fear-induced freezing. This is not likely to be a problem for studies only examining the conditioning phase (fear learning). However, if the experiment will include additional training sessions that examine fear extinction, bar-press training may be useful. Extinction training requires many CS trials (15–20 trials) in the absence of the US and if the duration of the inter-trial interval is long, the animals may fall asleep. Implementing bar-press training protects against the possibility that the animal might fall asleep. However, including bar-press training carries the disadvantage of introducing the possibility that motivation for reward may compete with the fear-circuitry behavior. This makes it challenging to compare results from studies that include bar-press training with those that do not.

3.6. Fear Conditioning in Humans

While the US always follows the CS presentation in rodents, this is not necessarily the case in human studies (70–72). Reinforcement schedules may be either *continuous*, whereby the US follows each CS+ presentation, or *partial* (or *intermittent*), whereby only some CS+ presentations are paired with the US. When partial reinforcement is used, the participant learns that a shock never will be received following the CS−. However, there is less certainty associated with the CS+ as to whether or not a shock will occur. In other words, the CS+ will always be present before every US administration, but a US does not necessarily follow the CS+. The introduction of the uncertainty component makes partial reinforcement more resistant to extinction than continuous reinforcement, even though continuous reinforcement might involve a greater number of CS–US pairings overall. The US presentations can be administered on a *fixed* or *random* schedule. A fixed schedule would mean a predetermined ratio; for example, a US presentation for every third CS+ presentation. With partial reinforcement, random schedules produce conditioned responses that are more resistant to extinction than fixed schedules.

Recently, some studies have begun to use more sophisticated conditioned stimuli in order to investigate the role of *context* in fear conditioning. Instead of simply displaying blue or red circles as CSs, a study might use images of more detailed environments.

For example, a cue that is to serve as the CS, such as a colored lamp, might be depicted inside one of two different contexts such as a living room or a study (40, 73, 74). The acquisition session might then show pictures of the lamp depicted in the living room context and present the US when the lamp is lit (CS+) but not when the lamp is unlit (CS−). Extinction of the CR would be accomplished by presenting the same pictures of lit and unlit lamps in the living room context without the US. Renewal effects could be examined by depicting the same lit and unlit lamps within a different context, i.e., the study. This procedure attempts to simulate animal conditioning models whereby acquisition, extinction, and renewal sessions are actually performed in different environments.

3.6.1. Virtual Reality

Virtual reality offers a new technology that is proving especially useful in treating clinical disorders that stem from conditioned fears. Effective methods for treating fear-related disorders in humans include exposure therapy and systematic desensitization. Often used to treat phobias, exposure therapy involves direct, often intense, repeated exposures to the stimulus which is thought to underlie the patient's fear. Historically, many studies in both humans and animals have used extinction sessions as a model for the processes involved in exposure therapy. More recently, some investigators have explored virtual reality as a way to realistically expose patients to their fear object or situation while maintaining control over the exposure process. For example, it has been found that patients suffering from acrophobia, or an extreme fear of heights, experience more anxiety when they are in a virtual reality environment that simulates *movement* in high places than if they were simply shown pictures of views from tall buildings (75). For this reason, virtual reality exposure therapy, or VRET, is seen as a viable alternative to exposure therapy because it can elicit fear and anxiety, therefore serving as a sound model for in vivo exposure to the object of one's phobia. Based on the fact that it occurs in the safety of a lab or therapy environment, it is often used as an intermediate step before live exposure, or as a substitute to imaginational exposure (76). Since the advent of VRET in 1996, most studies on acrophobia treatment have focused solely on the potential of this method (75).

Even more recently, researchers have begun to explore the use of VRET for the treatment of post-traumatic stress disorder (PTSD) and for active duty soldiers. System designers will create environments that seek to closely mimic those experienced by soldiers in combat. In one particular study, for example, soldiers with PTSD symptoms attended six 90-min sessions in either a convoy scenario or one made to look like a dismounted patrol in an Iraqi city (77). While undoubtedly less convenient than its older counterparts, VRET remains a promising new development in the treatment of PTSD, anxiety disorders, and specific phobias.

3.7. Measuring Skin Conductance Response (SCR) in Human Conditioning Studies

There are several approaches that have been used to measure SCR in humans. SCR is a relatively slow response that can be observed 1–3s after the onset of the CS. As such, investigators have used approaches in which the duration of the CS is divided into first interval response (FIR, first few seconds of the CS) and second interval response (SIR, the last few seconds of the CS). The FIR is generally viewed as an orienting response and not necessarily as a reflection of a conditioned response. As such, investigators using this approach typically focus on the SIR phase of the SCR. We have used a different approach that has also been well validated (40, 44). That is to subtract the skin conductance values prior to the presentation of the CS (usually the average of the last 2s prior to CS presentation) from the peak SCR during the CS presentation (the peak could be obtained at any point in time during the CS presentation). The advantage of this approach is any differences in skin conductance levels (SCL) between subjects could be cancelled out (or accounted for). Importantly, it has been recently shown that the subtraction method looking at the peak response of SCR is as effective as dividing the CS interval into FIR and SIR phases (80).

3.8. Conducting Fear Conditioning in Humans While in the fMRI: Issues to Consider

With the recent advancement of neuroimaging tools, investigators have already began to take advantage of neuroimaging to study the neural circuits of fear conditioning in the human brain, both in healthy humans and in populations with psychopathology (43, 78, 79). There were several concerns that were raised at the early stages that are now resolved. One of the major challenges is to conduct the conditioning protocol (delivery of electric shock and measuring SCR) in the scanner without any harm to the participants. The second major concern is that of noise artifact: noise that can be injected from the scanner onto the SCR data being recorded and noise being injected from the cables and electrodes attached to the subjects into the MRI images. All of the above issues and concerns have been resolved as there have been a number of published studies that conducted fear conditioning while acquiring fMRI data. The solutions included the use of MRI-safe electrodes that contain no magnetic material or metals such as iron. All cables that are passed into the scanner room are shielded to both protect the SCR data acquisition from scanner noise and vice versa. In addition, the cables carrying the electric shock into the scanner room and cables carrying SCR data can be passed through a batch board onto which radio frequency (rf) filters can be connected to. The rf filter will be able to remove most of electrical noise being carried into the fMRI environment. Employing all of the above has allowed investigators to successfully combine imaging tools with classic fear-conditioning paradigms that have recently provided us with valuable data to help translate what has been learned from the rodent brain into the human brain.

4. Summary

In this chapter, we provided a broad overview on the methodology and implementation of fear conditioning in rodents and in humans. As can be noted in this chapter, there are various ways in which the fear-conditioning paradigm can be carried out, including the type of conditioned stimulus (CS, i.e., auditory vs. visual), the number of trials used, the type of unconditioned stimulus (US), and the type of conditioning (cued vs. contextual). Also varies is the number of experimental manipulations and tests to follow the conditioning paradigm (e.g., conducting extinction sessions). The scientific question at hand will of course be an important factor on determining the specific parameters to be used and the type of follow-up sessions to be conducted. Regarding human conditioning, additional issues should be carefully considered. For example, ethical issues regarding the use of electric shock should be considered, especially when implementing this paradigm in patient populations or in younger populations (i.e., adolescents or children). Also, the use of fear-conditioning paradigms during functional imaging studies should be carefully implemented to ensure the safety of the participants and the quality of the data to be gathered.

References

1. Pavlov I (1927) Conditioned reflexes. Oxford University Press, London
2. Rescorla RA, Heth CD (1975) Reinstatement of fear to an extinguished conditioned stimulus. J Exp Psychol Anim Behav Process 1(1):88–96
3. Bouton ME, Bolles RC (1979) Role of conditioned contextual stimuli in reinstatement of extinguished fear. J Exp Psychol Anim Behav Process 5(4):368–378
4. Bouton ME, King DA (1983) Contextual control of the extinction of conditioned fear: tests for the associative value of the context. J Exp Psychol Anim Behav Process 9(3):248–265
5. Davidson PO, Payne RW, Sloane RB (1964) Introversion, neuroticism, and conditioning. J Abnorm Psychol 68:136–143
6. Kim JH, Richardson R (2009) Expression of renewal is dependent on the extinction–test interval rather than the acquisition–extinction interval. Behav Neurosci 123(3):641–649
7. Burgos-Robles A, Vidal-Gonzalez I, Quirk GJ (2009) Sustained conditioned responses in prelimbic prefrontal neurons are correlated with fear expression and extinction failure. J Neurosci 29(26):8474–8482
8. Quirk GJ, Russo GK, Barron JL, Lebron K (2000) The role of ventromedial prefrontal cortex in the recovery of extinguished fear. J Neurosci 20(16):6225–6231
9. Rabinak CA, Orsini CA, Zimmerman JM, Maren S (2009) The amygdala is not necessary for unconditioned stimulus inflation after Pavlovian fear conditioning in rats. Learn Mem 16(10):645–654
10. Morgan MA, Romanski LM, Ledoux JE (1993) Extinction of emotional learning:

contribution of medial prefrontal cortex. Neurosci Lett 163(1): 109–113
11. Knapska E, Maren S (2009) Reciprocal patterns of c-Fos expression in the medial prefrontal cortex and amygdala after extinction and renewal of conditioned fear. Learn Mem 16(8):486–493
12. Lebron K, Milad MR, Quirk GJ (2004) Delayed recall of fear extinction in rats with lesions of ventral medial prefrontal cortex. Learn Mem 11(5):544–548
13. Quinn JJ, Wied HM, Ma QD, Tinsley MR, Fanselow MS (2008) Dorsal hippocampus involvement in delay fear conditioning depends upon the strength of the tone-footshock association. Hippocampus 18(7):640–654
14. Joseph R, Gallagher RE (1980) Gender and early environmental influences on activity, overresponsiveness, and exploration. Dev Psychobiol 13(5):527–544
15. Santini E, Ge H, Ren K, Pena dO, Quirk GJ (2004) Consolidation of fear extinction requires protein synthesis in the medial prefrontal cortex. J Neurosci 24(25): 5704–5710
16. Walker DL, Ressler KJ, Lu KT, Davis M (2002) Facilitation of conditioned fear extinction by systemic administration or intra-amygdala infusions of D-cycloserine as assessed with fear-potentiated startle in rats. J Neurosci 22(6):2343–2351
17. Rogelj B, Hartmann CE, Yeo CH, Hunt SP, Giese KP (2003) Contextual fear conditioning regulates the expression of brain-specific small nucleolar RNAs in hippocampus. Eur J Neurosci 18(11):3089–3096
18. Anagnostaras SG, Maren S, Sage JR, Goodrich S, Fanselow MS (1999) Scopolamine and Pavlovian fear conditioning in rats: dose–effect analysis. Neuropsychopharmacology 21(6):731–744
19. Blair HT, Tinkelman A, Moita MA, Ledoux JE (2003) Associative plasticity in neurons of the lateral amygdala during auditory fear conditioning. Ann NY Acad Sci 985: 485–487
20. Corcoran KA, Quirk GJ (2007) Activity in prelimbic cortex is necessary for the expression of learned, but not innate, fears. J Neurosci 27(4):840–844
21. Milad MR, Quirk GJ (2002) Neurons in medial prefrontal cortex signal memory for fear extinction. Nature 420(6911):70–74
22. Rescorla RA (1973) Effect of US habituation following conditioning. J Comp Physiol Psychol 82(1):137–143
23. Bouton ME (1993) Context, time, and memory retrieval in the interference paradigms of Pavlovian learning. Psychol Bull 114(1):80–99
24. Pearce JM, Hall GA (1980) Model for Pavlovian learning: variations in the effectiveness of conditioned but not of unconditioned stimuli. Psychol Rev 87(6): 532–552
25. Rescorla RA, Wagner AD (1972) A theory of Pavlovian conditioning: variations in the effectiveness of reinforcement and nonreinforcement. In: Black AH, Prokasy WF (eds) Classical conditioning II. Appleton-Century-Crofts, New York, NY, pp 64–99
26. Stote DL, Fanselow MS (2004) NMDA receptor modulation of incidental learning in Pavlovian context conditioning. Behav Neurosci 118(1):253–257
27. Fanselow MS (2000) Contextual fear, gestalt memories, and the hippocampus. Behav Brain Res 110(1–2):73–81
28. Maren S, Tocco G, Chavanne F, Baudry M, Thompson RF, Mitchell D (1994) Emergence neophobia correlates with hippocampal and cortical glutamate receptor binding in rats. Behav Neural Biol 62(1):68–72
29. Hitchcock JM, Davis M (1987) Fear-potentiated startle using an auditory conditioned stimulus: effect of lesions of the amygdala. Physiol Behav 39(3):403–408
30. Davis M, Astrachan DI (1978) Conditioned fear and startle magnitude: effects of different footshock or backshock intensities used in training. J Exp Psychol Anim Behav Process 4(2):95–103
31. Stoddart CW, Noonan J, Martin-Iverson MT (2008) Stimulus quality affects expression of the acoustic startle response and prepulse inhibition in mice. Behav Neurosci 122(3):516–526
32. Blumenthal TD, Goode CT (1991) The startle eyeblink response to low intensity acoustic stimuli. Psychophysiology 28(3): 296–306
33. Davis M (1992) The role of the amygdala in fear-potentiated startle: implications for animal models of anxiety. Trends Pharmacol Sci 13(1):35–41
34. Davis M, Walker DL, Myers KM (2003) Role of the amygdala in fear extinction measured with potentiated startle. Ann NY Acad Sci 985:218–232
35. Orr SP, Lanzetta JT (1980) Facial expressions of emotion as conditioned stimuli for human autonomic responses. J Pers Soc Psychol 38(2):278–282
36. Ohman A, Dimberg U (1978) Facial expressions as conditioned stimuli for electrodermal responses: a case of "preparedness"? J Pers Soc Psychol 36(11):1251–1258

37. Hamm AO, Vaitl D, Lang PJ (1989) Fear conditioning, meaning, and belongingness: a selective association analysis. J Abnorm Psychol 98(4):395–406
38. Lang PJ, Bradley MM, Cuthbert BN (1998) Emotion and motivation: measuring affective perception. J Clin Neurophysiol 15(5):397–408
39. Jones T, Davey GC (1990) The effects of cued UCS rehearsal on the retention of differential 'fear' conditioning: an experimental analogue of the 'worry' process. Behav Res Ther 28(2):159–164
40. Milad MR, Orr SP, Pitman RK, Rauch SL (2005) Context modulation of memory for fear extinction in humans. Psychophysiology 42(4):456–464
41. Milad MR, Rauch SL, Pitman RK, Quirk GJ (2006) Fear extinction in rats: implications for human brain imaging and anxiety disorders. Biol Psychol 73:61–71
42. Milad MR, Orr SP, Lasko NB, Chang Y, Rauch SL, Pitman RK (2008) Presence and acquired origin of reduced recall for fear extinction in PTSD: results of a twin study. J Psychiatr Res 42:515–520
43. Phelps EA, Delgado MR, Nearing KI, Ledoux JE (2004) Extinction learning in humans: role of the amygdala and vmPFC. Neuron 43(6):897–905
44. Orr SP, Metzger LJ, Lasko NB, Macklin ML, Peri T, Pitman RK (2000) De novo conditioning in trauma-exposed individuals with and without posttraumatic stress disorder. J Abnorm Psychol 109(2):290–298
45. Milad MR, Pitman RK, Ellis CB, Gold AB, Shin LM, Lasko NB, Handwerger K, Orr SP, Rauch SL (2009) Neurobiological basis for failure to recall extinction memory in posttraumatic stress disorder. Biol Psychiatry 66(5):1075–1082
46. Birbaumer N, Veit R, Lotze M, Erb M, Hermann C, Grodd W, Flor H (2005) Deficient fear conditioning in psychopathy: a functional magnetic resonance imaging study. Arch Gen Psychiatry 62(7):799–805
47. Rabinak CA, Maren S (2008) Associative structure of fear memory after basolateral amygdala lesions in rats. Behav Neurosci 122(6):1284–1294
48. LaBar KS, Gatenby JC, Gore JC, Ledoux JE, Phelps EA (1998) Human amygdala activation during conditioned fear acquisition and extinction: a mixed-trial fMRI study. Neuron 20(5):937–945
49. Bechara A, Damasio H, Damasio AR, Lee GP (1999) Different contributions of the human amygdala and ventromedial prefrontal cortex to decision-making. J Neurosci 19(13):5473–5481
50. Hoefer M, Allison SC, Schauer GF, Neuhaus JM, Hall J, Dang JN, Weiner MW, Miller BL, Rosen HJ (2008) Fear conditioning in frontotemporal lobar degeneration and Alzheimer's disease. Brain 131(Pt 6):1646–1657
51. Beaver JD, Mogg K, Bradley BP (2005) Emotional conditioning to masked stimuli and modulation of visuospatial attention. Emotion 5(1):67–79
52. LaBar KS, Ledoux JE, Spencer DD, Phelps EA (1995) Impaired fear conditioning following unilateral temporal lobectomy in humans. J Neurosci 15(10):6846–6855
53. Armony JL, Dolan RJ (2001) Modulation of auditory neural responses by a visual context in human fear conditioning. Neuroreport 12(15):3407–3411
54. Grillon C, Davis M (1997) Fear-potentiated startle conditioning in humans: explicit and contextual cue conditioning following paired versus unpaired training. Psychophysiology 34(4):451–458
55. Jovanovic T, Keyes M, Fiallos A, Myers KM, Davis M, Duncan EJ (2005) Fear potentiation and fear inhibition in a human fear-potentiated startle paradigm. Biol Psychiatry 57(12):1559–1564
56. Norrholm SD, Jovanovic T, Vervliet B, Myers KM, Davis M, Rothbaum BO, Duncan EJ (2006) Conditioned fear extinction and reinstatement in a human fear-potentiated startle paradigm. Learn Mem 13(6):681–685
57. Dawson ME, Schell AM, Filion DL (2000) The electrodermal system. In: Cacioppo JT, Tassinary LG, Berntson GG (eds) Handbook of psychophysiology, 2nd edn. Cambridge University Press, New York, NY, pp 200–223
58. Orr SP, Metzger LJ, Lasko NB, Macklin ML, Hu FB, Shalev AY, Pitman RK (2003) Physiologic responses to sudden, loud tones in monozygotic twins discordant for combat exposure: association with posttraumatic stress disorder. Arch Gen Psychiatry 60(3):283–288
59. Hasegawa K, Fukuda M (1999) Emotional changes on blood pressure in classical fear conditioning. Jpn J Physiol 49(Supplement):S191
60. Davis M (1992) The role of the amygdala in fear-potentiated startle: implications for animal models of anxiety. Trends Pharmacol Sci 13(1):35–41
61. Sinha R, Lovallo WR, Parsons OA (1992) Cardiovascular differentiation of emotions. Psychosom Med 54(4):422–435

62. Hall G, Honey RC (1990) Context-specific conditioning in the conditioned-emotional-response procedure. J Exp Psychol Anim Behav Process 16(3):271–278
63. Bouton ME, Swartzentruber D (1986) Analysis of the associative and occasion setting properties of contexts participating in a Pavlovian discrimination. J Exp Psychol 12:333–350
64. Lovibond PF, Preston GC, Mackintosh NJ (1984) Context specificity of conditioning, extinction, and latent inhibition. J Exp Psychol Anim Behav Process 10:360–375
65. Kaye H, Mackintosh NJ (1990) A change of context can enhance performance of an aversive but not of an appetitive conditioned response. Q J Exp Psychol B 42(2):113–134
66. Chang CH, Maren S (2009) Early extinction after fear conditioning yields a context-independent and short-term suppression of conditional freezing in rats. Learn Mem 16(1):62–68
67. Schiller D, Levy I, Niv Y, Ledoux JE, Phelps EA (2008) From fear to safety and back: reversal of fear in the human brain. J Neurosci 28(45):11517–11525
68. Skinner BF (1937) The behavior of organisms. Appleton-Century-Crofts, New York, NY, p 277
69. Smith SG, Borgen LA, Davis WM, Pace HB (1971) Automatic magazine and bar-press training in the rat. J Exp Anal Behav 15(2):197–198
70. Dirikx T, Hermans D, Vansteenwegen D, Baeyens F, Eelen P (2004) Reinstatement of extinguished conditioned responses and negative stimulus valence as a pathway to return of fear in humans. Learn Mem 11(5):549–554
71. Hermans D, Craske MG, Mineka S, Lovibond PF (2006) Extinction in human fear conditioning. Biol Psychiatry 60(4):361–368
72. Vansteenwegen D, Hermans D, Vervliet B, Francken G, Beckers T, Baeyens F, Eelen P (2005) Return of fear in a human differential conditioning paradigm caused by a return to the original acquisition context. Behav Res Ther 43(3):323–336
73. Milad MR, Quinn BT, Pitman RK, Orr SP, Fischl B, Rauch SL (2005) Thickness of ventromedial prefrontal cortex in humans is correlated with extinction memory. Proc Natl Acad Sci USA 102(30):10706–10711
74. Rauch SL, Milad MR, Orr SP, Quinn BT, Fischl B, Pitman RK (2005) Orbitofrontal thickness, retention of fear extinction, and extraversion. Neuroreport 16(17):1909–1912
75. Coelho CM, Santos JA, Silva C, Wallis G, Tichon J, Hine TJ (2008) The role of self-motion in acrophobia treatment. Cyberpsychol Behav 11(6):723–725
76. Rothbaum BO, Hodges LF, Kooper R, Opdyke D, Williford JS, North M (1995) Effectiveness of computer-generated (virtual reality) graded exposure in the treatment of acrophobia. Am J Psychiatry 152(4):626–628
77. Reger GM, Gahm GA (2008) Virtual reality exposure therapy for active duty soldiers. J Clin Psychol 64(8):940–946
78. Pace-Schott EF, Milad MR, Orr SP, Rauch SL, Stickgold R, Pitman RK (2009) Sleep promotes generalization of extinction of conditioned fear. Sleep 32(1):19–26
79. Kalisch R, Korenfeld E, Stephan KE, Weiskopf N, Seymour B, Dolan RJ (2006) Context-dependent human extinction memory is mediated by a ventromedial prefrontal and hippocampal network. J Neurosci 26(37):9503–9511
80. An alternative scoring method for skin conductance responding in a differential fear conditioning paradigm with a long-duration conditioned stimulus. Pineles SL, Orr MR, Orr SP. Psychophysiology. 2009 46(5):984–995

Chapter 6

Conditioned Place Preference in Rodents and Humans

Devin Mueller and Harriet de Wit

Abstract

Place conditioning is among the most commonly used procedures to assess drug reward in animals. The procedure is used to study acquisition of conditioning, extinction, and reinstatement, to compare across drugs and doses of drugs, and to examine interactions between drugs and environmental or organismic variables. Studies using the procedure have provided a rich source of data regarding contextual conditioning in rodents, and most recently, in humans. Despite its widespread use, the place preference procedure has also raised theoretical and practical questions. Some of the questions are related to the procedural details and methods used: methodological variations on the procedure can affect the outcome and interpretation. In this review, we will examine some of the important methodological considerations in place conditioning with drugs and discuss how these have bearing on the results and conclusions. First, we will discuss what is being measured with place conditioning. Second, we will review the key phases of the procedure and methodological variations in the procedure that can influence the outcome. Third, we will describe place conditioning in humans and the unique methodological issues that arise in applying the procedure to humans. Finally, we will discuss potential limitations and future directions related to drug-induced place conditioning.

1. Introduction

1.1. What Is Measured in Place Conditioning?

Place conditioning procedures are designed to measure the rewarding properties of drugs. Because reward is a subjective experience, it can only be studied indirectly in non-humans, by the behavior that it elicits. Two main approaches have been used to assess reward-related behavior. One approach, the one used in the place preference procedure, is to measure elicited approach and contact responses. This tendency to approach rewarding stimuli is termed incentive salience and it is based on the principles of classical conditioning. The second approach, which is used in self-administration procedures, assesses the ability of the drugs or other stimuli to increase the probability of responses that precede them. This is the reinforcing feature of rewards and it is based on operant principles of learning. Thus, place conditioning assesses reward based on the principles of classical conditioning, and it essentially assesses the association between two stimuli, an unconditioned stimulus (UCS) consisting of the drug effect and a conditioned stimulus (CS) consisting of the environment in which the drug is experienced.

The place conditioning method is based on the observation that an animal will approach stimuli that have been previously paired with the rewarding effects of a drug. Thus, when an animal approaches and maintains contact with an environment where it has previously received a drug, then we infer that the drug was rewarding. In this situation, the rewarding property of the drug serves as a UCS, which is repeatedly paired with a previously neutral place (i.e., experimental chamber). During the course of conditioning, the chamber acquires secondary motivational properties, becoming the CS that can elicit approach and contact. During conditioning, animals are usually confined to one chamber of a two-chamber apparatus after drug administration and confined to the other chamber following a vehicle injection. In some studies the two chambers are separated by a central choice area. During the testing phase, the animals are allowed to explore the entire apparatus in a drug-free state, and the amount of time spent in the two conditioning chambers is recorded. Animals that spend more time in the drug-associated chamber are considered to exhibit a "conditioned place preference" (CPP) whereas animals that spend more time in the non-drug-associated chamber are considered to exhibit a "conditioned place aversion" (CPA).

Place conditioning was first used to explore the motivational effects of radiation on rats, which resulted in a CPA (1). Shortly after, Beach published the first report of a drug-induced CPP, showing that rats preferred the morphine-paired arm of a Y maze (2). Since then, a vast literature has accumulated using the place conditioning method (3–5). For example, just within the past 10 years, the number of publications based on this procedure has tripled (**Fig. 6.1**). Most drugs that serve as reinforcers (i.e., are self-administered) also induce a CPP, including morphine, amphetamine, methylphenidate, nicotine, and cocaine (4). Drug-induced CPP has been demonstrated with several species, including mice (6), hamsters (7), primates (8), birds (9), zebrafish (10), rats (3), and most recently, humans (11). Given the very widespread use of this procedure, it is not surprising that variations in the procedure have evolved, some of which affect the results and interpretation of the findings. Here, we will

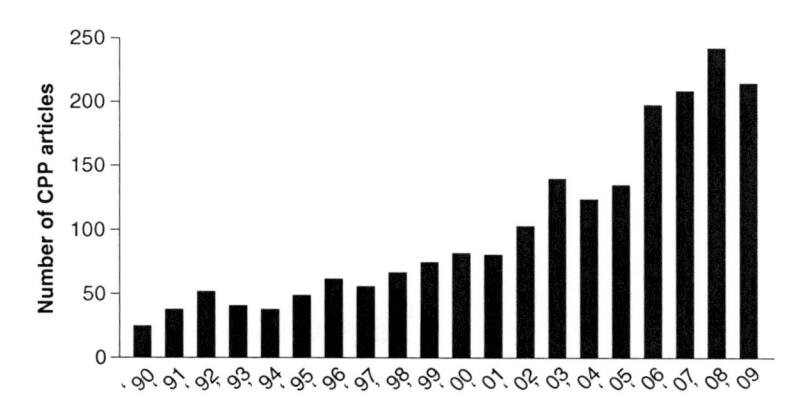

Fig. 6.1. Number of peer-reviewed articles retrieved from PubMed reporting the use of conditioned place preference in the past two decades by year.

discuss some of the key features of place conditioning procedures, including their applications to the study of extinction and reinstatement.

2. Key Phases of the Procedure

2.1. Pre-conditioning Test

Most place conditioning studies include a pre-conditioning exposure and test, in which animals are allowed to explore the apparatus before any drug administration to reduce the novelty of the procedure and to assess pre-existing preferences for the two conditioning chambers. Because animals sometimes avoid novel environments, a novel environment may mask a potential CPP for a drug, including a reliably rewarding drug such as cocaine (12). For this reason, animals are usually acclimated to the environment before conditioning. During the pre-conditioning test, the amount of time an animal spends in the two to-be-conditioned chambers is recorded. In most versions of the place conditioning procedure, this initial preference is used to evaluate the rewarding effects of the drug after conditioning, and in some versions of the procedure, this preference is used to assign which chamber is to be associated with drug or vehicle (see below).

2.2. Conditioning

The standard procedure for place conditioning with drugs is to pair one distinct chamber with a drug injection for one session, and pair a second chamber with vehicle in a separate session (*see* **Fig. 6.2**). The number of pairings used may vary from one to six, depending on the drug, the dose, and the route of administration (usually intraperitoneal (13)). Although the most common number of pairings is four, CPP has been demonstrated with just a single pairing with morphine (14), heroin (15), or β-endorphin

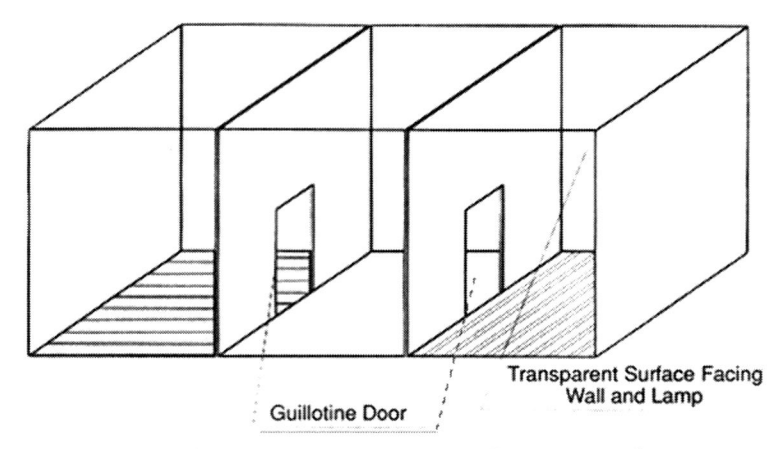

Fig. 6.2. Example of a place conditioning apparatus. Two distinctly different chambers are separated by a neutral center chamber or tunnel. During conditioning, the animal is confined to one chamber following a drug injection, and the opposite chamber following vehicle injection. At test, the animal is free to explore all three chambers.

(16). Morphine-induced CPP increases with the number of pairings, from two to four (17). The most common duration of each pairing is 30 min, although this has varied from 4 (18) to 120 min (19). The duration of the pairings is determined in part by the half-life of the drug used for conditioning: drugs with a shorter half-life are typically paired with a chamber for a shorter period (e.g., cocaine 20–30 min) whereas drugs with a longer half-life are paired with a chamber for a longer period (e.g., morphine 30–120 min). Interestingly, Parker and colleagues (19), using a three-choice apparatus to compare preference for morphine, cocaine, and vehicle, demonstrated that the relative CPP for morphine or cocaine-paired chambers depended on the length of the conditioning session; rats showed a stronger preference for the morphine-paired chamber than the cocaine-paired chamber when the pairing durations were 120 min, whereas they preferred the cocaine-paired chamber when the pairing durations were only 15 min. The authors argued that this shift in relative preference was the result of the faster onset of action of cocaine, compared to morphine.

Researchers have used two procedures to assign animals to the two to-be-conditioned chambers, the "biased" approach and the "unbiased" approach. With the "biased" technique, animals receive the drug conditioning trials in the least preferred side, as determined by the pre-conditioning test. With the "unbiased" technique, animals are assigned to receive drug or saline in either of the two chambers, based on counterbalanced assignments. There are advantages and disadvantages to each approach. The biased approach takes into account pre-existing preferences for one chamber over the other, which may confound the interpretation of a shift in preference. That is, the effect of the drug may interact with the factors that lead to the initial unconditioned preference, or avoidance, of one of the chambers, in ways that are not related to reward (e.g., fear reduction). On the other hand, with the unbiased procedure, animals spend an equal amount of time in the two chambers before conditioning which, some have argued, may allow a clearer measure of the rewarding properties of a drug (20). Although relatively few studies have directly compared the two techniques, the animals' initial preferences have been shown to affect the observed CPP with several different drugs. For example, cocaethylene induced a CPP when paired with the initially non-preferred side, but not when paired with the initially preferred side (21). Similar results were obtained with intraperitoneally (IP) administered cocaine (but not intravenous cocaine (12)), subcutaneously administered heroin (22), and for IP administered clonidine (23). [Leu]enkephalin, administered IP, produced a CPP when it was paired with the initially non-preferred chamber, but this drug produced a CPA when paired with the initially preferred chamber (6). The magnitude of CPP for amphetamine also reportedly depends on the baseline preference of the animals (24). These inconsistencies raise some concern about the interpretation of the results of the biased procedure, and indeed, some researchers now recommend the unbiased procedure (3, 4).

Several control conditions have been employed with the place preference procedure to ensure that the observed CPP is related specifically to the process of conditioning. First, some studies have included an explicitly unpaired group that receives the same exposure to the drug and to the chambers, but the drug and environments are not systematically paired. This rules out the possibility that any exposure to the drug, or to the environments, affects place preference. Although this control is not usually considered necessary for most behavioral studies, it is commonly used when examining learning-induced changes in neurochemistry and signaling molecules. For example, in one study, animals that had received explicit pairing between drug and environment exhibited less ERK2 activity in the frontal association cortex than animals that had received the drug in both chambers (25). In another study, explicit drug pairings, but not unpaired drug exposure, induced a significant increase in Fos/GAD67 co-labeling in the prelimbic cortex (26). Another control condition that is sometimes used is to administer vehicle prior to confinement in both chambers. This control is necessary to demonstrate that a CPP is dependent on explicit drug pairings rather than simply exposure to the apparatus and is also necessary when examining signaling molecules associated with drug-related learning (25). Overall, few CPP studies incorporate these control groups, and instead rely on a within-subjects design that allows for comparison of preferences between chambers, rather than between conditioning groups.

2.3. Testing

After drug conditioning trials are completed, animals are tested for preference for the two chambers in the drug-free state. During the preference test, they are given free access to the entire apparatus with all the barriers removed, and the primary outcome measure is the amount of time the animal spends in each of the two chambers. The most typical duration of this test is 15 min, but the times range from 10 to 45 min (13). The duration of the test must be sufficient to allow for a preference to emerge, as animals' preference for the drug-paired chamber may increase as the test progresses, at least up to 15 min (27). Mueller and Stewart (27) noted that a cocaine-induced CPP was not apparent during the first 3 min of the test, but emerged after 6 min, and then increased in magnitude throughout the 15-min test.

2.4. Extinction

Place preference procedures are sometimes used to study the extinction of drug-associated conditioning. Extinction in Pavlovian conditioning refers to the decline in the frequency or intensity of a conditioned response (CR) when the CS is presented without the UCS. Thus, extinction of a CPP is measured as a decline in the amount of time spent in the chamber previously paired with drug administration, to the point that the animal shows no preference for one chamber over the other, or reverts to its initial, pre-conditioning preference. Some researchers use as a criterion for extinction a significant decrease in time spent in the drug-paired chamber, whereas others examine time spent in the drug-paired chamber among drug-treated versus vehicle-treated animals (28), and yet others use a criterion of spending less than 55% of the total time in the drug-paired chamber on two consecutive days (29).

Two distinct extinction procedures have been used in the CPP paradigm. In one procedure, animals are given repeated CPP tests with access to the entire apparatus in the absence of the drug until the preference is diminished (see **Fig. 6.3**). In the other procedure, following an initial CPP test, animals receive vehicle in both chambers and are confined to each of the chambers in a similar manner to conditioning. After pairing each chamber with saline across several days, a subsequent CPP test is given to ensure that a CPP no longer is observed. Repeated testing results in a gradual decline in CPP across days, as shown with a cocaine-induced CPP (27, 30–32). Pairing both chambers with the vehicle results in the extinction of a CPP, including that induced by morphine (33), cocaine (27, 32), and ethanol (34). This second procedure does not provide information regarding extinction across

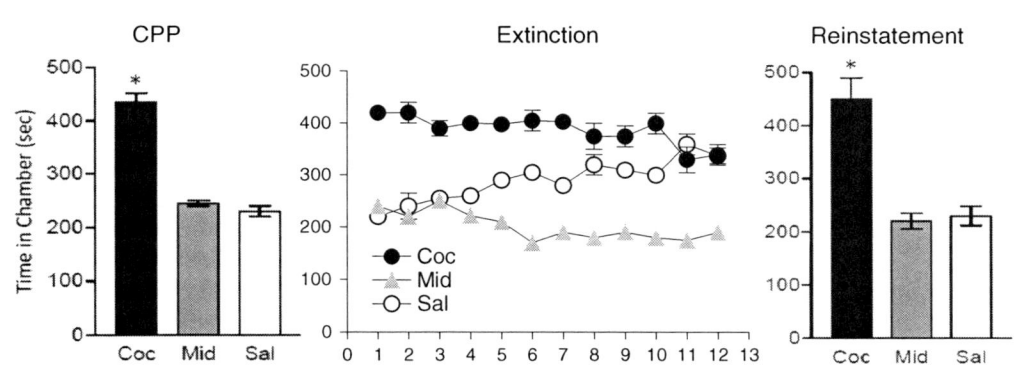

Fig. 6.3. Extinction–reinstatement procedure. Mean (± sem) time spent in each of three chambers in which rats (N = 12) received cocaine (Coc; 10 mg/kg, IP), saline (Sal), or no treatment (middle chamber, Mid). Rats were tested for a cocaine-induced CPP, then given extinction training consisting of repeated testing. When a side preference was no longer detectable, rats were given an injection of cocaine and tested for a preference. Cocaine induced a significant reinstatement of the CPP. Adapted from Mueller and Stewart (27). *$p < 0.05$.

days, but rather extinction is evaluated after multiple vehicle trials in a single test.

A drug-induced CPP can persist over time, and extinction of the CPP depends upon the number and timing of non-paired exposures to the environment. Drug-induced CPPs remain intact for long periods of time, in the absence of explicit extinction training. For example, a cocaine-induced CPP endured up to 4 weeks (27) and a morphine-induced CPP endured for at least 6 weeks (33) when there were no intervening experiences in the test environment. In other studies, CPP based on morphine conditioning and CPA based on naloxone remained 1 month following the last conditioning trial (17). The intervals between extinction trials also influence the rate of extinction of CPP. Animals that received repeated preference tests with long inter-test intervals of 2 weeks maintained robust preferences across the extinction trials, for both a cocaine-induced (27) and morphine-induced CPPs (33). These results suggest that tests given at sufficiently long intervals serve as reminders to maintain the significance of the drug-related stimuli. Thus, extinction procedures should ensure that extinction sessions are massed (i.e., daily), not spaced (i.e., weekly or longer), in time.

2.5. Reinstatement

Reinstatement refers to the return of a conditioned response after extinction, when an animal is exposed to the unconditioned stimulus. In the CPP paradigm, reinstatement is observed when the drug is administered prior to a standard CPP test (27) and is measured as an increase in time spent in the previously drug-paired chamber (*see* **Fig. 6.3**). Other, non-drug stimuli may also lead to reinstatement of a CPP, including exposure to stressors or conditioned stressors (35, 36).

Drug-induced reinstatement of a CPP has been demonstrated with several drugs and in several species. For example, injections of low doses of morphine can reinstate morphine-induced CPP after extinction in both rats (33, 37, 38) and mice (29, 39); heroin reinstates heroin-induced CPP (40), and cocaine reinstates cocaine-induced CPP in rats (27, 36, 41) and mice (42–44), and reinstatement has also been reported with amphetamine (28, 45), nicotine (46), ethanol (34, 47), and MDMA (48). Typically, the dose of drug used to reinstate CPP is lower than (e.g., half) the dose used to establish conditioning. This use of a lower dose is based on the idea that the reinstatement dose provides the animal with a "taste" of the drug while minimizing possible effects of the drug on locomotor and exploratory behavior.

Some studies have also demonstrated drug-induced reinstatement utilizing a drug other than the drug that was used during conditioning. For example, small doses of either cocaine or amphetamine can reinstate a morphine-induced CPP (49, 50), and cocaine-induced CPP can be reinstated by methylphenidate, methamphetamine, morphine, nicotine, and ethanol (42, 51). Finally, morphine and ethanol can also reinstate a nicotine-induced CPP after extinction (46, 52). These cross-drug reinstatement studies are especially valuable to characterize the generalizability of the reinstatement cue. Further, to the extent that reinstatement can be considered to be a model for relapse in humans (53, 54), the ability of other drugs to reinstate CPP may also provide an indication of which drugs pose a risk for relapse among abstinent drug users.

Other studies have shown that acute stress reinstates drug-induced CPP. Acute stress is implicated in relapse to drug use (55, 56), and it effectively reinstates operant responding for self-administered drugs (57). Thus, it is not surprising that stress can induce reinstatement of a CPP after extinction. Several stressors have been shown to reinstate morphine-induced CPP, including intermittent footshock (58), immobilization (59), tail-pinch (59), forced swim (60), conditioned withdrawal in morphine-dependent rats (38), and social defeat (59). Similarly, cocaine-induced CPP is reinstated by intermittent footshock (61), restraint stress (62), forced swim (63), and fear conditioned stimuli (36). Thus, drug-induced CPP can be readily reinstated by any of a number of stressful interventions, suggesting that the memory of the place conditioning remains robust even after an initial period of extinction. A standard rodent place conditioning procedure with extinction and reinstatement test is provided in **Table 6.1**.

Table 6.1
Rodent place conditioning procedure with extinction and reinstatement test

Pre-test (day 1)

1. Prepare all conditioning boxes with clean floors and drop pans
2. Weigh each rat/mouse and place into the center chamber of the box. Record activity
3. After the 15-min pre-test session is complete, remove the rat/mouse from the apparatus and return it to its home cage
4. Determine whether an unconditioned place preference exists. If so, assign the least preferred chamber as the drug conditioning chamber (biased procedure). If no preference, assign half of rats/mice to one chamber, and half to other in counterbalanced fashion (unbiased procedure). Next, assign half of rats/mice to conditioning (CS+ trials) on odd days, and half to even days

Conditioning (days 2–9)

5. Twenty-four hours after the pre-test session, initiate the first conditioning session
6. Prepare conditioning boxes with the appropriate floors
7. Weigh each rat/mouse and immediately inject IP with drug (if CS+ trial) or vehicle (if CS− trial). Place the rat/mouse into the assigned chamber. Record activity during trial (optional)
8. After the conditioning trial is complete (i.e., 20 min for cocaine, 45 min for morphine), remove the rat/mouse from the apparatus and return it to its home cage. These trials will occur at 48-h intervals (i.e., on days 2, 4, 6, and 8)
9. Twenty-four hours later, prepare the boxes with clean floors
10. Weigh each rat/mouse and immediately inject IP with drug (if CS+ trial) or vehicle (if CS− trial). Place the rat/mouse into the appropriate chamber. Record activity during trial (optional)
11. After the conditioning trial is complete, remove the rat/mouse from the apparatus and return it to its home cage. These trials will occur at 48-h intervals (i.e., days 3, 5, 7, and 9)

Note: We recommend a 2-day break between conditioning and the preference test. We have found that this break results in a greater conditioned place preference than if animals are tested the day after conditioning is complete

Preference test (day 12)

12. Seventy-two hours after the final conditioning session, prepare each conditioning box with clean floors
13. Weigh each rat/mouse and immediately and place into the center of the box. Record test activity
14. After the preference test is complete, remove the rat/mouse from the apparatus and return it to its home cage

Extinction training (days 13–21)

15. On days 13–20, either repeat preference test protocol (repeated testing procedure) or repeat conditioning protocol with vehicle *ONLY* in both chambers (vehicle pairing procedure)

Note: Repeated testing procedure may require additional days

16. On day 21, repeat preference test protocol

Reinstatement test (day 22)

17. Weigh each rat/mouse and immediately inject IP with drug (drug-induced reinstatement) or initiate stressor (stress-induced reinstatement). Place the rat/mouse into the center of the box. Record test activity

Note: For drug-induced reinstatement, the dose used should be low enough not to impede locomotor behavior (usually half of conditioning dose)

18. After the reinstatement test session is complete (typically 15 min), remove the rat/mouse from the apparatus and return it to its home cage

3. Physical Aspects of the Testing Environment

The apparatus most commonly used for place conditioning with rodents is a two-chamber or three-chamber (one neutral) box with removable doors to isolate each chamber during conditioning (*see* **Fig. 6.2**). The two-chamber apparatus (with or without a neutral chamber) is well suited for studies of place conditioning when only a single drug is being examined, whereas a multiple-chamber apparatus is typically used when researchers compare more than two drugs (e.g., cocaine, morphine, and vehicle (19)). Multiple-chamber units are useful to assess the relative rewarding effects of different drugs. Typically, the apparatus is designed so that the to-be-conditioned chamber dimensions are equal, but the chambers are maximally distinguishable to facilitate the discrimination (e.g., variations in flooring, lighting, and color).

Place conditioning has also been assessed using different procedures, most notably an open field separated into four quadrants (64). The open field version was designed to assess the degree to which animals will maintain contact with drug-paired floor quadrants. During conditioning, drug administration is paired with one type of flooring (e.g., wire mesh) that is inserted into all four quadrants of the open field. The following day, vehicle administration is paired with another type of flooring (e.g., steel bars) that covers all four quadrants of the open field. At test, the animal is placed in the open field, with some quadrants containing the drug-paired flooring and other quadrants containing the vehicle-paired flooring. The number of quadrants containing the drug-paired flooring can be varied from one to three to demonstrate the extent to which an animal will maintain contact with that flooring during a CPP test. Technically, in this case, the animal does not learn an association between a drug stimulus and a place, but rather between a drug stimulus and a floor texture. However, the outcome and interpretation is similar, that animals track stimuli that have been associated with a rewarding drug stimulus.

Natural variations in the physical features of the environment, such as lighting and odors, can influence the outcome of CPP procedures. This is especially true with stimuli for which animals have pre-existing preferences. Because rodents prefer darker chambers over lit chambers, differences in lighting in the two chambers may result in a chamber bias. The animals' bias toward darker chambers may also be relevant in the selection of colors or wall patterns to differentiate the two chambers. Sometimes researchers use this natural tendency deliberately, when using the "biased" procedure. Animals may also be differentially reactive to different flooring materials. Flooring features are readily discriminated by rodents, and the animals may have pre-existing preferences for certain materials (e.g., rough vs. smooth). To minimize pre-existing preferences while permitting a discrimination, some studies use two types of flooring made of similar material but varying in texture (e.g., stainless steel grid flooring with two vs. four squares per inch). Some studies have used odors to distinguish the two chambers, but these are difficult to control, and may easily be strong enough to be aversive to animals, resulting in a new source of bias. Overall, the physical stimuli associated with the chambers can influence unconditioned preferences, thus inadvertently influence the outcome, or require the investigator to utilize a "biased" procedure to use for conditioning. It should be noted that most commercially available apparatus have a preferred side, requiring the use of the "biased" procedure.

4. CPP in Humans

We have recently demonstrated that CPP also occurs in human volunteers, using oral doses of *d*-amphetamine as the drug (11). Although there is not as yet a standardized human CPP protocol, the methodological details of the one published study may serve as a guide for future studies. Healthy young adults participated in four 4-h conditioning sessions separated by at least 48 h, in which they received capsules containing placebo or *d*-amphetamine (20 mg), in alternating order. Half of the participants were assigned to a "paired" group, who received the drug on two sessions in one room and placebo on two sessions in another room, and half were assigned to an "unpaired" group, who received drug and placebo in random order in the two rooms. On the test session, we asked subjects to rate how much they liked each of the two rooms. Subjects in the paired group reported liking the amphetamine-associated room more than the placebo room, whereas the unpaired subjects were indifferent in their room-liking ratings (**Fig. 6.4**). Most importantly, within the paired group, individuals who reported experiencing the most pleasant effects from the drug on the conditioning sessions also reported greater liking of the drug-associated room (**Fig. 6.5**). This is consistent with the common, but untestable, assumption that place preference in non-humans is associated with "pleasant" internal effects of the drug. That is, the procedure provided an index of the relationship between place preference and the quality and magnitude of the mood-altering effects of the drug.

The human CPP study extends the animal studies in two important ways. First, it extends the evidence for CPP to another species, humans. More importantly, it allows investigators to investigate the relationship between subjective measures of

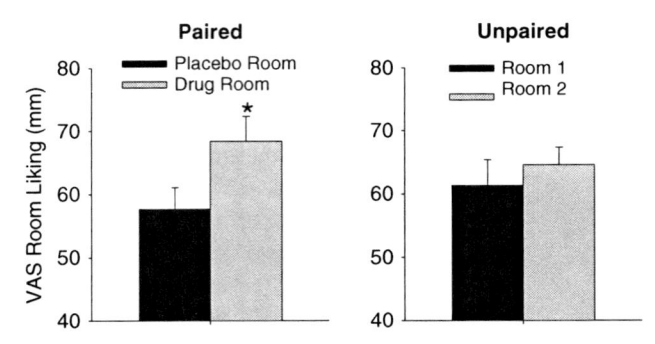

Fig. 6.4. Mean (± sem) ratings of liking of the room on a Visual Analogue Scale (VAS) in which subjects received d-amphetamine (20 mg) or placebo (paired group; N=19) or the two rooms (unpaired group; N=12). Subjects in the paired group reported liking the drug-associated room significantly more than the placebo room. Adapted from Childs and de Wit (11). *$p = 0.05$.

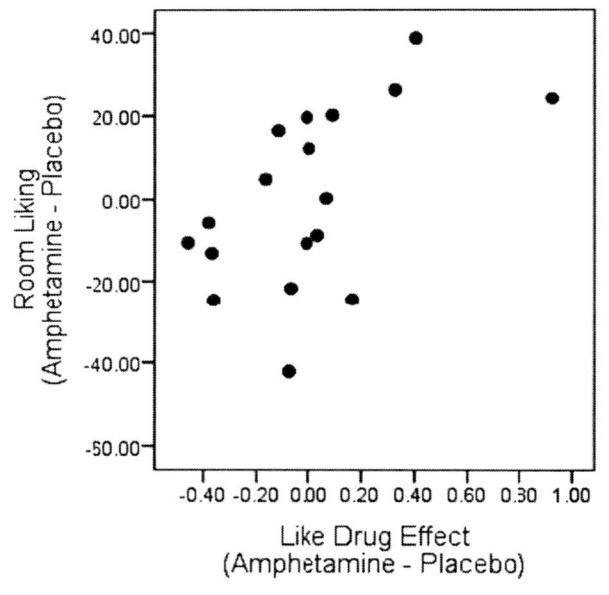

Fig. 6.5. Individual subjects' ratings of drug liking (amphetamine minus placebo) were positively correlated with room liking (amphetamine-paired room minus placebo-paired room) in the subjects in the paired group ($r^2 = 0.71$, $p = 0.001$). Subjects (N=19) received d-amphetamine (20 mg) in one room and placebo in the other room, on two occasions each. Drug-liking ratings were obtained after the conditioning sessions, and room-liking ratings were obtained on the test session. Adapted from Childs and de Wit (11).

"reward," i.e., subjective feelings of well-being, with the behavioral measure of place preference. Although researchers often implicitly infer that animals exhibit a CPP because they "like" the drug, this can only be measured in humans. Future empirical studies using this approach may yield novel and interesting relationships between mood effects of drugs and behavioral preferences. In general, we expect that drugs that produce "pleasurable" or affectively positive internal effects will induce a CPP. However, the interesting possibility also exists that certain drugs may induce a CPP *even without* producing these feelings of well-being. This latter possibility would have fundamentally important implications for our understanding of drug abuse.

4.1. Methodological Considerations for Human CPP

Certain methodological considerations with the human CPP procedure are unique to humans. These include the ethical issues related to administering drugs with potential for abuse to humans and the related constraints on the doses and frequency of dosing. Human volunteers must be carefully screened before participating, for both safety and scientific concerns. Pre-study screening typically includes both medical and psychiatric criteria, and subjects are also screened before each session for recent drug use and pregnancy. Instructions to participants must be carefully considered to minimize the influence of expectancies, and the drugs must be administered under double-blind conditions. Medical and psychiatric assistance must be readily available in the event of an adverse response. Many methodological aspects of the human CPP remain to be explored. For example, the relationship between the dose used during conditioning and preference, the possible influence of the physical characteristics and proximity of the two conditioning rooms, the number of conditioning sessions, the instructions to the subjects, the length of time spent in the conditioning rooms, and the time between conditioning sessions and test sessions are but a few of the unknown variables that influence conditioning. One important direction for future studies using the human place preference procedure is to identify the most appropriate and sensitive primary outcome measure. In our study, the outcome measure was ratings of liking of the rooms. In future studies with the proper physical configuration of rooms, it may be possible to assess time spent in the two rooms, parallel to the measure used in non-humans.

Several important questions arise in the human CPP procedure, some of which relate to similar issues that arise in the non-human studies, and some of which are unique to humans. Most importantly, it remains to be determined whether a biased procedure might yield a more sensitive measure of drug conditioning in humans. The Childs and de Wit Study (11) used an unbiased procedure and included both a paired and an unpaired group to control for unconditioned changes in preference. As in the studies with non-humans, however, the human subjects had a small but detectable preference for one of the two rooms, regardless of conditioning, which may have influenced the CPP. Thus, it will be worth determining, in future studies, whether a biased procedure provides more sensitive results, by determining the subjects' initial room preferences to assign the to-be-conditioned drug and placebo rooms.

As noted above, many methodological variables remain to be studied in human CPP procedures. Two methodological issues arise, however, that are unique to human studies. The first concerns the nature of the dependent measures. Although the primary outcome measure in animal studies is based on the amount of time spent in the two chambers, this is not a practical measure to use in humans. Typically, the conditioning sessions last several hours, during which the subjects are allowed to sit and read or relax. Thus, at the time of testing, human subjects are less likely to walk around and explore the two rooms, and more likely to simply choose one, and sit there. For this reason, in our first study, we used ratings of how much subjects liked the room, rather than a behavioral measure of time spent in each room. In future studies, it may be possible to devise behavioral measures that are more closely related to the non-human model, and less dependent on explicit, verbal ratings of room liking. This raises a second important question regarding the instructions that are given to subjects. In order to make a procedure credible to human volunteers, it is usually necessary to provide verbal instructions, including a rationale for the study. However, there is a risk that the instructions to subjects can influence the outcome. For this reason, in most human psychopharmacology studies the drug is administered under double-blind conditions to minimize the influence of expectancies. In the case of the human CPP procedure, it is also important to provide a reason for asking subjects to rate the rooms where they were tested. We have used two "cover" stories for this purpose: that we were considering redecorating the rooms and requested their assistance, and more recently, we informed subjects that we were testing the interactions between drug effects and physical environments. It is not known whether, or how, these instructional sets influence the outcome of the human CPP procedures. Clearly, this is a potentially valuable paradigm to be developed in humans. The procedure in humans may also have direct clinical relevance to improve our understanding of the strong place associations that drug-dependent individuals report experiencing when they return to former drug-using environments.

5. Possible Limitations of CPP Procedures

Place conditioning also has certain limitations (65). One limitation is the difficulty in obtaining clear dose-dependent effects with the procedure. At best, it is cumbersome to demonstrate that the CPP is dose related, using between groups designs. Even with group designs, it is sometimes noted that CPP is an "all-or-nothing" phenomenon, so that once a threshold dose is reached, there is a CPP and higher doses do not increase this effect further. This makes the procedure anomalous in the field of behavioral pharmacology in which dose responses are the norm, and it also makes it difficult to evaluate the efficacy and potency of various drugs or to compare across drugs or doses.

A second limitation is one of interpretation, and whether CPP is influenced by other factors such as novelty seeking or avoidance of a novel environment. Thus, while drugs may change the amount of time spent in the conditioned chamber, it is not certain that this effect is directly related to the rewarding effect of the drug or some other effect of the drug. The recent study in humans (11) addresses this concern to some degree, but it remains an issue to the extent that other, non-reward-related processes can affect time spent in an environment.

There are also some limitations of the CPP procedure as a model of drug seeking. First, drug is administered non-contingently (i.e., by the experimenter) such that the animal or person does not have any control over drug intake. In contrast, human drug users have control over their use of the drug, and studies with animal drug self-administration techniques have shown that drugs have markedly different effects if they are controlled by the animal or controlled by the experimenter (66, 67). Further, the total number of drug exposures in the CPP procedure is usually limited to the few conditioning trials. As such, this procedure may model initial preference, but does not model the compulsive and chronic use of drugs seen in addicts. Thus, the CPP procedure only provides an index of simple associative learning that occurs when the rewarding properties of a drug are associated with an environment.

One concern that has been raised about the use of the CPP reinstatement procedure is that reinstatement may reflect state-dependent learning. It is well known that the possibility of recalling previously learned information at the time of testing increases when the subject is in the same state as during the encoding phase (68–70). Thus, reinstatement of a CPP induced by a priming injection of the drug may be attributed to the state induced by the drug. That is, the drug produces an altered state that is present during conditioning and reinstatement tests, but not during extinction. Although state-dependent learning may contribute to reinstatement, it is unlikely to play a significant role. First, the initial CPP test is given under drug-free conditions indicating that the approach response is guided by the incentive salience of the cues and not state-dependent learning. Second, the fact that a CPP can be reinstated by drugs other than that used during conditioning (e.g., cocaine reinstates a morphine-induced CPP (49)) or by stress (50, 61) argues against a state-dependent retrieval mechanism.

6. Future Directions

Perhaps because of its relative simplicity, the CPP procedure is widely used to assess the rewarding effects of drugs. It has been used to address a wide range of questions pertaining to drug abuse, including questions about different drugs, doses, brain mechanisms, genetic factors, and behavioral manipulations. More recently, it has been used to study processes of both extinction and reinstatement, as well as other addiction-related processes. For instance, studies have used CPP procedures to examine interactions between drugs and social stimuli. In one study social interaction reduced the aversive properties of alcohol (71), while in another study social interaction enhanced preference for a low (subthreshold) dose of cocaine (72). These studies show that the CPP can be influenced by social interaction in the testing environment, both to dampen aversive effects and enhance positive effects. Place preference has also been used to study individual differences, or "traits" related to drug use, but the results of these studies have been mixed (65). In a widely cited series of studies with drug self-administration in rats, Piazza and colleagues (73, 74) suggested high levels of spontaneous activity and high levels of activity after a stimulant drug predicted a higher propensity to take drugs in operant self-administration paradigms. The idea that "high responder" (HR) and "low responder" (LR) animals differ in the rewarding effects of drugs has subsequently been tested using the place preference procedure, with mixed results. In some studies, HR animals exhibit a stronger CPP (75), but other studies have reported either no effect (76, 77) or the opposite relationship (78, 79). For example, in one study, rats that exhibited lower locomotor response to cocaine tended to develop stronger CPP (78), and similar results were reported in another study in mice (79). Several factors complicate the interpretation of these findings. One concern is that some strains of animals that differ on the HR and LR measures also score higher on behavioral measures of anxiety, and that these baseline levels of "anxiety" may confound the CPP results (75, 80). Another concern is that the HR and LR animals may differ in "novelty-seeking," and that it is this process, rather than the rewarding effects of the drug, that accounts for the original observation with drug self-administration (65). Additionally, the greater locomotor activity observed in HR animals may mask a CPP, as it has been shown that the overall CPP increases as locomotor activity decreases during the course of a preference test (27). Taken together, these studies show that the use of the CPP procedure to study individual differences or "traits" raises a number of interpretational problems that remain to be resolved.

We have reviewed some of the key features of the place preference procedure, as it has been used in rodents, and more recently, also in humans. The extension of the CPP procedure to humans provides a direct link between clinical observations and the rich existing literature on place preference in non-humans. The human place preference procedure may predict which drugs are rewarding or which individuals are likely to develop strong learned associations with drugs. Clinically, it could be used as a screening procedure to assess abuse liability of new compounds or to test the ability of a potential medication to block the positive effects of a known drug of abuse. Third, and most interesting, it could be used to study the conditioning, extinction, and reinstatement of drug-related memories in humans. This would be critically important to test potential medications that might interfere with the acquisition or facilitate the extinction of drug-related conditioning and memories involved in the relapse process.

Acknowledgments

Preparation of this chapter was supported by grants RO1DA02812 (HdW) and RO1DA09133 (HdW) from the National Institute on Drug Abuse, and by a grant to DM from the University of Wisconsin-Milwaukee Research Growth Initiative.

References

1. Garcia J, Kimeldorf DJ, Hunt EL (1957) Spatial avoidance in the rat as a result of exposure to ionizing radiation. Br J Radiol 30:318–321
2. Beach HD (1957) Morphine addiction in rats. Can J Psychol 11:104–112
3. Schechter MD, Calcagnetti DJ (1998) Continued trends in the conditioned place preference literature from 1992 to 1996, inclusive, with a cross-indexed bibliography. Neurosci Biobehav Rev 22:827–846
4. Tzschentke TM (2007) Measuring reward with the conditioned place preference (CPP) paradigm: update of the last decade. Addict Biol 12:227–462
5. Carlezon WA Jr (2003) Place conditioning to study drug reward and aversion. Methods Mol Med 84:243–249
6. Heinrichs SC, Martinez JL Jr (1986) Modification of place preference conditioning in mice by systemically administered [Leu]enkephalin. Behav Brain Res 22:249–255
7. Schnur P, Morrell J (1990) Morphine conditioned place preference in the hamster. Pharmacol Biochem Behav 37:383–385
8. Pomerantz AS, Wertz J, Hepner B, Walso L, Piazza J (1992) Cocaine-induced conditioned place preferences in rhesus monkeys. Soc Neurosci Abstr 18:1572
9. Hughes RA, Baker MR, Rettig KM (1995) Cocaine-conditioned place preference in young precocial domestic fowl. Exp Clin Psychopharmacol 3:105–111
10. Ninkovic J, Bally-Cuif L (2006) The zebrafish as a model system for assessing

the reinforcing properties of drugs of abuse. Methods 39:262–274
11. Childs E, de Wit H (2009) Amphetamine-induced place preference in humans. Biol Psychiatry 65:900–904
12. Nomikos GG, Spyraki C (1988) Cocaine-induced place conditioning: importance of route of administration and other procedural variables. Psychopharmacology (Berl) 94:119–125
13. Bardo MT, Rowlett JK, Harris MJ (1995) Conditioned place preference using opiate and stimulant drugs: a meta-analysis. Neurosci Biobehav Rev 19:39–51
14. Bardo MT, Neisewander JL (1986) Single-trial conditioned place preference using intravenous morphine. Pharmacol Biochem Behav 25:1101–1105
15. Bozarth MA, Wise RA (1982) Dissociation of the rewarding and physical dependence-producing properties of morphine. In: Harris LS (ed) Problems of drug dependence. National Institute on Drug Abuse, Rockville, MD, pp 171–177
16. Amalric M, Cline EJ, Martinez JL Jr, Bloom FE, Koob GF (1987) Rewarding properties of beta-endorphin as measured by conditioned place preference. Psychopharmacology (Berl) 91:14–19
17. Mucha RF, Iversen SD (1984) Reinforcing properties of morphine and naloxone revealed by conditioned place preferences: a procedural examination. Psychopharmacology (Berl) 82:241–247
18. Reid LD, Hunter GA, Beaman CM, Hubbell CL (1985) Toward understanding ethanol's capacity to be reinforcing: a conditioned place preference following injections of ethanol. Pharmacol Biochem Behav 22:483–487
19. Parker LA, Tomlinson T, Horn D, Erb SM (1994) Relative strength of place conditioning produced by cocaine and morphine assessed in a three-choice paradigm. Learn Motiv 25:83–94
20. Carr GD, Fibiger HC, Phillips AG (1989) Conditioned place preference as a measure of drug reward. In: Liebman JM, Cooper SJ (eds) The neuropharmacological basis of reward. Clarendon Press, Oxford, pp 264–319
21. Schechter MD (1995) Cocaethylene produces conditioned place preference in rats. Pharmacol Biochem Behav 51:549–552
22. Schenk S, Ellison F, Hunt T, Amit Z (1985) An examination of heroin conditioning in preferred and nonpreferred environments and in differentially housed mature and immature rats. Pharmacol Biochem Behav 22:215–220
23. Cervo L, Rossi C, Samanin R (1993) Clonidine-induced place preference is mediated by alpha 2-adrenoceptors outside the locus coeruleus. Eur J Pharmacol 238:201–207
24. Costello NL, Carlson JN, Glick SD, Bryda M (1989) Dose-dependent and baseline-dependent conditioning with d-amphetamine in the place conditioning paradigm. Psychopharmacology (Berl) 99:244–247
25. Li T, Yan CX, Hou Y, Cao W, Chen T, Zhu BF, Li SB (2008) Cue-elicited drug craving represses ERK activation in mice prefrontal association cortex. Neurosci Lett 448:99–104
26. Miller CA, Marshall JF (2004) Altered prelimbic cortex output during cue-elicited drug seeking. J Neurosci 24:6889–6897
27. Mueller D, Stewart J (2000) Cocaine-induced conditioned place preference: reinstatement by priming injections of cocaine after extinction. Behav Brain Res 115:39–47
28. Li SM, Ren YH, Zheng JW (2002) Effect of 7-nitroindazole on drug-priming reinstatement of D-methamphetamine-induced conditioned place preference. Eur J Pharmacol 443:205–206
29. Shoblock JR, Wichmann J, Maidment NT (2005) The effect of a systemically active ORL-1 agonist, Ro 64-6198, on the acquisition, expression, extinction, and reinstatement of morphine conditioned place preference. Neuropharmacology 49:439–446
30. Botreau F, Paolone G, Stewart J (2006) D-Cycloserine facilitates extinction of a cocaine-induced conditioned place preference. Behav Brain Res 172:173–178
31. Paolone G, Botreau F, Stewart J (2009) The facilitative effects of D-cycloserine on extinction of a cocaine-induced conditioned place preference can be long lasting and resistant to reinstatement. Psychopharmacology (Berl) 202:403–409
32. Malvaez M, Sanchis-Segura C, Vo D, Lattal KM, Wood MA (2010) Modulation of chromatin modification facilitates extinction of cocaine-induced conditioned place preference. Biol Psychiatry 67:36–43
33. Mueller D, Perdikaris D, Stewart J (2002) Persistence and drug-induced reinstatement of a morphine-induced conditioned place preference. Behav Brain Res 136:389–397
34. Font L, Miquel M, Aragon CM (2008) Involvement of brain catalase activity in the acquisition of ethanol-induced conditioned place preference. Physiol Behav 93:733–741

35. Leao RM, Cruz FC, Planeta CS (2009) Exposure to acute restraint stress reinstates nicotine-induced place preference in rats. Behav Pharmacol 20:109–113
36. Sanchez CJ, Sorg BA (2001) Conditioned fear stimuli reinstate cocaine-induced conditioned place preference. Brain Res 908:86–92
37. Parker LA, McDonald RV (2000) Reinstatement of both a conditioned place preference and a conditioned place aversion with drug primes. Pharmacol Biochem Behav 66:559–561
38. Lu L, Chen H, Su W, Ge X, Yue W, Su F, Ma L (2005) Role of withdrawal in reinstatement of morphine-conditioned place preference. Psychopharmacology (Berl) 181:90–100
39. Ribeiro Do Couto B, Aguilar MA, Manzanedo C, Rodriguez-Arias M, Minarro J (2003) Reinstatement of morphine-induced conditioned place preference in mice by priming injections. Neural Plast 10:279–290
40. Leri F, Rizos Z (2005) Reconditioning of drug-related cues: a potential contributor to relapse after drug reexposure. Pharmacol Biochem Behav 80:621–630
41. Zavala AR, Weber SM, Rice HJ, Alleweireldt AT, Neisewander JL (2003) Role of the prelimbic subregion of the medial prefrontal cortex in acquisition, extinction, and reinstatement of cocaine-conditioned place preference. Brain Res 990:157–164
42. Itzhak Y, Martin JL (2002) Cocaine-induced conditioned place preference in mice: induction, extinction and reinstatement by related psychostimulants. Neuropsychopharmacology 26:130–134
43. Szumlinski KK, Price KL, Frys KA, Middaugh LD (2002) Unconditioned and conditioned factors contribute to the 'reinstatement' of cocaine place conditioning following extinction in C57BL/6 mice. Behav Brain Res 136:151–160
44. Maldonado C, Rodriguez-Arias M, Castillo A, Aguilar MA, Minarro J (2007) Effect of memantine and CNQX in the acquisition, expression and reinstatement of cocaine-induced conditioned place preference. Prog Neuropsychopharmacol Biol Psychiatry 31:932–939
45. Cruz FC, Marin MT, Planeta CS (2008) The reinstatement of amphetamine-induced place preference is long-lasting and related to decreased expression of AMPA receptors in the nucleus accumbens. Neuroscience 151:313–319
46. Biala G, Budzynska B (2006) Reinstatement of nicotine-conditioned place preference by drug priming: effects of calcium channel antagonists. Eur J Pharmacol 537:85–93
47. Kuzmin A, Sandin J, Terenius L, Ogren SO (2003) Acquisition, expression, and reinstatement of ethanol-induced conditioned place preference in mice: effects of opioid receptor-like 1 receptor agonists and naloxone. J Pharmacol Exp Ther 304:310–318
48. Daza-Losada M, Rodriguez-Arias M, Aguilar MA, Minarro J (2009) Acquisition and reinstatement of MDMA-induced conditioned place preference in mice pre-treated with MDMA or cocaine during adolescence. Addict Biol 14:447–456
49. Do Ribeiro Couto B, Aguilar MA, Rodriguez-Arias M, Minarro J (2005) Cross-reinstatement by cocaine and amphetamine of morphine-induced place preference in mice. Behav Pharmacol 16:253–259
50. Wang B, Luo F, Zhang WT, Han JS (2000) Stress or drug priming induces reinstatement of extinguished conditioned place preference. Neuroreport 11:2781–2784
51. Romieu P, Meunier J, Garcia D, Zozime N, Martin-Fardon R, Bowen WD, Maurice T (2004) The sigma1 (sigma1) receptor activation is a key step for the reactivation of cocaine conditioned place preference by drug priming. Psychopharmacology (Berl) 175:154–162
52. Biala G, Budzynska B (2008) Calcium-dependent mechanisms of the reinstatement of nicotine-conditioned place preference by drug priming in rats. Pharmacol Biochem Behav 89:116–125
53. Katz JL, Higgins ST (2003) The validity of the reinstatement model of craving and relapse to drug use. Psychopharmacology (Berl) 168:21–30
54. Epstein DH, Preston KL, Stewart J, Shaham Y (2006) Toward a model of drug relapse: an assessment of the validity of the reinstatement procedure. Psychopharmacology (Berl) 189:1–16
55. Sinha R (2001) How does stress increase risk of drug abuse and relapse? Psychopharmacology (Berl) 158:343–359
56. Stewart J (2000) Pathways to relapse: the neurobiology of drug- and stress-induced relapse to drug-taking. J Psychiatry Neurosci 25:125–136
57. Shaham Y, Erb S, Stewart J (2000) Stress-induced relapse to heroin and cocaine seeking in rats: a review. Brain Res Brain Res Rev 33:13–33
58. Wang J, Fang Q, Liu Z, Lu L (2006) Region-specific effects of brain corticotropin-releasing factor receptor type 1 blockade on footshock-stress- or drug-priming-induced

reinstatement of morphine conditioned place preference in rats. Psychopharmacology (Berl) 185:19–28
59. Ribeiro Do Couto B, Aguilar MA, Manzanedo C, Rodriguez-Arias M, Armario A, Minarro J (2006) Social stress is as effective as physical stress in reinstating morphine-induced place preference in mice. Psychopharmacology (Berl) 185:459–470
60. Ma YY, Chu NN, Guo CY, Han JS, Cui CL (2007) NR2B-containing NMDA receptor is required for morphine – but not stress-induced reinstatement. Exp Neurol 203:309–319
61. Lu L, Zhang B, Liu Z, Zhang Z (2002) Reactivation of cocaine conditioned place preference induced by stress is reversed by cholecystokinin-B receptors antagonist in rats. Brain Res 954:132–140
62. Sanchez CJ, Bailie TM, Wu WR, Li N, Sorg BA (2003) Manipulation of dopamine $d1$-like receptor activation in the rat medial prefrontal cortex alters stress- and cocaine-induced reinstatement of conditioned place preference behavior. Neuroscience 119:497–505
63. Kreibich AS, Blendy JA (2004) cAMP response element-binding protein is required for stress but not cocaine-induced reinstatement. J Neurosci 24:6686–6692
64. Vezina P, Stewart J (1987) Conditioned locomotion and place preference elicited by tactile cues paired exclusively with morphine in an open field. Psychopharmacology (Berl) 91:375–380
65. Bardo MT, Bevins RA (2000) Conditioned place preference: what does it add to our preclinical understanding of drug reward? Psychopharmacology (Berl) 153:31–43
66. Mark GP, Hajnal A, Kinney AE, Keys AS (1999) Self-administration of cocaine increases the release of acetylcholine to a greater extent than response-independent cocaine in the nucleus accumbens of rats. Psychopharmacology (Berl) 143:47–53
67. Hemby SE, Co C, Koves TR, Smith JE, Dworkin SI (1997) Differences in extracellular dopamine concentrations in the nucleus accumbens during response-dependent and response-independent cocaine administration in the rat. Psychopharmacology (Berl) 133:7–16
68. Costa VC, Xavier GF (2007) Atropine-induced, state-dependent learning for spatial information, but not for visual cues. Behav Brain Res 179:229–238
69. Shulz DE, Sosnik R, Ego V, Haidarliu S, Ahissar E (2000) A neuronal analogue of state-dependent learning. Nature 403:549–553
70. Izquierdo I, Dias RD (1983) Memory as a state dependent phenomenon: role of ACTH and epinephrine. Behav Neural Biol 38:144–149
71. Gauvin DV, Briscoe RJ, Goulden KL, Holloway FA (1994) Aversive attributes of ethanol can be attenuated by dyadic social interaction in the rat. Alcohol 11:247–251
72. Thiel KJ, Okun AC, Neisewander JL (2008) Social reward-conditioned place preference: a model revealing an interaction between cocaine and social context rewards in rats. Drug Alcohol Depend 96:202–212
73. Piazza PV, Deroche-Gamonent V, Rouge-Pont F, Le Moal M (2000) Vertical shifts in self-administration dose–response functions predict a drug-vulnerable phenotype predisposed to addiction. J Neurosci 20:4226–4232
74. Piazza PV, Deminiere JM, Le Moal M, Simon H (1989) Factors that predict individual vulnerability to amphetamine self-administration. Science 245:1511–1513
75. Orsini C, Buchini F, Piazza PV, Puglisi-Allegra S, Cabib S (2004) Susceptibility to amphetamine-induced place preference is predicted by locomotor response to novelty and amphetamine in the mouse. Psychopharmacology (Berl) 172:264–270
76. Gong W, Neill DB, Justice JB Jr (1996) Locomotor response to novelty does not predict cocaine place preference conditioning in rats. Pharmacol Biochem Behav 53:191–196
77. Erb SM, Parker LA (1994) Individual differences in novelty-induced activity do not predict strength of amphetamine-induced place conditioning. Pharmacol Biochem Behav 48:581–586
78. Allen RM, Everett CV, Nelson AM, Gulley JM, Zahniser NR (2007) Low and high locomotor responsiveness to cocaine predicts intravenous cocaine conditioned place preference in male Sprague-Dawley rats. Pharmacol Biochem Behav 86:37–44
79. Shimosato K, Watanabe S (2003) Concurrent evaluation of locomotor response to novelty and propensity toward cocaine conditioned place preference in mice. J Neurosci Methods 128:103–110
80. Pelloux Y, Costentin J, Duterte-Boucher D (2009) Anxiety increases the place conditioning induced by cocaine in rats. Behav Brain Res 197:311–316

Chapter 7

Social Interactions in the Clinic and the Cage: Toward a More Valid Mouse Model of Autism

Garet P. Lahvis and Lois M. Black

Abstract

Autism spectrum disorders (ASDs) are characterized by impairments in communication, social interaction, and the presence of restricted and repetitive behaviors. Impairments in the capacity for social interaction include deficits in the use of nonverbal behaviors, failure to develop peer relationships, lack of motivation to share enjoyment with others, and lack of social reciprocity. This multidimensional diagnosis, both sophisticated and nuanced, contrasts sharply with our limited assessments of mouse "social behavior," which typically involve measures of approach behavior. Our objectives here are to examine the deficits in social interaction in ASD, highlighting the role of affective and nonverbal communication, and then to provide the theoretical and empirical basis for expanding translation-relevant measures of impaired social interaction in mouse models. We ask whether diminished approach toward another mouse, the current behavioral paradigm for mouse models of autism, is sufficient to capture the nuanced social impairments featured in the autism diagnosis. We conclude that these tests should be complemented by assessments of mouse social reward, vocal communication, and mouse abilities to respond to the emotional cues of others. General protocols for several of these behavioral tests are presented.

1. Introduction to the Autism Diagnosis: Social Interaction Impairments

Impairment in social interaction, as described in DSM-IV-TR (1), is manifested by at least two of the following:
- marked impairment in the use of multiple nonverbal behaviors such as eye-to-eye gaze, facial expression, body postures, and gestures to regulate social interaction;
- failure to develop peer relationships;

- a lack of spontaneous seeking to share enjoyment, interests, or achievements with other people (e.g., by a lack of showing, bringing, or pointing out objects of interest); and
- a lack of social or emotional reciprocity.

The ADOS, as a "gold standard" research tool in ASD, provides, through a series of semi-structured activities, opportunities to observe how a child behaves and interacts with a clinician–examiner in order to rate the child on core ASD symptomology (2–4). The clinician judges the child's use of both verbal and nonverbal forms of communication that support social interaction. Ratings include whether communications have reciprocal intent or are dominated by the child's own interests or focus; whether the child's social overtures and social responses are appropriate; whether the child shows a broad range of communicative affective expressions; and whether the child links verbal and nonverbal communications in a subtle and flexible way that supports reciprocal, emotional social interactions.

The ADOS, in combination with other diagnostic instruments, mitigates some of the inherent variability of clinical diagnosis of a human social disorder. Inevitably, as carried out by many clinician–examiners, ratings can be subject to frame of reference effects and the clinician's own level of insight and experience. These ratings require substantial experience, clinical training, and expertise. Moreover, ratings are susceptible to changes in cultural and clinical perspectives regarding normal and impaired levels of social expressivity and responsivity. Culturally, there are disparities in emotional expression across societies, ethnicity, social-economic status, rural versus urban upbringing, gender, age, and variations in parental personalities in response to societal expectations. Against these environmental influences, the child inherits a rich genetic variability and its own individual experiences that underlie a variety of social capacities and motivations that have nothing to do with ASD per se. A child can be shy, reclusive, introverted, or extroverted. A child can also have sensory and other impairments and psychological issues that influence social skills. For example, poor peripheral vision, internal pain, sensory hypersensitivity, and language impairment can all impair social functioning. Other contributing factors include abuse, poor coping skills, and poor social models at home or school.

In this regard, there is growing recognition that autism research with human subjects would benefit from automated tools to measure and analyze on-going verbal and nonverbal communication in ASD. Such tools would ideally employ thoroughly blind methods that allow for a precise focus on the behavioral feature of interest without interference from other features. Technologies are being developed that may ultimately provide better tools to identify atypical behaviors. For example, eye-tracking technology can help elucidate idiosyncratic nonsocial forms of attentional focus (5, 6), and machine learning/pattern recognition analyses can be employed to help automate assessment of facial affect and gestural movements (7, 8). Speech technology combined with statistical analysis methods can aid in exploration of vocalizations and automated assessment of prosody in autism and other disorders (9). Given the enormous progress that has been made in computer performance and algorithm sophistication, it is likely that these technologies only mark the beginning of a revolution in automated, highly sensitive behavioral assessment. *As many of these methods focus on nonverbal behavior, they raise the possibility of applying similar or even identical methods to the study of behavioral features of ASD in nonhuman species.*

Social interactions require that a participant detect a host of verbal and nonverbal cues associated with the other participant's emotions and intentions, that one participant has an internal emotion in response to these cues, and that the other participant then expresses his own emotional responses in a fashion that can be detected by the first individual. We might consider this interaction as a *reciprocity chain* that links emotion with expression and ensuing emotional response with expressive response. We can consider each link by its temporal and contextual features. For example, there could be a temporal progression from the first change in the emotion of an individual (1), to the outward expression of the emotion (2), to detection of that individual's expression by a companion (3), to changes in the companion's emotional state (4), which begins the cycle again. Among humans, this progression of expressed and perceived emotions is mediated primarily via nonverbal visual cues, such as eye gaze, facial expressions, bodily postures, gestures (10), as well as vocal prosody cues. Affective communication can be amplified when visual and auditory cues are spontaneously expressed in a coordinated fashion.

Communication and expression of emotions in the "reciprocity chain" is supported when an individual is attuned to the inter-subjective and situational context and can respond and express himself consistently across communication modalities. Consider, for example, a child who is talking to another in a flat tone, irrespective of the emotional import of the environmental context or unresponsive to the semantic content of the child's own or the other's communications. In this case, the flat-toned expression could index either a deficit in awareness of the other's emotions or mental state, and/or a deficit in the child's own self-expressivity. In either case, it may signal impairment in social–emotional reciprocity. The same concerns could be raised if a child were to talk in a singsong melody regardless of the content of what was said or the situational context. Thus, a lack of a dynamic, multimodal response to an emotionally salient event may identify an abnormal response or impairment in social–emotional reciprocity. To translate impairments of social interaction from the clinic to the lab bench and then back again, we might directly investigate the "stimulus–response" relationships between outward expressions of emotion.

1.1. Expressions of Emotion: Vocal Prosody

Let's now consider in more detail the deficits in vocal communication associated with ASD. Social interactions critically involve individuals to both vocally express their emotions and to understand and respond to the emotional expressions of others. The emotions experienced by an individual can be expressed through the semantic content of *what* is said. If I tell you that I feel angry, despondent, frustrated, defeated, overwhelmed, cagy, indignant, exhilarated, content, proud, or surprised, you have an idea of how I feel. Emotions are also critically conveyed by *how* words are said; that is, through vocal prosody. Prosody can be described at multiple levels. At the acoustic level, it refers to the usage of features, such as variations in the pitch of a phrase or sentence, the duration or loudness of a particular syllable, or vocal quality (e.g., tense, hoarse, breathy). At the mechanical–physiological level, prosody refers to how these acoustic features result from the complex interplay of all components of the speech production apparatus, ranging from the lips to the lungs. At the brain level, prosody refers to how these components, in turn, are controlled in a coordinated fashion by multiple brain systems. Finally, at the functional level, we can delineate the following classes of functions. *Affective prosody* refers to the expression of internal emotional experiences of joy, sadness, achievement, or anger. For example, a flat pitch contour can convey sadness, while significant dynamic changes in pitch and energy levels can indicate greater emotional salience. A tremble in a voice can indicate sadness or fear.

Grammatical prosody helps us to distinguish between meanings of words or sentences. Stress on a particular syllable can help us to differentiate a verb (pre-sent') from a noun (pre'-sent) and the different meanings of a word, while upward trend in pitch can help us identify a question from a string of statements. *Pragmatic prosody* is used to help us put emphasis on an important idea, sentence, or word in a sentence and thus reveals our focus and intentions. Enhanced duration and loudness (energy) can indicate the salient point of a statement. For example, "It wasn't the blue chair you spilled ink on, it was the NEW chair." Prosody allows us to express subtle or not so subtle attitudes. For instance, one can be sincere or sarcastic depending on *how* one says, "I *really* want to go to dinner with you." In another, and quite different function of pragmatic prosody, people speak to infants in "motherese" or "infant-directed talk" (11) which they do not use with older children. People speak differently to a policeman than they do to an intimate partner. Pragmatic prosody thus also signals the context, roles, and relationships between speakers. Although it is customary in human research to include role-related prosody in the category of pragmatic prosody, we will include it in the category of affective prosody for studies of rodent vocalizations. While role-related prosody is relevant for mice, its classification as "pragmatic prosody" would be inappropriate because of its strong verbal connotations and because, even in humans, role-related prosody typically has an affective component (e.g., motherese cannot be decoupled from the feelings toward the baby).

1.2. Prosody in ASD

Atypical prosodic expression has been considered a hallmark of autism since the disorder was first described (12). Prosody in ASD has been described as monotonous or flat (13), stilted, robotic, overly melodious, or singsong (14). Atypical features include aberrant stress, unusual pitch patterns, abnormalities of rate and volume, and problematic quality of voice (e.g., hypernasality) (15, 16). Children with ASD are more likely to express odd phrasing patterns and to exhibit abnormalities in their modulation of pitch, and in their rate and volume of speech compared to typically developing children (17). Unusual prosody in very young children, ages 12–24 months, has been found to be one of the "red flags" that distinguish ASD from other developmental disabilities (18).

Unusual prosody may interfere with the communicative effectiveness of speech, for example, by making it difficult to determine whether or not a statement is intended as a question, or what emotion is being conveyed. In fact, certain repetitive patterns of specific pitch, duration, or energy changes irrespective of verbal content may trigger an unintentional negative emotional response in the listener, which may then severely disrupt the reciprocity chain. In ASD, atypical prosody very much impacts communicative competence and reciprocal social interactions.

Using objective methods, expressive prosody issues have been documented in about 50% of a group of adolescents and young adults with ASD (16). Receptive prosody issues have been less frequently studied, but are thought to be closely associated with expressive deficits, both having been looked at in autism in studies that use controlled perceptual methods (17, 18). Automated methods for the study of expressive prosody are few and will likely expand on the labor-intensive approaches pioneered in seminal studies, such as Shriberg's PVSP (16, 19). Manual approaches unavoidably reduce the amount of data available for research (e.g., it is not unusual to analyze only the first 50–100 utterances of an individual's language sample, even when there may be many more relevant utterances in a spontaneous conversational dialogue across different topics or activities, for example, using the ADOS). Second, the use of the human ear in the PVSP and in Peppé's methods introduces additional variables, particularly for detection of more subtle cases of atypical prosody (9). Clinical judgment ratings on the ADOS for atypical prosody, for example, have such poor inter-rater reliability, that this item, despite its importance, has never been included in any version of the ADOS "algorithm," that subset of items most critical for a diagnosis.

Advances in speech technology, mathematical analyses, and algorithm development, however, are making it increasingly possible to perform automated analyses of expressive prosody. For example, preliminary findings using automated analysis of expressive prosody show that even when in ASD certain prosodic contrasts (e.g., focal stress) are expressed clearly; they are expressed with an atypical usage of prosodic features that are not detected by human observers (9). Speech technology can also create stimuli for the study of receptive prosody with carefully controlled or manipulated speech stimuli that cannot be produced by humans, such as affectively distinctive speech stimuli that have identical speaking rates, or stimuli in which certain prosodic features (e.g., pitch) are altered while keeping others (e.g., duration) invariant (20, 21). These speech manipulation methods can in principle be applied to nonhuman vocalization as well, for example, to determine which acoustic features are behaviorally relevant when studying different strains of mice in different situations.

1.3. Nonverbal Communication in ASD: Facial Expression, Eye Gaze, and Gesture

Nonverbal communication is thought to be a critical component of social interaction and communication of affect. Evidence exists that facial and gestural communication are impaired in ASD, both expressively and receptively, as is the use of eye gaze.

1.3.1. Facial Expression

Facial expressivity has been described as more constricted in range, more flat or neutral, and more idiosyncratic and ambiguous in children with ASD (22). Behavioral research on facial *expression* is itself extremely limited, however, due to the enormous amount of labor involved in the manual coding of facial expressions (e.g., using variants of the Facial Affect Coding Scheme (FACS) (23)) and the still early state of automated methods. Machine learning models of automatic affect recognition, although fast progressing, are not yet capable of analyzing facial affect in natural situations. Current methods, able to capture spontaneous affect, still require a nearly invariant frontal view of the individual and hence cannot easily be used in dynamic social situations (7).

Difficulties in the receptive processing of facial affect have been documented in ASD in numerous behavioral studies (24–29). Lindner and Rosén (30), using both static and dynamic (video) facial affect stimuli, vocal affect, and verbal content, as assessed with the Perception of Emotion Test or POET (31), showed that children with Asperger's (As) have more difficulty than TD children identifying emotions. Mazefsky and Oswald (32) using the DANVA (33), which measures facial affect and tone of voice unimodally at two levels of intensity, showed that children with As perform similarly to the standardization sample, whereas those with high functioning autism (HFA) perform significantly worse on both facial and vocal affect, and significantly worse than As at low levels of intensity. Thus differences emerge depending on the subtlety of the affects expressed, the behavioral measures used, the methodology employed, and who the subjects in the research groups are, which, too frequently, are not adequately characterized.

Studies with functional MRI indicate that children with autism have less activity in the fusiform gyrus, more activity in the precuneus, and reduced activity in the amygdala relative to TD children (34). Among adults, fMRI scans indicate that individuals with ASD take longer to habituate to repeated exposures to faces, accounting for what sometimes looks like conflicting findings of both under-arousal and hyperarousal of the amygdala in ASD; also shown was a significant correlation between reduced neural habituation and the severity of social deficits (35). Given that habituation is associated with learning and novelty detection, reduced neural habituation implies difficulty in extracting social information and adapting to the unfamiliar.

1.3.2. Eye Gaze

Averting eye gaze or atypical eye contact with others is a well-recognized aspect of ASD (17). Children with ASD have difficulties using eye gaze to initiate and maintain interactions, to reference objects beyond need-related requests as, for example, in pure social exchanges of sharing pleasure, feelings, uncertainties about the environment, and, in fact, people with ASD prefer to look at objects rather than other people or their faces (5). Notable is the deficit in the linkage of eye gaze with other forms of communication, especially gestures and vocalizations, in order to share attentional focus with another. Difficulties in initiating joint attention or responding to the bids of others in joint attention are among the most prominent deficits that lead to ASD diagnosis before the age of 3 (36–38).

1.3.3. Gesture

Gestures are communicative actions that make use of fingers, hands, arms, facial, and head movements, as well as body movements and postures. Limited use of gestures in children has also been considered a *red flag* for early identification of ASD in 12–24-month olds (18, 39–41) and a diagnostic marker for older children (4). Lack of communicative, descriptive, and emotional gestures appears, for example, on all ADOS algorithms for all modules. Gestures have a critical pre-linguistic communicative and interpersonal function in the very young. The earliest deictic gestures emerge between 7 and 12 months, with their meaning known from context (i.e., reaching to be picked up; pushing an object away for refusal; pointing to an object wanted; showing or giving an object to another for sharing) and make up nearly 80% of infants and toddlers' gestures (42). Representational and symbolic gestures, which have meaning in themselves and signify features of an object or action, start to appear around 12–24 months (e.g., fingers to mouth to represent "eating;" waving a hand to say "bye"). These gestures usually emerge within intersubjective contexts and social routines, such as play with caregivers and singing songs accompanied by gestures with peers (i.e., "itsy bitsy spider"), and are thought to depend on modeling by and observation of others.

Tools currently in use to assess quality and types of gestures typically involve impressionistic observations or videotaping activities structured to elicit gestures and then using manual coding schemes to rate them (e.g., 18, 42). Although not yet applied as an analysis tool for autism, there is progress in pattern recognition methods for gestural analysis in other fields. For example, technology tools have been developed to precisely capture expert surgeons' manual movements in order to implement robotic surgery (43). The possibility of applying similar methods to the analysis of gestural communication in autism may be on the horizon.

Weaknesses in understanding and using gestures are thought to relate to fundamental impairments in ASD in imitation (44, 45), possibly due, in part, to underlying motor deficits (46). In turn, imitation difficulties in ASD have been linked to neural dysfunction in the mirror neuron system (47, 48). Mirror neurons, which were first discovered in area F5 of the macaque, have been documented in humans, especially in the analogous area to F5, the pars opercularis of the inferior frontal cortex (Brodmann's area 47), as well as in parietal cortex (49, 50). Mirror neurons are said to fire when one *performs* a goal-directed action as well as when one *observes* a goal-directed action performed by someone else. Thus the mirror neuron system has been characterized as the neural mechanism whereby others' *actions*, and the motor plans for them, as well as others' *intentions and goals* are quickly or even automatically learned, even by just observing. In interconnection with other cortical areas, such as superior temporal sulcus and visual areas that respond to biological motion, the mirror neuron system likely plays a role in the human ability to imitate others. Via interconnection with the insula and limbic system, the mirror neuron system may support our understanding of others' minds and the ability to empathize (48, 51, 52). Using facial affect imitation and observation tasks in an fMRI study, Dapretto et al. (53) found significant differences in mirror neuron activity between children with ASD and TD (especially in the right inferior frontal lobe with links through the insula to the limbic system).

1.4. Empathy in ASD

ASD has been linked to impairments in the ability to perceive others' emotions, intentions, and mental states, as well as in the ability to perceive that the experience of another individual is different from one's own. These concepts are embodied in the notion of empathy. In the early twentieth century, empathy was coined from the German word, einfühlung, or "feeling into" (54). This term has since been expanded and refined by numerous authors (55, 56), but contemporary definitions of empathy continue to maintain a primary role for affective reactivity to others (57). Eisenberg (58) defined empathy as "an affective response that stems from the apprehension or comprehension of another's emotional state or condition, and which is similar to what the other person is feeling or would be expected to feel" (p. 72).

The mechanisms that underlie deficits in empathic ability in autism are far from well understood and are being researched at both neural and behavioral levels. Some studies focus on impaired and atypical social information processing at the behavioral level and the cascade of events that it can induce over time. These deficits may be tied to (both precursor and consequent) functional and structural brain abnormalities. There is research that relates many of the known social deficits in autism to dysfunction in the "social brain" (which includes, especially, the amygdala, fusiform gyrus, limbic system, and areas of prefrontal cortex) (59, 60). Other studies, as mentioned, are investigating dysfunction in the mirror neuron system and its interconnected circuitry, especially as related to deficits in imitation ability to understand others' intentions, and empathy (48). There has also been important evidence of uneven distribution of white matter and neural under-connectivity in recent research. For instance, studies have shown *functional under-connectivity*, measured using correlations between time series of pairs of regions of interest in fMRI (61), as well as *structural under-connectivity* (62). Uneven distribution and volumes of white matter in ASD may contribute to this reduced interconnectivity (63). White matter volumes may be significantly (25% or more) larger in certain brain areas (e.g., the frontal lobe) and smaller in other areas (e.g., the corpus callosum). The uneven distribution of white matter, together with atypically larger, but not functionally more productive areas, points to inefficient interconnectivity. In addition to these findings, there are additional features at the microscopic level that may also have implications, such as increased numbers of minicolumns (vertical columns through the cortex layer) with greater cell dispersion in autism (e.g., Casanova 64).

Implications of under-connectivity include impairments in the integration of complex social information. Understanding of others' emotions and the processing of other relevant nonverbal social cues may itself be thwarted by the lack of exposure and focus in autism on the relevant information. For example, Klin et al. (5) employed eye-tracking technology to better understand how individuals with ASD view the world. In a series of experiments in which participants watched video clips from Edward Albee's *Who's Afraid of Virginia Woolf*, individuals with autism were found to focus significantly less time on the eye region of the actors' faces and significantly more time on the mouth, body, and object regions in the film compared to normal controls. Eye-tracking studies in toddlers with ASD have shown, similarly, a preference to look at the mouth rather than eyes, objects over people, and to orient to physical rather than social contingencies (6). Because the nonsocial focus of visual pursuit takes place at a very young age, children then miss out on much socially relevant information and social learning that would allow for greater understanding of interpersonal contexts and improved social adaptation. However, it is unknown to what degree such biases and lack of input is causal, leading to further neural and neurodevelopmental changes, or themselves caused by fundamental neural impairments in brain circuitry.

Under-connectivity may also result in poor cross-modal integration in ASD, which has been documented via behavioral, anatomic, electrophysiological, and neuroimaging methods. At the behavioral level, Wetherby et al. (18), for example, used the systematic observation of red flags (SORF), rated by two people with good inter-rater reliability ($k=0.94$) to confirm the presence of six red flags that distinguish children with ASD during the second year of life from those with DD (developmental delay) and TD. One of those flags was the lack of typical "coordination of gesture with eye gaze, facial expression, or vocalization" (with at least three modalities coordinated) and was found in 100% of the children with ASD.

As with the above explanation underlining the lack of vital social input early on, there is strong plausibility for poor cross-modal integration being a major obstacle to empathic ability, since empathy requires a near-automatic integration of many subtle behavioral cues in multiple modalities simultaneously. Although behavioral observations have testified often enough to the lack of linkage in multiple verbal and nonverbal modalities, most of these studies have used subjective judgment or manual coding methods. There are, however, studies underway to examine this in an objective behavioral paradigm and to automate detection of affective communication in individual modalities (i.e., facial, gestural, postural, verbal, and vocal prosody) and analyze, via mathematical algorithms, their temporal coordination and synchrony.

As we transition here to a discussion about how we assess mouse models that may be relevant to autism, we can see the tremendous translational benefit of analysis tools that assess gestures and vocalizations and behavioral responses to these social signals. Analogous diagnostics can be developed, in theory and in practice, for studying laboratory mice. Do mice share these nuanced social abilities? In the subsequent sections, we will explore the social repertoire of the feral and laboratory mouse, including a broad assessment of the expressive and receptive aspects of mouse communication.

1.5. The Social Aptitude of the Laboratory Mouse

When a genetic linkage to autism is found, mouse studies can help to discern whether there is a causal relationship between the mutation and changes in brain function or behavior relevant to autism. Mice are useful species for genetics research because they have a mammalian brain and a relatively short generation time (10 weeks from conception to reproductive maturity). The entire mouse genome has been sequenced, mouse genes can be deleted, gene expression can be changed (65), and human genes can be inserted into the mouse genome (66, 67). With these tools, relationships between genetic mutations and social behavior can be examined. After mutations in *fmr1* were linked to fragile-X disorder (68), a knockout mouse strain (*fmr1KO*) was generated to model the disorder (69). The fmr1KO was then studied for relevant cognitive, social, and seizure-related deficits (70–73). As a result, this mouse model is now being used to explore pharmacological and genetic treatments for the symptoms of fragile X (74–76). Other findings of genetic linkages to autism have resulted in knockout mice with disruptions in WNT2 (77), MECP2 (78), engrailed 2 (79), CAPS2 (80), and neuroligin-4 (81).

Are mice capable of social interactions that would make the study of mouse behavior relevant to autism? Feral mice have been found at population densities of up to 4,000 mice per acre (82, 83). In the wild, both males and females establish territories and social hierarchies (84–86) and compete for food, territory, and mates (87). Dominant males can distinguish mice that belong in their territory and the degree of relatedness of mice in adjacent territories (86). In natural environments, social arrangements are dynamic, fluctuating with season and food availability (88). Aggressive behavior is more common at high population densities, while gregarious social groups form when the resources for each mouse are more plentiful (89, 90). Social arrangements are also dynamic, changing with the birth of new litters, with movements of mice into and out of geographic areas, with the death of dominant males, and changing food and mate availability.

One could imagine that a mouse with an impaired ability to recognize expressions of dominance, submissiveness, or the changing needs of offspring might have difficulties surviving in a natural population. For instance, a mother is behaviorally responsive to the wriggling calls (91) and distress calls (92) of her pups. One could envision that a nonresponsive mother would be less likely to produce viable litters. Inability to recognize some of the cues that dictate social hierarchies might also result in forced expulsion from the territory of a dominant male or female. A mouse that avoids social interaction might remain ignorant of social cues for safe foods to eat, safe times and places to forage, and mate availability. Taken as a whole, the natural environment likely imposes strong selective pressures for abilities to detect fear or pain in another mouse and recognize the conventions of social hierarchies (for an enlightened overview, read Crowcroft 93).

1.6. Mouse Social Approach Tests

Against this backdrop of mouse social capacities in natural environments, we can consider the most useful and widely employed approach for examining the social capabilities of a laboratory mouse; the social approach test. The social approach test has been variously attributed as measure of mouse "sociability," "sociality," "reciprocal sociality," or "social interaction." This social interaction test has been used extensively to examine mice with targeted alleles relevant to autism (*see* Section 2). Social approach behavior of a "test" mouse is measured typically after a period of social isolation. During isolation, the test mouse becomes habituated to a cage where it resides. A *stimulus* mouse is then introduced to the test mouse's cage. During the test, the stimulus mouse explores the cage and the test mouse follows or "approaches" the stimulus mouse. "Sociality" in this test is defined as a composite measure of social approach; the fraction of time that the nose of the test mouse is within a short distance of the head, flank, or anogenital region of the stimulus mouse. Social approach is most vigorous after extended periods of social isolation and can vary with the time of day when social isolation occurs (94). This test has various permutations. Among juveniles, it can indicate "social interaction" (95). Among adult males or nursing females confronted with a visitor, this resident–intruder test measures aggressive behaviors (96).

It should be noted here that some very insightful information are gained about the social approach behaviors of a laboratory mouse by simply observing the home cage. For instance, dvl1 knockout mice do not huddle together or barber the whiskers of their cagemates like their wild type controls (97). This particular study highlights the importance of simply observing the home cage, in the colony room during periods when the white lights are on and also under dim red light (see details on mouse husbandry at the end of this chapter).

There are also automated versions of social approach tests, to rapidly determine whether a test mouse is more likely to approach an unfamiliar mouse, versus a familiar mouse, or versus an inanimate object (98). Approach behavior is indicated when the test mouse breaks an infrared beam and enters a room that contains a cage wherein the stimulus mouse resides. A more detailed description of this test is provided at the end of the chapter.

Just as social behaviors and motivations of children are different from those of adults, mouse behaviors likewise change with development. For instance, social approach behaviors of early adolescent mice can be very sensitive to genetic background and insensitive to the gender of the test mouse or the stimulus mouse. As mice approach reproductive maturity, social approach becomes more stereotypical: males approach females more than females approach males. Genetic influences of social approach behaviors may be marked by gender differences that emerge with reproductive maturity (99). There are several excellent references on the ontogeny of mouse behavior that should be considered (100–104). A common mistake is to consider a juvenile mouse as a smaller version of a juvenile rat, which is a very different rodent (105–108). For instance, social play among young mice is very infrequent and not at all robust (106), whereas young rats express "rough and tumble" play (108, 109). Further, mouse motor behaviors do not indicate strong "reciprocal" social interactions in this test. The stimulus mouse typically walks around the perimeter of the cage while the test mouse follows the stimulus mouse. These behaviors do not indicate play or reciprocity. It is possible that there are reciprocal vocal exchanges, but this aspect of their social exchange is currently unknown.

It can also be difficult to interpret differences in social approach response. Diminished approach toward another mouse could indicate shyness, social anxiety, reduced preference to engage in a social encounter, or an increased preference for social isolation. Shyness and social anxiety are not core features of autism. Preference for social isolation may have more to do with autism, at least as originally described and emphasized by Kanner (110) who characterized "autistics" as "aloof," "indifferent," and "isolated." Currently, however, the spectrum of autism disorders is much broader and involves a more heterogeneous group of children for whom aloofness or lack of social interest may not be descriptive. Reduced social motivation is not mentioned as a core symptom of the DSM-IV, but it does characterize the broader autism phenotype as described by Dawson et al. (111).

Moreover, in humans "impaired social interaction" has a much more inter-subjective connotation than does diminished social approach, sociability, or "sociality." *A person with autism can be sociable and still have a severe impairment in social interaction.* Social interaction has to do with the behavior of one individual impacting another individual, of individuals in *interaction*. At the very least, impaired social interaction assumes that social behavior is "inappropriate" in the means, timing, or manner in which it is done with respect to both norms (e.g., age, sex, cultural, contextual, social) and with respect to the individual to or for whom the behavior is directed. A question then arises, does the social approach test paradigm capture core symptoms of

1.7. Interpreting Social Approach Through the Lens of Reward Theory

"impaired social interaction" in autism? Is it reasonable to think that it could, given some modifications to the paradigm?

Can we be sure that a diminished social approach behavior indicates a lack of social motivation? Approach behaviors of any sort toward a stimulus might suggest that the stimulus induces a psychological expectation of reward or pleasure, but this inference is far from guaranteed. For instance, an amoeba, which lacks a nervous system altogether, swims up a chemical gradient. Having a brain allows an animal to navigate through its environment to gain access to rewards and avoid painful stimuli, through the cues that are associated with these states. Such cues can be contextual (berries to eat or thorns to avoid) or temporal (a sting following a buzzing sound). This capacity to make associations allows an animal to adapt to dynamic conditions. For example, an animal can learn that a novel red berry is not sweet. By contrast, an amoeba cannot modify its behavioral strategy to obtain food. Its movement is dictated by membrane receptors, directed by fixed stimulus–response action patterns.

But how do we infer whether a mouse finds a stimulus rewarding or aversive? The question of how to infer animal motivation from animal behavior is a hard problem but it has been extensively studied. Through conditioning experiments (described below), we can infer animal capacities for fear and fear learning (112–114) and for the anticipation and consumption of rewards (115, 116). These capacities are supported by a highly differentiated limbic anatomy (117) and physiological systems that include glucocorticoids, dopamine, serotonin, and endogenous opiates (118). Drugs of abuse can co-opt these natural reward systems (119–123). Thus, mice respond to their environment, in part, based upon associations between particular contextual or temporal cues and underlying affective states, such as feelings of reward or fear.

How can we explore the relationship between social approach and these underlying experiences of reward or pain? The conditioned place preference (CPP) test is particularly valuable in this regard. CPP tests were developed in the context of a rich theoretical framework (120, 124–127). If a particular stimulus (such as access to morphine, cocaine, or a highly palatable food) is rewarding, then a test animal will choose to spend its time in an environment that it associates with access to this putative reward. In the CPP experiment, the test mouse is exposed to two environments discernable by distinct but unimportant qualities, such as particular beddings or odors. One environment is repeatedly paired with the presence of the putative reward and the other environment is conditioned by the absence of the reward. During the test, the mouse is allowed to roam between both of the conditioned environments. *No reward is provided during the test period,* but the mouse is expected to spend time in the environment formerly paired with the putative reward and less time in an environment where it either did not experience the reward (for example, saline or standard lab chow) or where it previously experienced something aversive. CPP tests show that exposures to morphine, cocaine, and highly palatable foods can be rewarding for rodents (120). CPP tests have also demonstrated that different kinds of social encounters, including play (126, 127), sexual interactions (128, 129), juvenile social interactions (130), mother–infant bonding (131), and even aggression (132), are rewarding to mice.

Within the context of the social CPP, we can ask whether social approach can be associated with social reward. The experimental framework of a social CPP is depicted in **Fig.7.1**. BALB/cJ mice (BALB) express levels of much lower social approach than C57Bl/6 (B6) mice, consistent with previous studies (133, 134). The B6 strain shows a strong place preference for environments paired with social access. By contrast, the BALB strain shows no conditioned respond to the different environments; they do not find access to other juveniles rewarding. Comparison of these two strains indicates that social approach can be associated with social reward (130). A more detailed description of the social CPP test is described at the end of the chapter.

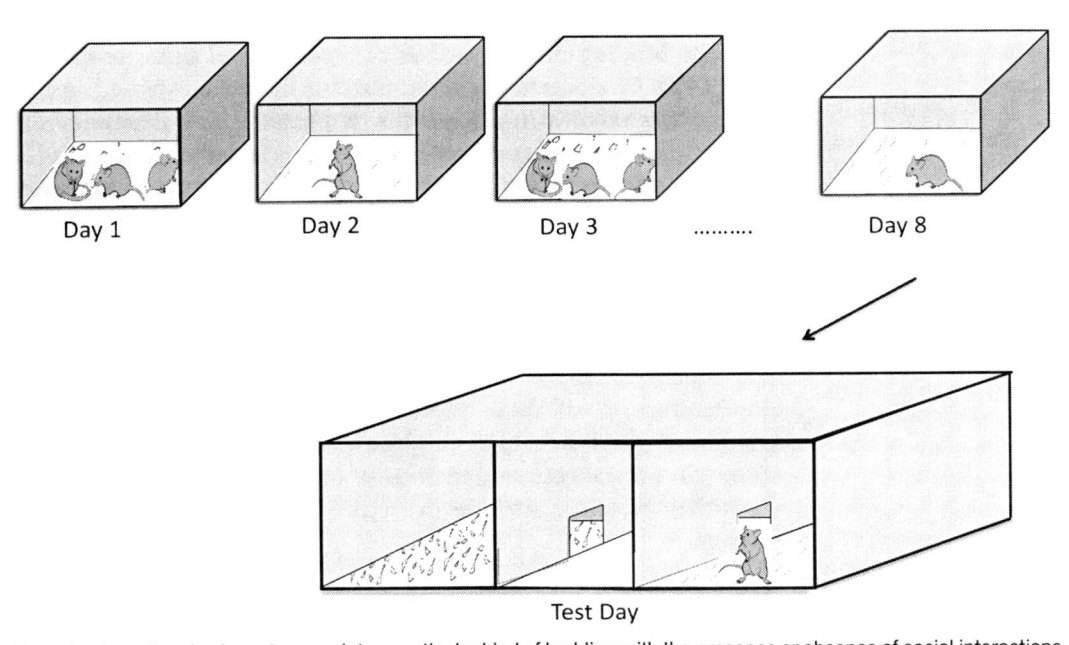

Fig. 7.1. Juvenile mice learn to associate a particular kind of bedding with the presence or absence of social interactions. Mice are placed in clean bedding material each day that alternates between aspen shavings and recycled paper and either the presence of other juvenile mice or social isolation. After 10 days of conditioning, mice are tested for their preference of the bedding materials in the absence of other mice. Typical juvenile mice, such as the B6 mouse, prefer the beddings that they experienced when they had access to social encounter versus beddings associated with social isolation. BALB mice do not show this preference. This is a measure of social reward.

We can also envision situations in which the association between social approach and social reward does not exist. As mentioned earlier, a mouse strain might have a strong motivation for social novelty but an aversion to group housing. A mouse might avoid social approach for other reasons, such as shyness or social anxiety, yet prefer social housing to isolation. Social reward can be multiplicitous and ephemeral. Nursing dams, when they must choose between environments paired with access to newborn offspring versus access to cocaine, choose environments paired with offspring during the first week after birth, but when they have older pups, they prefer environments paired with access to cocaine (131).

Social reward is likely mediated, at least in part, by endogenous opioids, dopamine, and their receptors within the brain. These endogenous molecules are distributed within regions involved in emotional regulation and mediate responses to pain and stress, reward, and social bonding (135, 123). Activation of opioid receptors promotes maternal behavior in mothers and social play among juvenile animals (136–138) and can be modulated by agonists and antagonists of opioid receptors (139, 140). Separation distress, exhibited by archetypal behavior and calls in most mammals and birds, is reduced by opioid agonists and increased by opioid antagonists (141). Access to social encounter can, in turn, influence opioid physiology. For instance, social play and short-term social isolation alter opioid receptor binding in the nucleus accumbens and other regions in rats (142). Dopamine is also important in social interactions. In the nucleus accumbens, dopamine appears to be a necessary signal for pair-bond formation in male prairie voles (143). Social stress or social isolation promotes long-term changes in mesolimbic dopamine (144), and social isolation can render animals more sensitive to dopamine-releasing drugs of abuse (145, 146). There are also important roles for oxytocin and vasopressin in social approach behaviors (147, 148), which influence dopaminergic reward circuits (149). While endogenous opioids and dopamine can moderate social approach and response to social access, the interaction of these physiological mechanisms with social encounter does not rule out the possibility that mice, under some circumstances, might engage in or avoid social approach for other reasons, such as novelty-seeking, shyness, or social anxiety.

1.8. Modeling Impairments in Social Interaction in Mice

How can impairments in social interaction be examined in a way that captures mouse social capacities and has optimal translational value for autism research? Let us first consider the mouse social approach test within the context of the DSM IV-TR diagnosis. In what follows, we review the defining features of "qualitative impairment in social interaction," as listed in the DSM IV, and ask within this set of criteria whether existing or new behavioral tests can be employed to interrogate behaviors relevant to autism. Let us look at a few of these manifestations and how they could be captured in our mouse models. Importantly, in considering these various tests, it is important to maintain a mouse colony and a behavioral testing environment that minimizes variation in mouse social behavior.

1.8.1. Impairment in the Use of Multiple Nonverbal Behaviors to Regulate Social Interaction

Could a mouse express nonverbal behaviors that fail to regulate social interaction? Just as there are proscribed behavioral patterns for bee waggle dances, fruit fly courtship behaviors, or the hunting sequence for members of a wolf pack, mouse social interactions that are bound by norms for the species. What would be the modalities of these interactions? Feral mice shelter in burrows and forage at night, so their natural environment typically occurs under low levels of light. In this environment, social interactions are mediated by vocalizations, visual and tactile cues, and odors. Mouse use of vision, relative to their other senses, is diminished. We will briefly review how mice communicate and sense each other via their auditory, olfactory, and visual senses, because these modalities form the basic components of social interaction.

Vocalizations: During social interactions, rats and mice often emit ultrasonic vocalizations (USVs) at frequencies above the limit of human hearing (15,000 cycles/s (hertz) = 15 kHz). Rats emit high frequency (50 kHz) upward modulated calls when they anticipate or experience reward and lower frequency (22 kHz) USVs under aversive conditions (150, 151). Consistent with these observations, rats actually self-administer playbacks of recordings of high frequency calls, but not low frequency USVs (150). When rats hear recordings of high versus low frequency calls, there is a differential induction of *c-fos* gene expression patterns in the brain (152).

Vocalizations among mice can provide information throughout development. When infant mice, from post-natal day (PD) 1 to 8, are displaced from the nest they emit distress calls at very high frequencies (> 90 kHz) that solicit the dam to return the pup to the nest (153). "Wriggling" calls (about 35 kHz) are emitted by pups within the nest and engender the dam to free the pup from an uncomfortable position under the dam or relative to other pups (92, 154). Call rates of pup USVs and maternal responses to these calls are sensitive to drugs that target affective states, such as opiates, dopamine, and serotonin (92, 155, 156).

Juvenile mice emit more complex USV patterns between weaning (PD 21) and adulthood. When a juvenile mouse is reunited with a conspecific after a period of social isolation, it can express an elevated level of social approach and emit a rapid rate of vocalizations. These USVs can be classified as syllables (vocal sounds separated by silence from other USVs) according to how their fundamental frequencies are modulated (such as upward, downward, chevrons, and punctuated calls). These vocalizations can be remarkably complex interactions (99) and vary with mouse strain. For instance, juvenile mice of the B6 strain show a vigorous social approach response and high number of downward modulated and punctuated calls relative to juvenile BALB mice. BALB mice express a higher rate of upward modulated calls during social union interactions (67). Interestingly, the pitch jumps between frequencies of punctuated syllables are more stereotyped among juvenile BALB mice and more varied among age-matched B6 (99), in line with the concept that BALB/c mice have many behavioral features relevant to autism (133, 134). Among adults, these complex call structures can also be sensitive to specific genetic mutations, such as BTBR (157). Like rat vocalizations, adult mouse USVs are responsive to mating access (158, 159), the odors of conspecifics (160, 161), and exposure to drugs, such as amphetamine (162) and alcohol (163).

Odors: Mouse scents are also powerful social cues for mice. Information about age, sex, genetic background (164, 165), degree of kinship (166), and parasitic infection (167, 168) can be conveyed through odor. A mouse "marks" its urinary scent within its environment depending upon its own age, sex, and genetic background and depending upon the age, sex, and genetic background of another stimulus mouse in the vicinity of the test mouse (164). There are multiple behavioral responses to scent marking. Female urine applied to juvenile mice inhibits adult male aggression toward juveniles (169). When males are conditioned to express low levels of aggression, odors in their urine evoke lower levels of aggression from males and approach responses from females (170). Mice also avoid the urine of conspecifics that are distressed by shock (171) and parasitic infection (167). Interestingly, mice lacking a functional oxytocin gene (that modulates social affect among humans and pair bonding among voles) cannot detect parasitic infections of other mice (148), suggesting the possibility that social motivation can influence the extent that a mouse can derive information from social interactions. Mouse social behaviors are even sensitive to odors that indicate whether another mouse was previously housed in a social or isolate environment (172). Odors of female mice also elicit complex patterns of ultrasonic calls from male conspecifics (173, 174).

Movements and gesture: Visual cues can be useful for mouse social interactions. Visual cues are used in imitation, which has been studied in mice and is highly relevant to autism. In an ethological-relevant experiment, mice were allowed to observe other "demonstrator" mice respond to biting flies. The "observer" mice were then examined for their abilities to learn from their visual observations of demonstrator mice with biting flies. The demonstrator mice were then tested in response to biting flies that were surgically altered so that they were no longer capable of biting. After observation of the demonstrators, observer mice displayed both conditioned hypoalgesia and active burying responses in response to the altered flies. Thus, increased pain tolerance of observer mice in this paradigm was dependent on the experience of demonstrators being bitten. The learned active avoidance (burying) response was contingent on visual detection of a demonstrator actively burying itself in bedding to escape the biting flies (176).

Mice can perform additional imitation tests, such as learning manipulation of a puzzle box (177) and how to swing a door open to obtain a food reward (178). In the latter experiment, male and female mice observed demonstrator mice open a swinging door that was hinged at the top and could swing either to the right or to the left to open. Observer mice showed lower latencies to open the door relative to naïve controls. Further, male mice reliably were able to open the door in the same direction (to the right or to the left) as the demonstrator condition. An opaque barrier was sufficient to block this learning response (178). In these two measures of procedural learning, the ability to detect emotional cues is unlikely to play an important role.

"Emotional contagion" among mice can be expressed when individual members of a dyad are subjected to different levels of paw irritation. Mice express degrees of paw licking that is directly correlated with the concentration of irritant injected into the paw. When two mice are injected with different concentrations of the irritant, they experience different levels of pain and lick themselves at different levels. Importantly, when these individuals are placed adjacent to each other, the response of one of the mice influences the response of the other. This bidirectional response to differences in pain response indicates an ability to adjust individual responses according to the emotional cues of a distressed conspecific (175).

Cross-modal integration: The ability to regulate social interactions or to develop rewarding social interactions may require an appropriate mix and timing of specific motor patterns, vocalizations, scent markings, and other expressive cues. ASD likely involves not only deficits in the use of individual modalities of nonverbal cues, but also in their expressive cross-modal integration; that is, deficits in their synchronous, linked usage. Temporal relationships and linkages between movements, vocalizations, motor, and spatial patterns of approach/withdrawal behaviors might also be impaired in mouse models of ASD.

To determine if a test mouse appropriately regulates its social interactions, we could ask *how mice* respond to *the test mouse's* quality *of approach or withdrawal*. For instance, mice that lack a functional fragile-X gene (*fmr1*) can show enhanced (72, 179) or diminished (180) social approach relative to wild-type controls, depending upon experiments approach (for overview, *see* (181)). These apparent differences might all be indications that the target gene (*fmr1*) allows a mouse to regulate its social interactions with others. It may be possible that poorly regulated social behaviors are aversive to a "reference" mouse. In this case, the behavior of the reference mouse in response to the test mouse might be a very useful behavioral measure of the test mouse. We can then examine whether the test mouse expresses an atypical pattern of social cues toward the reference animal.

1.8.2. Failure to Develop Peer Relationships

An important deficit in ASD is the failure to develop peer relationships. While this process is obviously complex in humans, use of the social conditioned place preference (CPP) test can be used to determine the extent that a mouse values a peer relationship. If a mouse finds a social interaction rewarding, it would bring forth, via association, a preference for the environment paired with access to other juveniles. Juvenile B6 mice express a social preference in a CPP test, whereas BALB/cJ appear indifferent (**Fig. 7.1**). For example, by changing test conditions, we can ask whether a mouse prefers to spend time in environments associated with social interaction or avoids environments paired with social isolation. We can ask whether a mouse of a specific genetic background, or lack of a functional allele, values access to other similar mice or mice of a different reference strain. If the test strain shows no preference for environments paired with social access to other members of the test strain, we can ask more targeted questions to determine whether the test strain lacks a preference for social encounter or whether the experience of housing with the test strain is not rewarding. For example, individual test mice might find access to mice of another genetic background or targeted allele rewarding. Conversely, reference mice that express strong social reward (130) might avoid environments associated with the presence of the test mouse. By utilizing social groups that contain mixed strains, we can dissociate these effects. As with all behavioral experiments, these measures require several control experiments to rule out more general cognitive or motivation deficits.

1.8.3. Lack of Social or Emotional Reciprocity

Reciprocity is a more nuanced aspect of social interaction but aspects of emotional and social reciprocity are measurable in children with ASD and might be assessed in mice. There are numbers of ways to clinically assess reciprocity: One diagnostic approach for young children is to feign distress and ask whether the child orients toward the clinician "distressed" by, for example, feigning to burn her fingers with a match as she lights a candle. The clinician makes an exclamation that indicates surprise or pain and an expectation of a response from the child (see modules 1 and 2 of the ADOS) (182). A typically developing child looks toward the clinician, usually with a nonverbal or vocal empathic response. Children with ASD often fail to express a response to feigned distress (183, 184).

Are there analogous measures of the feigned distress test for mice? A modification of the fear-conditioning procedure can be used to assess whether observer mice orient toward shock-startled mice and whether these observer mice can subsequently become fearful of cues that predict this shock based upon their experience of a conspecific undergoing fear conditioning (185). By using a mouse model of cue-conditioned fear, a series of experiments can be conducted to determine whether exposure to conspecific distress influences how a mouse subsequently responds to environmental cues that predict this distress. In a standard fear-conditioning procedure, a neutral stimulus, such as a tone (conditioned stimulus, CS), is presented forward paired with an aversive stimulus, such as an electrical shock (unconditioned stimulus, UCS). Upon repeated administration of the paired CS–UCS, the mouse expresses a fear response (they freeze).

Using this approach, we asked whether B6 and BALB strains express differences in the abilities to detect distress in others, whether they show a differential fear responses to the tone after they observe other "demonstrator" mice experience the tone–shock contingency (*see* **Fig. 7.2**). Test mice were placed in the rooms adjacent to the fear-conditioning chambers, separated by metal bars, after they had themselves experienced shock in the demonstrator chambers. The demonstrator mice were then

fear conditioned mouse   observer mouse

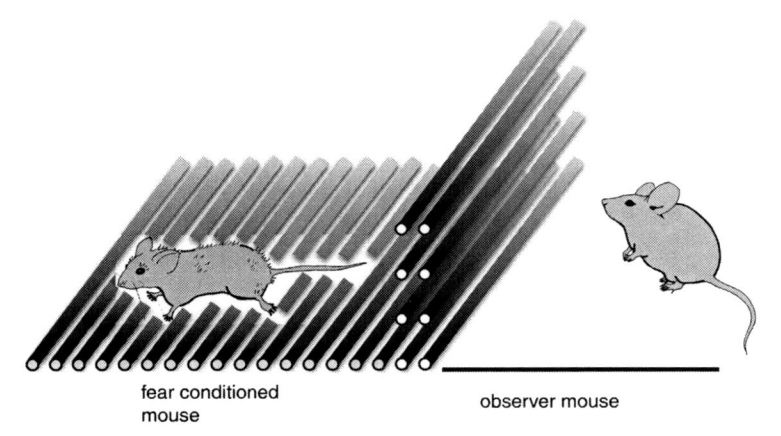

Fig. 7.2. An observer "subject" mouse experiences a demonstrator "object" mouse undergo fear-conditioning trials that involve a 30-s tone that ends with a 2-s shock, repeated at 2-min intervals. Mice are separated by two sets of metal bars. All bars lining the object mouse chamber are live whereas bars lining the subject chamber did not carry electric current. Both subjects and objects were exposed to the CS (tone), but only objects directly experienced the UCS (shock) during conditioning trials. Subjects had access to object distress cues that were emitted as a consequence of receiving the UCS.

exposed to a 30-s tone that ended with a 2-s mild shock. After 90 s, this tone–shock contingency was repeated. Demonstrator mice might be considered analogous to a clinician feigning distress. When B6 and BALB mice observed other mice undergo fear conditioning, both strains oriented toward the demonstrator mice while they were shocked. When placed back into the shock chambers, only B6 mice expressed a freezing response in response to hearing the tone-only. In other words, B6 mice expressed a fear response to a tone that was conditioned by administering a mild shock to other conspecifics. BALB mice, though they oriented toward the demonstrator mice during shock administration, did not acquire a fear response to the cue that predicted this shock.

The enhanced fear learning response in B6 mice could also be reproduced by playbacks of recorded vocalizations of mice distressed by shock concurrent with hearing the 30-s tone. By playing back distressed vocalizations of the demonstrator mice, we found that these vocalizations were sufficient to condition the tone for B6 mice. The lack of a BALB response could be due to inherent emotional insensitivity to vocalizations of distressed mice. This test has considerable translational validity, as heart rate deceleration was also expressed in B6, but not in BALB mice. The depression of B6 heart rate that accompanies the playbacks of vocalizations from distressed mice also occurs when children detect distress among others (58) allowing us to make inferences to the emotional state of this strain.

1.8.4. Lack of Spontaneous Seeking to Share Enjoyment with Other People

There may not be precedent for behavioral studies of a mouses's ability to indicate attention towards objects of interest. It is entirely possible that mice do not express vocalizations or gestures that engender sharing of enjoyment. However, some manipulations might be informative. For example, mice may emit different vocalizations when they are introduced to one another after an individual has been placed in a novel-enriched environment relative to the other mouse. Perhaps experiments in which individuals are differentially exposed to a novel stimulus and then reintroduced will have valuable effects.

1.9. Conclusion

We argue here for greater awareness of the potential to observe a much wider range of behavioral features in mice; features that correspond to the many documented behavioral features in ASD – than is addressed with conventional methodologies. In this regard, we hope to capitalize on human diagnostic tools that do not directly involve perceptual and clinical judgment, but utilize more automated forms of assessment. Such tools would augment both clinical assessments and accelerate translational research efforts. Just as urgent, such "*automated bridge tools*" should have a foundation in clinical expertise. They should also target behaviors that are truly comparable in mice and humans and reflect upon core issues of the disorder. For instance, mouse deficits in "sociality" may be easy to observe and measure, but they are not equivalent to human deficits in "reciprocal social interaction." This looseness in translation can lead to substantial problems. Clearly, mice are capable of many social behaviors that are relevant to autism, which can complement and help us to understand measures of social approach behaviors. Ongoing dialogue between clinicians and research scientists will provide an essential translational stepping stone to further advances in autism research, particularly in the area of expression of and receptivity to emotion.

2. Methods

2.1. Social Interaction (SI) Test

Purpose: This test examines whether a mouse expresses normal patterns of approach toward a mouse that has been introduced to its cage.

Test arena: The standard housing cage where the isolated test mouse resides in social isolation in a fresh cage with fresh bedding.

Preparation: The test mouse is socially isolated prior to testing to engender a vigorous social approach response toward a stimulus mouse. The test mouse is isolated from its social group into a clean cage containing fresh bedding without nesting material. SI testing is conducted during the dark phase under dim red illumination in a sound-dampened room.

Social isolation should not exceed 24 h. The duration of social isolation can be as short as 6 h if isolation is experienced during the dark phase (*see* (95)).

Test procedure: Near the end of the social isolation period, mouse cages are transported from the colony room to the procedure room, maintained under dimly lit or dark conditions. This transfer should occur at least 30 min prior to testing.

This test measures the response of the "test" mouse to a "stimulus" mouse that is introduced to its cage. Juvenile mice are collected from weaned social groupings that contained fixed sex ratios of four animals each. Two mice from each group are randomly designated as test mice, while the remaining individuals serve as stimulus mice. Then both test mice and stimulus mice are isolated for 24 h in this test. Stimulus mice should not be reused for multiple tests, unless it is with the same test mouse on different days. To determine whether the test mouse expresses normal social regulation, the responses of the stimulus mouse toward the test mouse can also be assessed.

The top of the cage containing a test mouse is replaced with transparent Plexiglas® 5–10 min before testing begins. The overhead video camera is turned on and the video capture software (such as Windows Movie Maker) is activated. The stimulus mouse is then removed from its own cage and added to the cage where the test mouse resides. The amount of social investigation that the test mouse directs toward the stimulus mouse is recorded during a 5-min period. Behaviors are video-recorded (we prefer 3CCD mini-DV cameras with a firewire port) and transferred via firewire to a computer for additional analysis.

Ultrasonic vocalizations (USVs) can also be recorded. We use an ultrasound microphone (UltraSoundGate model CM16, Avisoft Bioacoustics, Berlin, Germany) with a 10–180-kHz flat frequency range. The microphone is lowered to the plane of the cage top, where there is a small opening (30 mm diameter) centrally located within the Plexiglas® cage cover. USVs are collected with an UltraSoundGate 116 acquisition system and the Avisoft-Recorder v.2.97 (Avisoft Bioacoustics) and stored as "wav" files for subsequent analysis.

Video analysis: Behavioral variables include
- sniffing or snout contact with the head/neck/mouth area,
- sniffing or snout contact with the flank area,
- direct contact with the anogenital area,
- social pursuit within one body length as the stimulus mouse moves around the cage, and
- social grooming.

These variables are typically highly correlated and combined into a composite measure of SI. However, they should be reviewed first as individual variables to identify possible behavioral anomalies. Additional features of social interaction that can be recorded include social proximity (i.e., mice within one body length of each other without movement or direct contact) and "jerk-and-run," a play-like behavior but these behaviors are infrequent and highly variable among pairs of mice and therefore not considered in the comparisons of SI.

All behaviors are scored in duplicate with the aid of computer-assisted analysis software that tabulates the duration, frequency, and order of depression of keys on a computer keyboard (we use ButtonBox v.5.0, Behavioral Research Solutions, Madison, WI, USA). A trained observer punches keys during the test (or recording of the test) to tally the occurrence of different social behaviors. In all cases, behaviors are scored twice, during a subsequent "off-line" analysis session by a different observer blind to the age and gender of the interacting mice. The presentation of all SI data and statistical outcomes are based on an average of two independent measurements (inter-rater reliability, as determined by Pearson's correlation, should exceed 0.90).

The social behaviors of stimulus mice can also be included in the behavioral analyses.

Audio analysis: Spectrograms can contain a tremendous amount of ultrasonic vocalizations (USVs) when juvenile mice are reunited after separation. These USVs are complex. The ultimate utility of these calls for survival and reproduction purposes has been shown in elegant studies of male courtship vocalizations and the female responses to these calls. Mouse vocalizations have proximate value in conveyance of emotional information, indicated by their associations with social reward and fear learning and their responsiveness to drugs that modulate affective states (186).

Small segments of vocal recordings can be manually assessed, such as the first 10-s interval of each minute during an SI test for all pairs of mice tested. Quantitative analysis can be conducted with various programs. We use SASLab Pro (Avisoft Bioacoustics, Germany). A 40-kHz band-pass filter can help minimize background noise during recordings, though most "wav" files still contain a considerable amount of "non-USV" signal that compromised the accuracy of the automated parameter measurement functions available within the SASLab Pro software format. Thus, extraneous noise is identified and removed from all of the sonograms.

When a rater finds an ultrasound signal that is difficult to interpret, the call can be evaluated by a minimum of one additional, trained observer and identification required a consensus by all raters. Each spectrogram is then evaluated with a series of automated parameter measurements that tally the total number of USVs produced, USV duration, the mean dominant frequency of a USV, and the inter-vocalization interval (*see* (100)). Interpretation of vocal signals can involve assessments of mouse behavior in response to playbacks (186, 187).

SI controls: It is always important to consider alternative explanations for an atypical level of social approach behaviors.

Locomotion: An alternative explanation for a mouse that shows diminished social approach behaviors toward a stimulus mouse exploring the test cage is that it has impaired locomotor abilities. Locomotion should be independently assessed to consider this possibility in an open field under red-light conditions.

Approach toward a change in environment: One possibility is that the test mouse under consideration differs from a control mouse because it approaches all changes in environmental context, not only changes in social context, with added or diminished vigor.

One suggested test is to examine a mouse response to a novel olfactory stimulus. In such a test, the test mouse is socially isolated for the same duration of time used for the social interaction test, but rather than present the mouse with a stimulus mouse, the test mouse is presented with a scented cotton ball (for example, with 500 μl lemon extract). Then, olfactory investigation, including contact, active manipulation or sniffing (<10 mm) the cotton ball, is measured.

Consumption rate of rewards: Another consideration is that the test mouse differs from a control mouse because it consumes all rewards differently, not just social rewards, differently than the reference mouse. For instance, there may be a difference between the vigor with which a knockout and wild-type mice eat food or approach and sniff another young mouse. Two days after testing for investigation of a novel olfactory stimulus, mice can be re-isolated into a clean cage and food-deprived during a 24-h social isolation period. Then, each mouse is provided with a single pellet of standard lab chow on the floor of its cage and the total amount consumed within a 10-min period can be measured.

Aberrant nonautistic-like social behaviors: It is possible that mice could express atypical social phenotypes that are not relevant to autism, but are relevant to social impairments that accompany shyness, depression, or attention deficit disorder. A social approach test cannot distinguish between these possibilities. The social reward test provides some insight into the question of whether motivations for social interaction are valuable here. Finally, a core phenotype of autism social interaction is the inability to regulate its social interactions. In this case, it can be informative to examine the social behaviors of the stimulus mouse for aggressive or avoidance behaviors.

It is essential to recognize that this test, like all of the tests described in this chapter, should be used as one of a panel of measures, to avoid the possibility of interpreting the lack of an atypical mouse response as a lack of social deficit. For instance, if a mouse strain does not express an abnormal response to this test, it may have still express anomalous patterns in sociability, social reward, or social transmission of fear.

2.2. Sociability and Preference for Social Novelty Tests

These tests are useful for identifying mice that have abnormalities in their approach behaviors toward either (a) other mice versus an object (*sociability*) or (b) toward familiar versus unfamiliar conspecifics (*preference for social novelty*).

Purpose: These tests can rapidly assess two measures of mouse social behavior with minimal preparation (short habitation and no conditioning) of the test mouse and by automated tallies of mouse movements, providing a valuable screen or comparison of multiple mouse strains for atypical social behaviors.

Test arena: Mice move freely through a three-chambered structure through doorways that contain embedded photocells that detect directional movement. Stimulus mice are confined within small cages that are placed within the outside chambers.

Habituation: An empty wire cage is placed in both side chambers. A weight is placed on top of each wire cage to prevent a mouse from climbing them. The test mouse is placed in the middle chamber and allowed to explore the test arena.

Test procedure: After habituation, the test mouse is enclosed in the center compartment of the test arena. Doors between chambers are opened, allowing the test mouse to explore the test arena. In the *sociability* test, the test mouse is presented with an unfamiliar mouse placed in one chamber versus a novel object (an empty wire cage) in the opposing chamber. The position of these stimuli is counterbalanced across tests. In the *social novelty test*, the test mouse chooses between familiar and unfamiliar mice placed within opposing chambers.

Analysis: The automated measure in this test is the duration that a test mouse occupies each chamber and the number of entries to each chamber. Other measures that can be obtained from videotape include time spent sniffing cages. Controls for olfaction are necessary. See original papers for greater detail (188–190).

2.3. Social Conditioned Place Preference (SCPP) Test

Purpose: To determine whether a test mouse that shows diminished or enhanced social approach indicates differences in their experience of social reward or isolation aversion.

Conditioning: Every 24 h, the conditioning procedure entails a predictable alternation of the home cage living situation with respect to its social and nonsocial stimulus characteristics.

The first 24-h period of conditioning follows weaning, in which four mice (two per gender) are housed together in a standard home cage that contains one of two sets of novel environmental cues that are discernable by bedding and the presence of distinct kinds of PVC couplers (threaded and smooth) that are available at hardware stores. One conditioning environment ("paper") includes pelleted paper bedding, two schedule 40 1″ polyvinylchloride (PVC) couplers, and nesting material. The second conditioning environment ("aspen") included aspen shavings, two schedule 40 1″ PVC threaded couplers, and nesting material. The particular bedding used, whether corn-cob, paper, wood shavings, or something else, is not important so long as the mouse does not have a strong preference for either of the two beddings after conditioning for 10 days with no social contingency: as groups moving back and forth between the beddings.

Some animal facilities routinely use corn-cob beddings for mouse housing. Others use paper. However, it is crucial that:

- beddings are used to establish SCPP, providing a comfortable environment to establish social bonds,
- beddings used for the SCPP test are novel to the test mice, not experienced during rearing,
- mice do not have a strong preference for either bedding (control 1) but can discern them (control 2),
- tunnels are placed in the conditioning and test environments to provide a more complex environment.

During the second 24-h period, which follows 24 h of social housing, mice are socially isolated in a second novel home cage environment. During the third 24-h period, mice are returned to their social group, again placed in the bedding associated with social housing during the first day of conditioning. This pattern of alternating housing and social context every 24 h is repeated for a total of 10 days. The conditioning context (social or isolate housing) is always counterbalanced relative to its pairing with the home cage environment (aspen or paper bedding). Mice always began the conditioning phase of the experiment in a social group, making social isolation the default housing condition for the 24-h period prior to social conditioned place preference (SCPP) testing.

Very importantly, clean bedding, nesting material, and PVC tubes are provided for each conditioning session.

A 24-h duration of conditioning session produces a strong social conditioning response in juvenile mice. Use of shorter social conditioning sessions (e.g., 30–60 min) may result in smaller and more variable conditioning effects. The decision not to use shorter conditioning sessions also eliminates potentially complicating factors that result from maintaining mice in continuous isolate housing outside of the conditioning environments.

On the final day of conditioning, individual mice are allowed to freely explore the three-compartment testing arena ($300 \times 150 \times 150$ mm/compartment) for a 15-min habituation period with no conditioning cues present (habituations took place at 1,400–1,600 under dim red light).

Test procedure: On the test day (PD 30–35), an individual mouse is placed in the central compartment of the testing arena (where no conditioning cues were present) and clean beddings of the alternate environments are available in the opposing side chambers. Its movement throughout the arena is videotaped for a 30-min period. The spatial location and locomotor activity of individual mice are monitored. Litter size, sex bias within each litter, and maternal experience (primiparous vs. multiparous) are noted and their relationships to SCPP responses are also assessed.

Analysis: The time spent in each compartment (peripheral compartments contained the socially paired and isolation-paired beddings, respectively) and the number of transitions made between each compartment were quantified during a subsequent off-line analysis. Preference scores are calculated as the duration a mouse spent in the aspen bedding-lined compartment *minus* the duration spent in paper bedding-lined compartment. Videos are rated at 2× speed and then converted back to real time.

Controls: baseline preferences for the novel environments (i.e., no conditioning): To evaluate whether juvenile mice developed a preference for a particular bedding (such as aspen or paper) irrespective of a contingency after multiple conditioning sessions, a place preference test can be run after mice experience each environment without associated changes in social environment. Mice are handled according to the CPP procedure described above, but remain together in a social group as they are moved daily between the aspen and paper home cage environments. Environments on the first day of conditioning are counterbalanced across all groups to control for the possibility that environmental preferences of mice are sensitive to the environment experienced on the day prior to place preference testing. In other words, environment preferences for mice tested following 24 h in aspen are compared with those of mice that had spent 24 h in paper prior to testing.

Abilities to distinguish the bedding environments: This experiment is designed to control for the possibility that strain-dependent differences in SCPP are attributable to a more general difference in the ability of juvenile mice to establish a contextual association between the home cage environment and a rewarding or aversive stimulus. Following the conditioning protocol described above, groups of mice experience the two different environments every 24 h, paired with either ad libitum access to standard lab chow or with complete food deprivation. Mice begin the conditioning phase of this control experiment in the food-paired environment, so that CPP testing always occurs after 24 h of food deprivation. Weights for all food-deprived mice are monitored and compared with the weights of mice that are maintained under free-feeding conditions. Weights should not fall below 85% of free-feeding weights.

See original paper for greater detail (130).

2.4. Receptivity to Fear (Empathy) Test

Purpose: To determine whether a mouse can become fearful of a specific environmental cue that has been conditioned by its association with the distress vocalizations of another mouse. In this test, we utilize the terms "object" and "subject" mice. These terms are used extensively in the empathy literature and refer, respectively, to the individual receiving the distress (object) and the individual being tested (subject) for the ability to detect or respond to the distressed object.

Test arena: The fear-conditioning arena contains a "demonstration" compartment (130×165×150 mm) and two adjacent "observation" compartments (130×82.5×150 mm per observation compartment) see Fig. 7.2. This arena can be fabricated from ABS plastic and Plexiglas® or purchased from Cleversys, Inc., Reston, VA. The floor and one wall of the demonstration compartment are lined with a shock grid composed of stainless steel dowels (3.2 mm diameter) spaced 9.6 mm on center. The wall separating the observation and demonstration compartments consists of two sets of horizontal steel dowels that extend vertically 75 mm from the floor. This wall allows subjects to smell and hear objects, but eliminates the possibility of direct contact with the objects or the shock. Scrambled current is provided to the dowels lining the demonstration compartment, but current is not provided to steel dowels lining the observation compartments. Within the observation compartments, floors are also lined with inactive stainless steel dowels. An opaque plastic wall separates each observation compartment. The conditioned stimulus (CS) is a tone that can be delivered through computer speakers. The unconditioned stimulus (UCS) is a scrambled electrical shock that can be delivered via various devices produced for mouse and rat studies.

General considerations: All aspects of the behavioral test (habituation, conditioning and testing) are conducted under dim red illumination (30–40 lx) during the dark phase of the light–dark cycle. Cages containing the subjects and objects are transported from the mouse colony under dimly lit or dark conditions at least 30 min prior to the beginning of all phases of the conditioning protocol.

Habituation: For most experiments, subject mice lived together prior to conditioning. Individual subjects are habituated to the fear-conditioning arena for a 5-min period followed by presentation of a single 2-s shock. Mice are then returned singly into a clean cage.

Conditioning: After completion of the habituation sessions for two subject mice, two age-matched objects are randomly selected from a cage and placed together in the demonstration compartment. Two subjects that include same-sex individuals of the test animal and control conditions are then placed separately into the observation compartments. Home cages that contain the remaining subjects and objects are removed from the procedure room while their cage mates are pre-exposed to object mice receiving conditioning.

Object mice receive ten consecutive 120-s trials under one of several conditioning schedules while subjects remain in adjacent observation compartments where they do not receive the UCS. The conditioning apparatus is always cleaned very thoroughly with 70% ethanol before introducing new subjects/objects to the conditioning apparatus. Enzymes to remove urine are also recommended.

See **Fig. 7.3** for conditioning schedules.

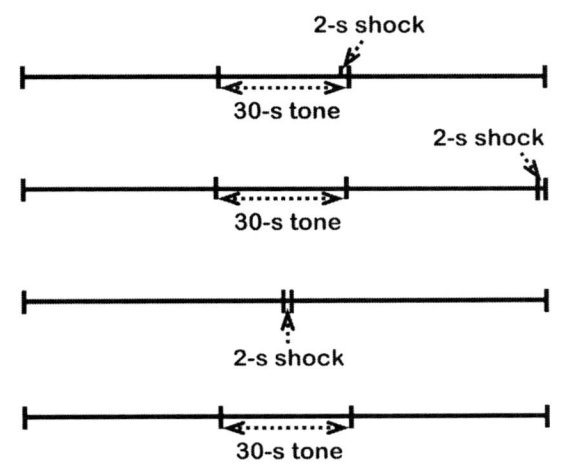

Fig. 7.3. The conditioning schedule for demonstrator mice includes three controls. The experimental condition is that the CS is forward paired with the UCS. Controls include an unpaired CS and USC, CS-only, and UCS-only. Mice that are receptive to the fear of others will show a strong response to the paired CS–UCS contingency, but will also likely show a greater fear response after experiencing object mice exposed to the UCS under the control conditions.

Following the first pre-exposure session, subjects and objects are re-grouped within their respective home cage. The next day subjects received another pre-exposure session as objects undergo another series of ten conditioning trials and then they are removed from the procedure room for 15 min.

Test procedure: Approximately 15 min after the second conditioning session, subjects are evaluated in the fear-conditioning arena. Each subject is placed singly into a clean cage and one subject is tested in the procedure chamber at a time. Test and control subjects are tested in a random order. Testing entails placing an individual subject in the fear-conditioning arena and then the freezing behavior of the subject is measured in response to 9 consecutive CS-only (tone) presentations followed by 11 consecutive presentations of the CS forward paired with the UCS (shock). Thus, freezing during test trials 1–10 occurs in response to presentation of the tone-only, whereas freezing on test trials 10–20 occurs in the context of direct presentation of the tone–shock contingency. All testing sessions are videotaped and transferred via a firewire cable directly to a computer for additional analysis. The conditioning apparatus is always cleaned thoroughly with several washes of 70% ethanol before introducing new subjects/objects to the conditioning apparatus. The duration of freezing behavior can be assessed by a trained observer with a stopwatch or by automated approaches.

Controls: To assess how exposure to social distress subsequently influences freezing behavior, it is essential to evaluate the freezing responses of test mice in response to direct presentation of the CS and the paired CS–UCS. Specifically, test mice should be placed individually in a fear-conditioning arena and presented with the CS-only (the CS was 30-s tone) for ten consecutive trials. Ideally, their freezing responses should be minimal; indicating that presentation of the CS is not salient without conditioning.

To assess whether test mice are responsive to fear conditioning, they should be assessed by measuring their freezing responses to CS forward paired with the UCS (a 2-s shock). Across successive conditioning trials of ten consecutive trials, test mice should express longer freezing responses.

To assess whether the test subject mice detect cues in the demonstration compartment, head orientations and freezing responses of subjects can be assessed during exposure to object distress. To obtain estimates of subject head orientations, freeze-frames of video recordings can be compared in the 1 s immediately prior to the presentation of the tone to objects versus immediately after the presentation of the tone and the shock to the objects. The longitudinal axis of the subject's head, running parallel with the sagittal midline, can be referenced at 15° increments. A 0° head orientation is defined by the longitudinal axis of the subject's head forming a right angle with the wall separating the observation and demonstration compartments.

Freezing responses of subjects during object distress can also be evaluated if the observation compartment is of sufficient size (such as 130×165×150 mm). Subjects are habituated to this observation compartment for 10 min (without objects present) on the day prior to testing. On the day of testing, two object mice are placed in the demonstration compartment and 1 subject mouse is placed in the observation compartment. Object mice then receive a 2-s shock every 120 s.

To assess whether the vocalizations are sufficient to engender a freezing response to the conditioned cue, recordings of object vocalizations during UCS presentation can be reproduced for subjects through an ultrasound-capable speaker (such as Ultra-

SoundGate ScanSpeak, Avisoft Bioacoustics) situated in or adjacent to, the demonstration compartment. To obtain recordings of vocalizations without enclosure-induced distortions, demonstrator mice should be tethered to an open shock grid (or be placed on a grid with low walls) and exposed to the UCS. Several recordings should be sampled. During the pre-exposure sessions, subjects are then exposed to ten consecutive CS-vocalization forward-pairings per session (randomly selected vocalizations are paired with the CS without substitution during each pre-exposure session). Vocalizations should be played back at 85–92 dB. In this case, subjects are exposed to the distress vocalizations of object mice, but not to objects directly. Following the second pre-exposure session, subjects are assessed as previously described. See original paper for greater detail (185).

Acknowledgments

GPL is grateful to Marsha Seltzer of the Waisman Center at the University of Wisconsin for her ongoing facilitation of collaborations with the autism clinical research community. He also wishes to thank Tina Iyama of the Waisman Center for promoting conversations with the families of autistic children. Both authors wish to acknowledge Jan Van Santen of Oregon Health and Science University for his thoughtful insights to prosodic communication. This work is a culmination of research experience support from several funding sources, including NIH grants (T32 MH018931, R03 HD046716, R01 DA022543, R01 DC007129, and P30 HD03352) and Autism Speaks.

References

1. American Psychiatric Association (2004) Diagnostic and statistical manual of mental disorders, 4th edn, Text Revision (DSM-IV-TR). American Psychiatric Press, Inc., Washington, DC
2. Lord C, Risi S, Lambrecht L, Cook EH, Leventhal BL, DiLavore PC, Pickles A, Rutter M (2000) The autism diagnostic observation schedule – generic: a standard measure of social and communication deficits associated with the spectrum of autism. J Autism Dev Disord 30(3):205–223
3. Lord C, Rutter M, Le Couteur A (1994) Autism diagnostic interview-revised. J Autism Dev Disord 24:659–686
4. Berument SK, Rutter M, Lord C, Pickles A, Bailey A (1999) Autism screening questionnaire: diagnostic validity. Br J Psychiatry 175:444–451
5. Klin A, Jones W, Schultz R, Volkmar F, Cohen D (2002) Visual fixation patterns during viewing of naturalistic social situations as predictors of social competence in individuals with autism. Arch Gen Psychiatry 59:809–816
6. Klin A, Lin D, Gorrindo P, Ramsay G, Jones W (2009) Two-year-olds with autism orient to non-social contingencies rather than biological motion. Nature 459:257–263
7. Zeng Z, Pantic M, Roisman G, Huang T (2009) A survey of affect recognition

methods: audio, visual, and spontaneous expressions. IEEE Trans Pattern Anal Mach Intell 31(1):39–58
8. Bartlett M, Littlewort M, Frank M, Lainscsek C, Fasel I, Movellan J (2006) Fully automatic facial action recognition in spontaneous behavior. Proceeding of the IEEE international conference, Automatic Face and Gesture Recognition (AFGR'06), Southampton, UK, pp 223–230
9. van Santen J, Tucker-Prud'hommeaux E, Black L (2009) Automated assessment of prosody production. Speech Commun 51:1082–1097
10. Schore AN (2010) The right brain implicit self: A central Mechanism of the psychotherapy change process. In: J. Petrucelli (Ed.) *Knowing, Not knowing and Sort of Knowing: Psychoanalysis and the Experience of Uncertainity* (pp. 22-45). London: Karnac Books.
11. Fernald A (1989) Intonation and communicative intent in mothers' speech to infants: is the melody the message? Child Dev 60(6):1497–1510
12. Kanner L (1943) Autistic disturbances of affective content. Nerv Child 2:217–250
13. Lord C, Rutter M (1994) Autism and pervasive developmental disorders. In: Rutter M , Hersov L, Taylor E (eds) Child and adolescent psychiatry: modern approaches, 3rd edn. Blackwell Scientific, Oxford, pp 569–593
14. Fay W, Schuler A (1980) Emerging language in autistic children. University Park Press, Baltimore, MD
15. Lord C, Paul R (1997) Communication. In: Cohen D, Volkmar F (eds) Handbook of autism and pervasive developmental disorders, 2nd edn. Wiley, New York, NY, pp 195–225
16. Shriberg L, Paul R, McSweeney J, Klin A, Cohen D, Volkmar F (2001) Speech and prosody characteristics of adolescents and adults with high functioning autism and Asperger syndrome. J Speech Lang Hear Res 44:1097–1115
17. Lord C, Rutter M, DiLavore PC, Risi S (2002) Autism diagnostic observation schedule. Western Psychological Services, Los Angeles, CA
18. Wetherby AM, Woods J, Allen L, Cleary J, Dickinson H (2004) Early indicators of autism spectrum disorders in the second year of life. J Autism Dev Disord 34:473–493
19. Shriberg L, Kwiatkowski J, Rasmussen C (1990) Prosody-voice screening profile. Communication Skillbuilders, Tuscon, AZ
20. Black L, van Santen J, Coulston R, Paul R, de Villiers J (2008) Effects of enhanced prosody on narrative recall in children with autism. Proceeding of the international meeting for autism research 2008, London
21. Black L, van Santen J, Coulston R, de Villiers J (2009) Vocal prosody in autism: understanding the effects of enhancing vocal prosody on children's comprehension and retention of story narratives. Proceedings of the international meeting for autism research 2009, Chicago, IL
22. Yirmiya N, Kasaari C, Sigman M, Mundy P (1989) Facial expressions of affect in autistic, mentally retarded, and normal children. J Child Psychol Psychiatr 30(5):725–735
23. Ekman P, Friesen W (1978) Facial action coding system. Consulting Psychologists Press, Palo Alto, CA
24. Hobson R (1986a) The autistic child's appraisal of expressions of emotion. J Child Psychol Psychiatr 27(3):321–342
25. Hobson R (1986b) The autistic child's appraisal of expressions of emotion: a further study. J Child Psychol Psychiatry 27(5):671–680
26. Fein D, Pennington B, Markowitz P, Braverman M, Waterhouse L (1986) Toward a neuropsychological model of infantile autism: are the social deficits primary? J Am Acad Child Psychiatry 25:198–212
27. Braverman M, Fein D, Lucci D, Waterhouse L (1989) Affect comprehension in children with pervasive developmental disorders. J Autism Devl Disord 19:301–315
28. Ozonoff S, Pennington B, Rogers S (1990) Are there emotion perception deficits in young autistic children? J Child Psychol Psychiatry 31:343–361
29. Schultz R (2005) Developmental deficits in social perception in autism: the role of the amygdala and fusiform face area. Int J Dev Neurosci 23:125–141
30. Lindner J, Rosén L (2006) Decoding of emotion through facial expression, prosody and verbal content in children and adolescents with Asperger's syndrome. J Autism Dev Disord 36:769–777
31. Egan G (1989) Assessment of emotional processing in right and left hemisphere stroke patients: a validation study of the perception of emotion test. Unpublished doctoral dissertation, Georgia State University
32. Mazefsky C, Oswald D (2007) Emotion perception in Asperger's and high-functioning autism: the importance of diagnostic criteria and cue intensity. J Autism Dev Disord 37:1086–1095

33. Nowicki S, Duke MP (1994) Individual differences in the nonverbal communication of affect: the diagnostic analysis of nonverbal accuracy scale. J Nonverbal Behav 18(1):9–35
34. Wang A, Dapretto M, Hariri AR, Sigman M, Bookheimer SY (2004) Neural correlates of facial affect processing in children and adolescents with autism spectrum disorder. J Am Acad Child Adolesc Psychiatry 43(4):481–490
35. Kleinhaans NM, Johnson LC, Richards T, Mahurin R, Greenson J, Dawson G, Aulward E (2009) Reduced neural habituation in the amygdala and social impairments in autism spectrum disorders. Am J Psychiatry 166:467–475
36. Mundy P (1995) Joint attention and social–emotional approach behavior in children with autism. Dev Psychopathol 7:63–82
37. Stone WL, Ousley OY, Yodar PJ, Hogan KL, Hepburn SL (1997) Nonverbal communication in two- and three-year old children with autism. J Autism Dev Disord 27:677–696
38. Chiang C-H, Soong W-T, Lin T-S, Rogers SJ (2008) Nonverbal communication skills in young children with autism. J Autism Dev Disord 38:1898–1906
39. Watson LR, Baranek GT, Dilavore PC (2003) Toddlers with autism developmental perspectives. Infants Young Children 16(3):201–214
40. Baranek GT (1999) Autism during infancy: a retrospective video analysis of sensory-motor and social behaviors at 9–12 months. J Autism Dev Disord 29:213–224
41. Zwaigenbaum L, Bryson S, Rogers T, Roberts W, Brian J, Szatmari P (2005) Behavioral markers of autism in the first year of life. Inter J Dev Neurosci 23:143–152
42. Crais ER, Watson LR, Baranek GT (2009) Use of gesture development in profiling children's prelinguistic communication skills. Am J Speech-Lang Pathol 18:95–108
43. Lin H, Shafran I, Murphy TE, Okamura AM, Yuh DD, Hager GD (2005) Automatic detection and segmentation of robot-assisted surgical motions. Proceedings of the 8th international conference on medical image computing and computer assisted intervention (MICCAI), vol I, Palm Springs, CA, pp 802–810
44. Rogers SJ, Hepburn SL, Stackhouse T, Wehner E (2003) Imitation performance in toddlers with autism and those with other developmental disorders. J Child Psychol Psychiatry 44(5):763–781
45. Rogers SJ, Williams JH (2006) Imitation and the social mind: autism and typical development. Guilford Press, New York, NY
46. Dziuk M, Larson J, Apostu A, Mahone E, Denckla M, Mostofsky S (2007) Dyspraxia in autism: association with motor, social, and communicative deficits. Dev Med Child Neurol 49:734–739
47. Iacoboni M, Woods RP, Brass M, Bekkering H, Mazziotta JC, Rizzolatti G (1999) Cortical mechanisms of human imitation. Science 286:2526–2528
48. Iacoboni M (2009) Imitation, empathy, and mirror neurons. Annu Rev Pyschol 60:653–670
49. Rizzolatti G, Craighero L (2004) The mirror neuron system. Annu Rev Neurosci 27:169–192
50. Rizzolatti G, Fadiga L, Gallese V, Fgassi L (1996) Premotor cortex and the recognition of motor actions. Brain Res Cogn Brain Res 3:131–141
51. Gallese V (2007) Before and below theory of mind: embodied simulation and the neural correlates of social cognition. Philos Trans R Soc Lond B Biol Sci 362:659–669
52. Pfeifer JH, Iacoboni M, Mazziotta JC, Dapretto M (2008) Mirroring others' emotions relates to empathy and interpersonal competence in children. Neuroimage 39:2076–2085
53. Dapretto M, Davies N, Pfeifer J, Scott A, Sigman M, Bookheimer S, Iacoboni M (2006) Understanding emotions in others: mirror neuron dysfunction in children with autism spectrum disorders. Nat Neurosci 9(1):28–30
54. Lipps T (1903) Einfühlung, innere nachahmung, und Organepfindungen. Arch Gesamte Psychol 1:185–204
55. Preston SD, de Waal FBM (2002) Empathy: its ultimate and proximate bases. Behav Brain Sci 25:1–72
56. de Waal FBM (2008) Putting the altruism back into altruism: the evolution of empathy. Annu Rev Psychol 59:279–300
57. Hoffman M (2000) Empathy and moral development: implications for caring and justice. Cambridge University Press, New York, NY
58. Eisenberg N, Fabes RA, Spinrad TL (2006) Prosocial development. In: Eisenberg N (eds) Handbook of child psychology, vol. 3. John Wiley and Sons, Inc., New Jersey, NJ, pp 646–718
59. Adolphs R (2009) The social brain: neural basis of social knowledge. Annu Rev Pyschol 60:693–716

60. Frith CD (2007) The social brain? Philos Trans R Soc Lond B Biol Sci 362: 671–678
61. Just MA, Cherkassky VL, Keller TA, Kana RK, Minshew NJ (2007) Functional and anatomical cortical underconnectivity in autism: evidence from an fMRI study of an executive function task and corpus callosum morphometry. Cereb Cortex 17: 951–961
62. Just MA, Cherkassky VL, Keller TA, Minshew NJ (2004) Cortical activation and synchronization during sentence comprehension in high-functioning autism: evidence of underconnectivity. Brain J Neurol 127:1811–1821
63. Herbert MR, Ziegler DA, Makris N, Filipek PA, Kemper TL, Normandin JJ, Sanders HA, Kennedy DN, Caviness VS Jr (2004) Localization of white matter volume increase in autism and developmental language disorder. Ann Neurol 55:530–540
64. Casanova M (2006) Neuropathological and genetic findings in autism: the significance of a putative minicolumnopathy. NeuroScript 12(5):435–441
65. Kurian JR, Bychowski ME, Forbes-Lorman RM, Auger CJ, Auger AP (2008) Mecp2 organizes juvenile social behavior in a sex-specific manner. J Neurosci 28: 7137–7142
66. Turakainen H, Saarimaki-Vire J, Sinjushina N, Partanen J, Savilahti H (2009) Transposition-based method for the rapid generation of gene-targeting vectors to produce Cre/Flp-modifiable conditional knockout mice. PLoS ONE [Electronic Resource] 4:e4341
67. Tabuchi K, Blundell J, Etherton MR, Hammer RE, Liu X, Powell CM, Sudhof TC (2007) A neuroligin-3 mutation implicated in autism increases inhibitory synaptic transmission in mice [see comment]. Science 318:71–76
68. Hagerman RJ, Jackson AW III, Levitas A, Rimland B, Braden M (1986) An analysis of autism in fifty males with the fragile X syndrome. Am J Med Genet 23:359–374
69. Comery TA, Harris JB, Willems PJ, Oostra BA, Irwin SA, Weiler IJ, Greenough WT (1997) Abnormal dendritic spines in fragile X knockout mice: maturation and pruning deficits. Proc Natl Acad Sci USA 94: 5401–5404
70. Kooy RF, D'Hooge R, Reyniers E, Bakker CE, Nagels G, De Boulle K, Storm K, Clincke G et al (1996) Transgenic mouse model for the fragile X syndrome. Am J Med Genet 64:241–245
71. McNaughton CH, Moon J, Strawderman MS, Maclean KN, Evans J, Strupp BJ (2008) Evidence for social anxiety and impaired social cognition in a mouse model of fragile X syndrome [see comment]. Behav Neurosci 122:293–300
72. Spencer CM, Graham DF, Yuva-Paylor LA, Nelson DL, Paylor R (2008) Social behavior in Fmr1 knockout mice carrying a human FMR1 transgene. Behav Neurosci 122: 710–715
73. Brodkin ES (2008) Social behavior phenotypes in fragile X syndrome, autism, and the Fmr1 knockout mouse: theoretical comment on McNaughton et al (2008) [see comment]. Behav Neurosci 122: 483–489
74. Hayashi ML, Rao BSS, Seo J-S, Choi H-S, Dolan BM, Choi S-Y, Chattarji S, Tonegawa S (2007) Inhibition of p21-activated kinase rescues symptoms of fragile X syndrome in mice. Proc Natl Acad Sci USA 104: 11489–11494
75. Yan QJ, Rammal M, Tranfaglia M, Bauchwitz RP (2005) Suppression of two major Fragile X syndrome mouse model phenotypes by the mGluR5 antagonist MPEP. Neuropharmacology 49:1053–1066
76. de Vrij FMS, Levenga J, van der Linde HC, Koekkoek SK, De Zeeuw CI, Nelson DL, Oostra BA, Willemsen R (2008) Rescue of behavioral phenotype and neuronal protrusion morphology in Fmr1 KO mice. Neurobiol Dis 31:127–132
77. Wassink TH, Piven J, Vieland VJ, Huang J, Swiderski RE, Pietila J, Braun T, Beck G et al (2001) Evidence supporting *WNT2* as an autism susceptibility gene. Am J Med Genet 105:406–413
78. Moretti P, Bouwknecht JA, Teague R, Paylor R, Zoghbi HY (2005) Abnormalities of social interactions and home-cage behavior in a mouse model of Rett syndrome. Hum Mol Genet 14:205–220
79. Cheh MA, Millonig JH, Roselli LM, Ming X, Jacobsen E, Kamdar S, Wagner GC (2006) En2 knockout mice display neurobehavioral and neurochemical alterations relevant to autism spectrum disorder. Brain Res 1: 166–176
80. Sadakata T, Washida M, Iwayama Y, Shoji S, Sato Y, Ohkura T, Katoh-Semba R, Nakajima M et al (2007) Autistic-like phenotypes in Cadps2-knockout mice and aberrant CADPS2 splicing in autistic patients. J Clin Investig 117:931–943
81. Jamain S, Radyushkin K, Hammerschmidt K, Granon S, Boretius S, Varoqueaux F, Ramanantsoa N, Gallego J et al (2008)

Reduced social interaction and ultrasonic communication in a mouse model of monogenic heritable autism. Proc Natl Acad Sci USA 105:1710–1715

82. Anderson PK (1961) Density, social structure, and nonsocial environment of housemouse populations and the implications for regulation of numbers. Trans NY Acad Sci 23:447–451

83. DeLong KT (1967) Population ecology of feral house mice. Ecology 48:611–634

84. Noyes RF, Barrett GW, Taylor DH (1982) Social structure of feral house mouse (*Mus musculus* L.) populations: effects of resource partitioning. Behav Ecol Sociobiol 10:157–163

85. Desjardins C, Maruniak JA, Bronson FH (1973) Social rank in house mice: differentiation revealed by ultraviolet visualization of urinary marking patterns. Science 182:939–941

86. Hurst JL, Barnard CJ (1992) Kinship and social behavior in wild house mice: effects of social group membership and relatedness on the responses of dominant males toward juveniles. Behav Ecol 3:196–206

87. Wolff RJ (1985) Mating behavior and female choice: their relation to social structure in wild caught house mice (*Mus musculus*) housed in a semi-natural environment. J Zool Ser A 207:43–51

88. Bronson FH (1983) Chemical communication in house mice and deer mice: functional roles in reproduction of wild populations. Am Soc Mammalogists Spec Publ 7:198–238

89. Krebs C, Chitty D, Singleton G, Boonstra R (1995) Can changes in social behaviour help to explain house mouse plagues in Australia? Oikos 73:429–434

90. Chambers LK, Singleton GR, Krebs CJ (2000) Movements and social organization of wild house mice (*Mus domesticus*) in the wheatlands of northwestern Victoria, Australia. J Mammal 81:59–69

91. Branchi I, Santucci D, Alleva E (2001) Ultrasonic vocalisation emitted by infant rodents: a tool for assessment of neurobehavioural development. Behav Brain Res 125:49–56

92. D'Amato FR, Scalera E, Sarli C, Moles A (2005) Pups call mothers rush: does maternal responsiveness affect the amount of ultrasonic vocalizations in mouse pups? Behav Genet 35:103–112

93. Crowcroft P (1966) Mice all over. Foulis, London

94. Panksepp JB, Wong JC, Kennedy BC, Lahvis GP (2008) Differential entrainment of a social rhythm in adolescent mice. Behav Brain Res 195:239–245

95. Van den Berg CL, Kitchen I, Gerrits MA, Spruijt BM, Van Ree JM (1999) Morphine treatment during juvenile isolation increases social activity and opioid peptides release in the adult rat. Brain Res 830:16–23

96. Miczek KA, Maxson SC, Fish EW, Faccidomo S (2001) Aggressive behavioral phenotypes in mice. Behav Brain Res 125:167–181

97. Lijam N, Paylor R, McDonald MP, Crawley JN, Deng CX, Herrup K, Stevens KE, Maccaferri G et al (1997) Social interaction and sensorimotor gating abnormalities in mice lacking Dvl1. Cell 90:895–905

98. Moy SS, Nadler JJ, Perez A, Barbaro RP, Johns JM, Magnuson TR, Piven J, Crawley JN (2004) Sociability and preference for social novelty in five inbred strains: an approach to assess autistic-like behavior in mice. Genes Brain Behav 3:287–302

99. Panksepp JB, Jochman K, Kim JU, Koy JJ, Wilson ED, Chen Q, Wilson CR, Lahvis GP (2007) Affiliative behavior, ultrasonic communication and social reward are influenced by genetic variation in adolescent mice. PLoS ONE [Electronic Resource] 2. doi:10.1371/journal.pone0000351

100. Terranova ML, Laviola G, Alleva E (1993) Ontogeny of amicable social behavior in the mouse: gender differences and ongoing isolation outcomes. Dev Psychobiol 26:467–481

101. Terranova ML, Laviola G, de Acetis L, Alleva E (1998) A description of the ontogeny of mouse agonistic behavior. J Comp Psychol 112:3–12

102. Tirelli E, Laviola G, Adriani W (2003) Ontogenesis of behavioral sensitization and conditioned place preference induced by psychostimulants in laboratory rodents. Neurosci Biobehav Rev 27:163–178

103. Spear L (2000) Modeling adolescent development and alcohol use in animals. Alcohol Res Health: J Nat Inst Alcohol Abuse Alcoholism 24:115–123

104. Laviola G, Adriani W, Terranova ML, Gerra G (2000) Psychobiologic risk factors and vulnerability to psychostimulants in adolescents and animal models. Annali Dell'Istituto Superiore di Sanita 36:47–62

105. Pellis SM, Iwaniuk AN (1999) The roles of phylogeny and sociality in the evolution of social play in muroid rodents. Anim Behav 58:361–373

106. Pellis SM, Pasztor TJ (1999) The developmental onset of a rudimentary form of play fighting in C57 mice. Dev Psychobiol 34:175–182

107. Pellis SM, Pellis VC (1997) The prejuvenile onset of play fighting in laboratory rats (*Rattus norvegicus*). Dev Psychobiol 31: 193–205
108. Siviy SM, Panksepp J (1985) Dorsomedial diencephalic involvement in the juvenile play of rats. Behav Neurosci 99:1103–1113
109. Ikemoto S, Panksepp J (1992) The effects of early social isolation on the motivation for social play in juvenile rats. Dev Psychobiol 25:261–274
110. Kanner L (1943) Autistic disturbances of affective contact. Nerv Child 2:217–250
111. Dawson G, Toth K, Abbott R, Osterling J, Munson J, Estes A, Liaw J (2004) Early social attention impairments in autism: social orienting, joint attention, and attention to distress. Dev Psychol 40:271–283
112. LeDoux JE, Iwata J, Cicchetti P, Reis DJ (1988) Different projections of the central amygdaloid nucleus mediate autonomic and behavioral correlates of conditioned fear. J Neurosci 8:2517–2529
113. Falls WA, Carlson S, Turner JG, Willott JF (1997) Fear-potentiated startle in two strains of inbred mice. Behav Neurosci August 111:855–861
114. Paylor R, Tracy R, Wehner J, Rudy JW (1994) DBA/2 and C57BL/6 mice differ in contextual fear but not auditory fear conditioning. Behav Neurosci 108:810–817
115. Kelley AE, Bakshi VP, Haber SN, Steininger TL, Will MJ, Zhang M (2002) Opioid modulation of taste hedonics within the ventral striatum. Physiol Behav 76:365–377
116. Moles A, Kieffer BL, D'Amato FR (1983) Deficit in attachment behavior in mice lacking the mu-opioid receptor gene [see comment]. Science 304:1983–1986
117. MacLean PD (1990) The triune brain in evolution: role in paleocerebral functions. Plenum Press, New York, NY
118. Panksepp J (1998) Affective neuroscience. Oxford University Press, New York, NY
119. Kelley AE, Berridge KC (2002) The neuroscience of natural rewards: relevance to addictive drugs. J Neurosci 22:3306–3311
120. Bardo MT, Bevins RA (2000) Conditioned place preference: what does it add to our preclinical understanding of drug reward? Psychopharmacology 153:31–43
121. Reith ME, Selmeci G (1992) Cocaine binding sites in mouse striatum, dopamine autoreceptors, and cocaine-induced locomotion. Pharmacol Biochem Behav 41:227–230
122. Abarca C, Albrecht U, Spanagel R (2002) Cocaine sensitization and reward are under the influence of circadian genes and rhythm. Proc Natl Acad Sci USA 99:9026–9030
123. van Ree JM, Gerrits MA, Vanderschuren LJ (1999) Opioids, reward and addiction: an encounter of biology, psychology, and medicine. Pharmacol Rev 51:341–396
124. Tzschentke TM (2007) Measuring reward with the conditioned place preference (CPP) paradigm: update of the last decade. Addict Biol 12:227–462
125. Berridge KC, Robinson TE (2003) Parsing reward. Trends Neurosci 26:507–513
126. Douglas LA, Varlinskaya EI, Spear LP (2004) Rewarding properties of social interactions in adolescent and adult male and female rats: impact of social versus isolate housing of subjects and partners. Dev Psychobiol 45: 153–162
127. Calcagnetti DJ, Schechter MD (1992) Place conditioning reveals the rewarding aspect of social interaction in juvenile rats. Physiol Behav 51:667–672
128. Camacho F, Sandoval C, Paredes R (2004) Sexual experience and conditioned place preference in male rats. Pharmacol Biochem Behav 78:419–425
129. Jenkins WJ, Becker JB (2003) Female rats develop conditioned place preferences for sex at their preferred interval. Horm Behav 43:503–507
130. Panksepp JB, Lahvis GP (2007) Social reward among juvenile mice. Genes Brain Behav 6:661–671
131. Mattson BJ, Williams S, Rosenblatt JS, Morrell JI (2001) Comparison of two positive reinforcing stimuli: pups and cocaine throughout the postpartum period. Behav Neurosci 115:683–694
132. Martinez M, Guillen-Salazar F, Salvador A, Simon VM (1995) Successful intermale aggression and conditioned place preference in mice. Physiol Behav 58:323–328
133. Sankoorikal GM, Kaercher KA, Boon CJ, Lee JK, Brodkin ES (2006) A mouse model system for genetic analysis of sociability: C57BL/6 J versus BALB/cJ inbred mouse strains. Biol Psychiatry 59:415–423
134. Brodkin ES (2007) BALB/c mice: low sociability and other phenotypes that may be relevant to autism. Behav Brain Res 176: 53–65
135. Shippenberg TS, Elmer GI (1998) The neurobiology of opiate reinforcement. Crit Rev Neurobiol 12:267–303
136. Panksepp J, Herman BH, Vilberg T, Bishop P, DeEskinazi FG (1980) Endogenous opioids and social behavior. Neurosci Biobehav Rev 4:473–487
137. Panksepp J, Nelson E, Siviy S (1994) Brain opioids and mother–infant social motivation. Acta Paediatr Suppl 397:40–46

138. Kalin NH, Shelton SE, Barksdale CM (1988) Opiate modulation of separation-induced distress in non-human primates. Brain Res 440:285–292
139. Benton D, Brain S, Brain PF (1984) Comparison of the influence of the opiate delta receptor antagonist, ICI 154,129, and naloxone on social interaction and behaviour in an open field. Neuropharmacology 23:13–17
140. Vanderschuren LJ, Niesink RJ, Spruijt BM, Van Ree JM (1995) Effects of morphine on different aspects of social play in juvenile rats. Psychopharmacology 117:225–231
141. Herman BH, Panksepp J (1981) Ascending endorphin inhibition of distress vocalization. Science 211:1060–1062
142. Vanderschuren LJ, Stein EA, Wiegant VM, Van Ree JM (1995) Social play alters regional brain opioid receptor binding in juvenile rats. Brain Res 680:148–156
143. Aragona BJ, Liu Y, Curtis JT, Stephan FK, Wang Z (2003) A critical role for nucleus accumbens dopamine in partner-preference formation in male prairie voles. J Neurosci 23:3483–3490
144. Lucas LR, Celen Z, Tamashiro KL, Blanchard RJ, Blanchard DC, Markham C, Sakai RR, McEwen BS (2004) Repeated exposure to social stress has long-term effects on indirect markers of dopaminergic activity in brain regions associated with motivated behavior. Neuroscience 124:449–457
145. Kosten TA, Zhang XY, Kehoe P (2005) Neurochemical and behavioral responses to cocaine in adult male rats with neonatal isolation experience. J Pharmacol Exp Ther 314:661–667
146. Kosten TA, Zhang XY, Kehoe P (2006) Heightened cocaine and food self-administration in female rats with neonatal isolation experience. Neuropsychopharmacology 31:70–76
147. Ferguson JN, Young LJ, Insel TR (2002) The neuroendocrine basis of social recognition. Front Neuroendocrinol 23:200–224
148. Kavaliers M, Choleris E, Agmo A, Muglia LJ, Ogawa S, Pfaff DW (2005) Involvement of the oxytocin gene in the recognition and avoidance of parasitized males by female mice. Anim Behav 70:693–702
149. Skuse DH, Gallagher L (2009) Dopaminergic-neuropeptide interactions in the social brain. Trends Cogn Sci 13:27–35
150. Burgdorf J, Kroes RA, Moskal JR, Pfaus JG, Brudzynski SM, Panksepp J (2008) Ultrasonic vocalizations of rats (*Rattus norvegicus*) during mating, play, and aggression: behavioral concomitants, relationship to reward, and self-administration of playback. J Comp Psychol 122:357–367
151. Knutson B, Burgdorf J, Panksepp J (1999) High-frequency ultrasonic vocalizations index conditioned pharmacological reward in rats. Physiol Behav 66:639–643
152. Sadananda M, Wohr M, Schwarting RKW (2008) Playback of 22-kHz and 50-kHz ultrasonic vocalizations induces differential c-fos expression in rat brain. Neurosci Lett 435:17–23
153. Branchi I, Santucci D, Vitale A, Alleva E (1998) Ultrasonic vocalizations by infant laboratory mice: a preliminary spectrographic characterization under different conditions. Dev Psychobiol 33:249–256
154. Ehret G, Bernecker C (1986) Low-frequency sound communication by mouse pups (*Mus musculus*): Wriggling calls release maternal behavior. Anim Behav 34:821–830
155. Dastur FN, McGregor IS, Brown RE (1999) Dopaminergic modulation of rat pup ultrasonic vocalizations. Eur J Pharmacol 382:53–67
156. Moles A, Kieffer BL, D'Amato FR (2004) Deficit in attachment behavior in mice lacking the mu-opioid receptor gene [see comment]. Science 304:1983–1986
157. Scattoni ML, Gandhy SU, Ricceri L, Crawley JN (2008) Unusual repertoire of vocalizations in the BTBR T+tf/J mouse model of autism. PLoS ONE [Electronic Resource] 3:e3067
158. Nyby J (1983) Ultrasonic vocalizations during sex behavior of male house mice (*Mus musculus*): a description. Behav Neural Biol 39:128–134
159. White NR, Prasad M, Barfield RJ, Nyby JG (1998) 40- and 70-kHz vocalizations of mice (*Mus musculus*) during copulation. Physiol Behav 63:467–473
160. Elwood RW, Kennedy HF, Blakely HM (1990) Responses of infant mice to odors of urine from infanticidal, noninfanticidal, and paternal male mice. Dev Psychobiol 23:309–317
161. Nyby J, Wysocki CJ, Whitney G, Dizinno G, Schneider J, Nunez AA (1981) Stimuli for male mouse (*Mus musculus*) ultrasonic courtship vocalizations: presence of female chemosignals and/or absence of male chemosignals. J Com Physiol Psychol 95:623–629
162. Wang H, Liang S, Burgdorf J, Wess J, Yeomans J (2008) Ultrasonic vocalizations induced by sex and amphetamine in M2, M4, M5 muscarinic and D2 dopamine receptor knockout mice. PLoS ONE [Electronic Resource] 3:e1893

163. Cabral A, Isoardi N, Salum C, Macedo CE, Nobre MJ, Molina VA, Brandao ML (2006) Fear state induced by ethanol withdrawal may be due to the sensitization of the neural substrates of aversion in the dPAG. Exp Neurol 200:200–208

164. Arakawa H, Arakawa K, Blanchard D, Blanchard RJ (2007) Scent marking behavior in male C57BL/6 J mice: sexual and developmental determination. Behav Brain Res 182:73–79

165. Eggert F, Luszyk D, Ferstl R, Muller-Ruchholtz W (1989) Changes in strain-specific urine odors of mice due to bone marrow transplantations. Neuropsychobiology 22:57–60

166. Gilder P, Slater P (1978) Interest of mice in conspecific male odours is influenced by degree of kinship. Nature 274:364–365

167. Kavaliers M, Colwell DD (1995) Odours of parasitized males induce aversive responses in female mice. Anim Behav 50:1161–1169

168. Kavaliers M, Choleris E, Pfaff DW (2005) Recognition and avoidance of the odors of parasitized conspecifics and predators: differential genomic correlates. Neurosci Biobehav Rev 29:1347–1359

169. Dixon A, Mackintosh J (1976) Olfactory mechanisms affording from attack to juvenile mice (Mus musculus L.). Z Tierpsychol 41:225–234

170. Sandnabba N (1986) Changes in male odours and urinary marking patterns due to inhibition of aggression in male mice. Behav Process 12:349–361

171. Rottman SJ, Snowdon CT (1972) Demonstration and analysis of an alarm pheromone in mice. J Com Physiol Psychol 81:483–490

172. Arakawa H, Arakawa K, Blanchard D, Blanchard RJ (2009) Social features of scent-donor mice modulate scent marking of c57bl/6j recipient males. Behav Brain Res 205:138–145

173. Nyby J, Wysocki CJ, Whitney G, Dizinno G (1977) Pheromonal regulation of male mouse ultrasonic courtship (Mus musculus). Anim Behav 25:333–341

174. Holy TE, Guo Z (2005) Ultrasonic songs of male mice. Plos Biol 3:e386

175. Langford DJ, Crager SE, Shehzad Z, Smith SB, Sotocinal SG, Levenstadt JS, Chanda ML, Levitin DJ, Mogil JS (2006) Social modulation of pain as evidence for empathy in mice. Science 312:1967–1970

176. Kavaliers M, Choleris E, Colwell DD (2001) Learning from others to cope with biting flies: social learning of fear-induced conditioned analgesia and active avoidance. Behav Neurosci 115:661–674

177. Carlier P, Jamon M (2006) Observational learning in C57BL/6j mice. Behav Brain Res 174:125–131

178. Collins RL (1988) Observational learning of a left–right behavioral asymmetry in mice (Mus musculus). J Com Psychol 102:222–224

179. Spencer CM, Alekseyenko O, Serysheva E, Yuva-Paylor LA, Paylor R (2005) Altered anxiety-related and social behaviors in the Fmr1 knockout mouse model of fragile X syndrome. Genes Brain Behav 4:420–430

180. Mineur YS, Sluyter F, de Wit S, Oostra BA, Crusio WE (2002) Behavioral and neuroanatomical characterization of the Fmr1 knockout mouse. Hippocampus 12:39–46

181. Bernardet M, Crusio WE (2006) Fmr1 KO mice as a possible model of autistic features. Sci World J 6:1164–1176

182. Zahn-Waxler C, Radke-Yarrow M (1990) The origins of empathic concern. Motiv Emot 14:107–130

183. Bacon A, Fein D, Morris R, Waterhouse L, Allen D (1998) The responses of autistic children to the distress of others. J Autism Dev Disord 28(2):129–142

184. Dawson G, Toth K, Abbott R, Osterling J, Munson J, Estes A, Liaw J (2004) Early social attention impairments in autism: social orienting, joint attention, and attention to distress. Dev Psychol 40(2):271–283

185. Chen Q, Panksepp JB, Lahvis GP (2009) Empathy is moderated by genetic background in mice. PLoS ONE [Electronic Resource] 4:1–14

186. Lahvis GP, Alleva E, Scattoni M-L (2010) Translating mouse vocalizations: prosody and frequency modulation. Genes Brain Behav no. doi: 10.1111/j.1601-183x.2010.00603.x, pp. 1–13

187. Hammerschmidt K, Radyushkin K, Ehrenreich H, Fischer J (2009) Female mice respond to male ultrasonic 'songs' with approach. Biol Lett

188. Moy S, Nadler J, Young N, Nonneman R, Grossman A, Murphy D, D'Ercole A, Crawley J et al (2009) Social approach in genetically engineered mouse lines relevant to autism. Genes Brain Behav 8:129–142

189. Moy S, Nadler J, Perez A, Barbaro R, Johns J, Magnuson T, Piven J, Crawley J (2004) Sociability and preference for social novelty in five inbred strains: an approach to assess autistic-like behavior in mice. Genes Brain Behav 3:287–302

190. Nadler J, Moy S, Dold G, Trang D, Simmons N, Perez A, Young N, Barbaro R et al (2004) Automated apparatus for quantitation of social approach behaviors in mice. Genes Brain Behav 3:303–314

Chapter 8

The Experimental Manipulation of Uncertainty

Dominik R. Bach, Christopher R. Pryce, and Erich Seifritz

Abstract

Uncertainty is an important concept in neuroscience: due to its relevance in everyday life, because of theoretical significance for neurocomputational models, and clinical implications. A body of empirical research has tackled fundamental questions about how uncertainty is represented in the brain and what impact it has on behaviour. In this chapter, we review how uncertainty on different variables can be studied in isolation and how it can be quantified. Building on theoretical and empirical work that has been carried out so far, we propose rigorous experimental designs that should help in testing and understanding uncertainty and its translational relevance to adaptive behaviour and affective disorders.

1. Introduction

Research about uncertainty has received increasing interest over the last decades. Indeed, uncertainty is one of the main constants in life. For any organism, nothing is ever certain. Any perception, any representation of external states, or any computation bears imprecision. Humans often have a conscious notion of this uncertainty, humans and animals behave as if they know about it, and behaviour has been shown to be more optimal when uncertainty is taken into account. This has raised the question of whether there is a specific representation of uncertainty in the brain and how and where this is organised. Such research questions embrace a clinical perspective which posits that, for example, disorders in the schizophrenic spectrum are characterised by a fundamental misrepresentation of uncertainty, arising through increased stochastic noise in neural circuits (*see*, e.g. 1), and on a higher cognitive level, maladaptive styles of coping with uncertainty could be crucial, for example, in generalised anxiety disorder (*see*, e.g. 2).

The bulk of uncertainty literature, starting in the 1950s and 1960s, takes the form of research into the effects of a situation where, e.g. aversive events are uncertain, or unpredictable, as opposed to another situation where they are certain, or predictable. This view posits discrete psychological states associated with the presence, or absence, of uncertainty and has advanced our understanding of clinical conditions such as depression and anxiety. It does, however, not address the fact that uncertainty is ubiquitous and its fundamental role in guiding (and possibly misguiding) behaviour.

This role is acknowledged by more recent research topics which evolved from areas as diverse as behavioural economics, theoretical neuroscience and machine learning. They try to pin down *how* uncertainty on many different quantities is detected, represented in the brain and used to guide optimal behaviour. These approaches require a finer understanding, and more careful experimental manipulation, of uncertainty, where the fundamental issue is its definition and quantification.

An exhaustive theoretical treatment of uncertainty is beyond the scope of this chapter. Here, we seek to summarise existing approaches, and make them available to interested researchers, by giving a brief introduction into fundamental concepts of uncertainty from probability theory, reinforcement learning, perceptual decision-making and economics. Some of the distinctions and definitions we make may be perceived as subtle, but they pertain to very fundamental issues of brain function and can be exciting to study.

Several computational theories of brain function are based on probabilistic concepts and thus naturally account for uncertainty on many levels at the same time. They are based, for example, on the framework of predictive coding or on empirical Bayes methods (*see* sensorimotor control theory (e.g. 3), models of visual cognition (e.g. 4), the free energy principle (5), and others). Drawing on such theories, one can freely vary sensory inputs and estimate uncertainty on different quantities. Such estimates can then be related to measured brain functions or behaviour. However, this strategy rests on the assumption that the model one is using accurately describes brain function, which is necessarily speculative. This motivates the need for the isolated investigation of different forms of uncertainty. In this approach, one seeks to vary uncertainty on one quantity and carefully keep it constant on other quantities. This is the approach that we propose here. It revolves around two central premises:

(1) There is no unitary account of uncertainty – uncertainty is always *about* something, and it can be about different things. *What* we are uncertain about is important because it shapes the way behaviour is organised. For example, the occurrence of electric shocks with uncertain onset might

make it optimal to move as little as possible in order to save energy (i.e. freezing in context conditioning) (6). On the other hand, the presence of a shadow at night with uncertain significance might require approaching it and having a closer look (i.e. risk assessment behaviour) (7). On a neuronal level, it has been proposed that uncertainty is coded in a common way for different variables we are uncertain about: population coding (i.e. uncertainty is represented in the pattern of a neural population) (8), fixed-form coding (i.e. uncertainty is directly represented as a discrete quantity) or implicit coding (i.e. uncertainty is calculated from the mean of some quantity) (5). However, if such a common principle exists, it will probably be implemented in different brain areas and neurotransmitter systems, and bear different behavioural consequences, depending on what we are uncertain about.

(2) Uncertainty can be quantitatively measured – therefore experiments on uncertainty can and should continuously manipulate uncertainty along a continuum from low to high, instead of contrasting it with certainty. Although the latter approach has many merits and forms the bulk of literature in many areas, uncertainty often is a psychologically salient feature such that a direct contrast of uncertainty and certainty encompasses processes that are theoretically unrelated to uncertainty. As an example, we have recently shown that brain responses that appeared to be caused by uncertainty about outcome contingencies (9) were not strongest when uncertainty was highest. Instead, they only occurred when it was made clear to subjects that the (uncertain) outcome rule was predetermined and hidden, such that it was known with certainty to the experimenter, perhaps implying that it was thus potentially knowable for the subject as well. This points to social factors and qualitative behavioural strategies as key sources for this brain activation and demonstrates that the direct contrast of uncertainty with certainty can be difficult to interpret (10, see for a discussion (11)). Recent attempts have been made to disambiguate a categorical manipulation of uncertainty from confounding variables (12), however, it seems more promising to investigate uncertainty by experimentally varying it continuously. Not only does this provide a dose–response relationship; but it also forces the researcher to conceptualise *why* uncertainty should be represented in the brain and by which statistical rule. This provides for a deeper understanding of the underlying processes than a merely phenomenological approach.

2. Uncertainty in a Hierarchical Model of Action

For every quantity represented anywhere in the brain, we can define and investigate its uncertainty. In order to make this chapter tractable, we group related uncertainty manipulations into four levels of a decision-making process; the idea is to give a heuristic framework while acknowledging that these groups do not form exclusive or coherent processing levels.

If we imagine an action episode, uncertainty can arise at various stages. In the first place, perceptual information needs to be processed to provide relevant information. At night, we might be uncertain about what we see. Such perceptual uncertainty thus arises, for example, from incomplete sensory information or lack of attention. Next, perceptual information is used to infer facts about the situation we are in. Are we on a busy road in London, or in New York or at a friend's party? Uncertainty about the state (or context) can put us into uncomfortable situations, as the context often prescribes certain rules according to which we behave. But are we certain about these rules? This might be the first party we attend in a town we recently moved to, and even though we know where we are, that does not imply that we know precisely how to behave. Uncertainty about rules is a typical initial condition in reinforcement learning. But then, at the end of learning, we might still not precisely know what is going to happen when action outcomes are probabilistic. Even if we know that a coin flip will result in tails half of the time, we do not know what the result of the next coin flip will be. Such outcome uncertainty is commonplace in everyday life.

Thus, a real-life stream of actions can often be broken up into small episodes to which the hierarchical model presented here can be applied. The central idea is that uncertainty on different levels of the hierarchy might be implemented by different neural mechanisms, such as different brain areas, networks or transmitter systems; but that related quantities might be represented in a similar manner. Therefore, it seems plausible to group similar forms of uncertainty and present them together: (1) perceptual uncertainty, i.e. uncertainty in the sensory information or its interpretation; (2) state uncertainty, i.e. uncertainty about which of several sets of (known) rules apply in a given moment, including uncertainty about the context and about state transitions; (3) rule uncertainty, i.e. uncertainty about stimulus–outcome or stimulus–action–outcome rules as such; (4) uncertainty about probabilistic outcomes, where outcome is defined as a motivationally salient event.

3. Quantification of Uncertainty

How can we quantify uncertainty? Imagine we could win £5 or £15 from heads or tails, respectively, at a coin flip, or £60 from a six on the dice and nothing for the numbers one to five. Are we more uncertain about how much we earn from the coin or from the dice? **Figure 8.1a** and **b** shows our expectations of winning the money from coin and dice. We do not have a deterministic expectation for either, so our expectations consist of a number of discrete probabilities. Panel **c** shows another case where we win the amount of money shown by the number on the dice, plus £6.50, and one can generalise this idea to situations with completely continuous outcome possibilities (panel **d**). What we

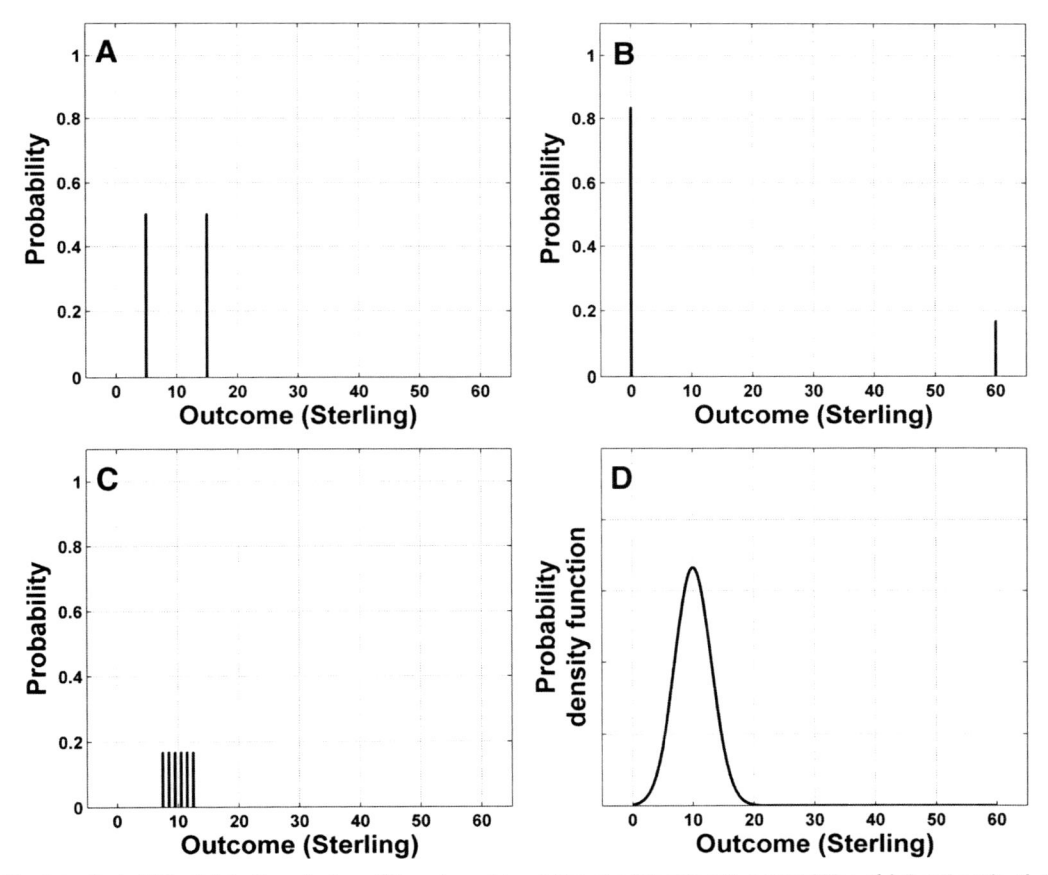

Fig. 8.1. Probability distributions for four different gambles. **a** Flip of a fair coin with probabilities of 0.5 and a win of £5 or £15 from heads or tails, respectively. **b** Toss of a fair dice with a win of £60 from a six and nothing for the numbers one to five. **c** Tossing a fair dice again, winning the amount of money shown by the number on the dice, plus £6.50. **d** Probability density function for an unspecified example of continuous outcome possibilities.

have to find is a statistic that captures the amount of uncertainty embedded in each of these probability distributions. In the following, we give an overview of different statistics that are in widespread use, pointing out how they differ in their quantifications.

Entropy (or Shannon entropy) is a measure from information theory (13) that is often used when outcomes are defined as a set of nominal and discrete events, one of which can occur at a time. Entropy takes only the probabilities for the occurrence of each event into account. That is, our uncertainty estimate does not depend on how much we win – it only depends on the probabilities of winning for each possible outcome. Entropy measures how much information we have about the upcoming event: we have some information about what will happen from throwing the dice, because we are not very likely to win. We have no information about the coin flip because both scenarios (head and tail) are equally likely. So, we should be more uncertain about the coin. That is reflected in entropy H, defined as the negative product of each probability with its (natural) logarithm, summed up over all possible outcomes:

$$H = -\sum_{i=1}^{n} p_i \cdot \log p_i.$$

For the three cases above, we derive

$$H(A) = -[0.5 \cdot \log(0.5) + 0.5 \cdot \log(0.5)] = 0.69$$
$$H(B) = -\left[\frac{1}{6} \cdot \log\left(\frac{1}{6}\right) + \frac{5}{6} \cdot \log\left(\frac{5}{6}\right)\right] = 0.45.$$
$$H(C) = \sum_{i=1}^{6} \frac{1}{6} \cdot \log \frac{1}{6} = 1.79$$

Thus, entropy prescribes in fact that we are more uncertain about the coin (A) than about the first dice throw (B) where we have two possible outcomes, one of which is much more likely than the other one. We are even much more uncertain about the outcome from the second roll of the dice (case C) where we have six possible outcomes, all with equal probability. However, it does not matter how much we earn, a perspective that economists and reinforcement learning theorists would probably disagree with. Does not the prospect of earning £5 or £15 from a coin flip feel much more uncertain than earning £9 or £11? The entropy is the same in both cases; however, their variance will differ.

Variance (and its square root, *standard deviation*) is an intuitive statistic that most researchers will be familiar with. It is commonly used to describe asset risk in finance (14) and hence has influenced behavioural economics and neuroeconomics where outcomes are defined in a continuous event space. Here, the magnitude of each possible outcome is taken into account. In the discrete case (*A–C*) variance is simply the squared deviation of each outcome from the expected (mean) outcome, weighted by the probability of this outcome and summed up over all outcomes (and similarly it can be generalised to continuous probability distributions as in case *D*). The expected earning in all three lotteries is £10, but the deviations from the mean are different. In case *A*, both possible outcomes (£5 and £15) deviate by £5 from the mean and have a probability of $p_i = 0.5$, such that the variance (**Note 1**) is

$$Var(A) = 0.5 \cdot (5-10)^2 + 0.5 \cdot (15-10)^2 = 25,$$

and analogous for the other two cases:

$$Var(B) = 0.833 \cdot (60-10)^2 + 0.167 \cdot (0-10)^2 = 500,$$

$$Var(C) = \sum_{i=1}^{n} 0.167 \cdot ((i+6.5)-10)^2 = 2.9.$$

This time, we are most uncertain about *B*. This is because the earning from the six deviates so much from the average earning of the dice, and also from the earnings in the other two lotteries *A* and *C*. This conforms to our intuition about earning money, or goods: our earnings are more uncertain when the possible earnings have wider spread. Therefore, variance is a good statistic when values can be assigned to different outcomes (**Note 2**). If we change the earnings from the coin flip to £0 or £20, the variance increases to

$$Var(A) = 0.5 \cdot (0-10)^2 + 0.5 \cdot (20-10)^2 = 100.$$

However, adding a constant to both possibilities does not change the variance, which is possibly problematic. Earning £490 or £510 from a coin flip entails the same variance as earning £0 or £20. This might not seem very intuitive: many people would probably be more indifferent towards earning one of two amounts that are already quite large than towards earning either £20 or nothing. To reconcile this intuition with the idea of an uncertainty estimate, the *coefficient of variation* has been proposed (e.g. (15)) which is simply the standard deviation of outcomes, divided by the expected (average) outcome. This prescribes that we are more uncertain about our earnings when the same deviation occurs at a smaller average earning. In the example above, the coefficient of variation would be much smaller when the expected outcome is £500 than when it is £10.

We will see that entropy and variance (and its derivates) are the most commonly used measures of experimentally manipulated uncertainty in neuroscience. In some areas, proxy measures are being applied which we will discuss in detail below. All of these aim at quantifying the *objective* uncertainty given in the experimental situation. Of course it is possible that subjects have a different estimate of the uncertainty, for example, because they have prior assumptions which were generated outside the experimental context. An interesting question is therefore how one can quantify the uncertainty of actual neural representations. Naturally, prior knowledge and assumptions are usually beyond experimental control and can only be inferred from subjects' behavioural or physiological responses. Therefore, we focus here on quantifying objective uncertainty but acknowledge that the actual uncertainty in neural representations might be slightly different.

4. Uncertainty About Sensory Information

4.1. Concepts

Sensory processing often involves categorising or quantifying incoming information. Uncertainty can then arise from noise in the sensory information or from uncertain representation of category templates, category boundaries or quantitative mappings. How can this be studied? Many experiments in this field are interested in how decisions are made when perceptual information is noisy and use decision-making models to describe the processes associated with the task. This will most often be some form of temporal integration model. We will therefore first give a brief overview of this class of decision-making models, then go on to show how measures from these models can be used as a proxy for the objective uncertainty.

Temporal integration models (also termed *sequential sampling*, *drift diffusion* or *bounded integrator models*) assume that sensory information is accumulated over time (see for overviews, e.g. (17, 18)). Imagine, for example, an array of dots; 80% of dots are moving randomly to the right or to the left, the remaining 20% are consistently moving to the left (**Fig. 8.2a**). Our task is to decide whether the net movement is to the left or to the right. At each moment in time, the majority of dots could be – just by chance – moving to the right or to the left. Temporal integration models assume that we gather some evidence over time: at each moment in time, the number of dots we see moving to the right side causes a signal that is depicted in **Fig. 8.2b**. This could, for example, be a neuronal signal, achieved by summing up input from all neurons that detected a dot moving to the right, minus the signal from all neurons that detected a dot moving to the left. At each moment in time, the random dot motion causes this signal to fluctuate,

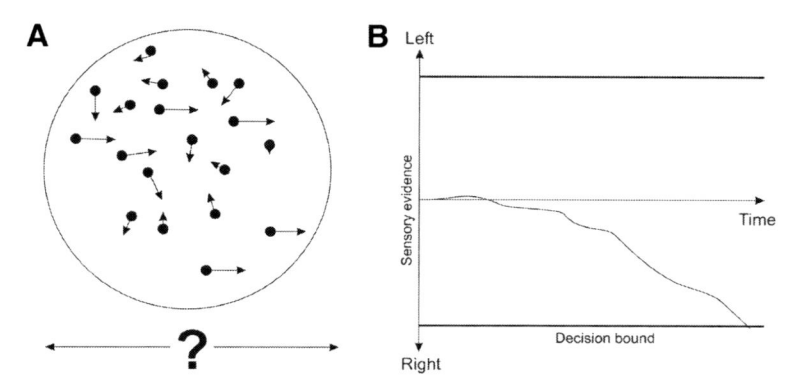

Fig. 8.2. Random dot motion task. **a** An array of dots is moving into random directions, with a proportion consistently moving left or right. The task is to detect the consistent motion direction. **b** Temporal integration model: Over time, evidence is accumulated for both possible motion directions; a decision is made when the evidence reaches a certain threshold.

but over time the consistent dot motion will cause the signal to increase (more or less) slowly. When this signal reaches a certain bound (criterion), we decide that the true movement direction is to the right. The opposite mechanism can account for a net movement to the left. The more random dots we add to this task, the shallower will the slope of this evidence curve be (the so-called drift rate), and the longer we will need to make a decision. At the same time, if we are forced to make a quick decision, we will be less accurate when there is more noise (18, 19).

Is there any place for uncertainty in this model? One might be tempted to use the drift rate as a surrogate for uncertainty, because the more uncertain the sensory evidence is, the smaller will the drift rate be. One can infer the drift rate from subjects' behaviour: with a lower drift rate, reaction times would presumably be longer, and when there is an incentive for speedy responses, accuracy will be lower. However, the reverse relation is not necessarily given: shorter reaction times do not imply less certainty. In fact, reaction times can be biphasic such that people respond very quickly when there is little sensory evidence (20). Also, cerebral responses to reaction times and to accuracy dissociate although both should in theory depend on the drift rate (20); and cerebral responses to accuracy and objective task difficulty dissociate (21, 22). In addition, such temporal integration models have recently been challenged (16).

Beyond such measures for subjective uncertainty, there seem to be two principles of quantifying objective uncertainty in these tasks. A model-based approach to a categorisation task would measure the physical evidence for the different categories and quantify the uncertainty of the ensuing probability distribution over categories, for example, as entropy. On the other hand, one might measure physical similarity to a template or physical distance from a category boundary as a proxy for uncertainty. So far, most experimental paradigms use either accuracy or objective physical characteristics to quantify uncertainty. We will review three studies here that use these measures, but emphasise the need for a model-based approach in order to clearly define uncertainty.

4.2. Examples

4.2.1. Animal Experiments

Kepecs et al. (23) used an odour discrimination task in rats. Two pure odours (caproic acid and 1-hexanol) were mixed in six different ratios (0/100, 32/68, 44/56%). The rat's task was to categorise the odour according to the major ingredient, with a reward for correct responses. Uncertainty was not quantified; instead the odour mixtures were treated as discrete levels: if the odour ratio was closer to 0.5, the perceptual uncertainty was assumed to be higher. The authors then investigated brain areas where neuron firing conformed to a U-shape or inverted U-shape when plotted according to the odour ratios, thus implying lower or higher firing with higher uncertainty. While this is a very elegant design, it shares a problem with most variations of perceptual uncertainty in reward-based decision-making. Because accuracy is lower with higher uncertainty, rats will learn to expect less reward with the intermediate odour categories. In order to disentangle neuronal responses to expected reward and to uncertainty, the authors performed a model-based analysis, comparing models based on either of the two factors.

Kiani and Shadlen (24) gave monkeys the random dot motion task described above, where the monkey had to make a saccade in order to indicate the net dot motion. Harvesting the fact that a more uncertain stimulus leads to a lower expected reward, they gave the monkey on half of the trials a chance to "opt out" and make a saccade to a third target which would yield a smaller, but fixed reward. Thus, they could measure the monkey's estimate of his expected reward, corresponding to the confidence that his decision would be correct which can then be used as a behavioural measure of uncertainty in analysis of neuronal firing patterns. As in the example above, uncertainty is again correlated with expected reward, such that higher neural firing could be due to high uncertainty or low reward. Kiani and Shadlen avoid this confound by focusing on neurons that show either high or low firing when uncertainty is low, and intermediate firing when the uncertainty is high, and the monkey opts out. This corresponds to the temporal integration model shown in **Fig. 8.1** where a fictive neuron that collects evidence for one motion direction would have high firing when the dots move into this direction, low firing when they move into the opposite direction and intermediate firing when the direction uncertain.

4.2.2. Human Experiments

Grinband et al. (25) used an elegant method to introduce uncertainty into a very simple perceptual categorisation task. Remember that in the aforementioned task setups, uncertainty is not inherent to the stimulus – it arises during categorisation of the stimulus. If we were, for example, tasked to indicate the amount of random dots in a random dot display, we would be more certain, the more random dots are in the display. If we are asked to indicate the net motion, we are less certain when more random dots are in the display. Uncertainty is thus a function of the categorisation task. Grinband et al. harvested this fact and introduced uncertainty by having subjects learn a categorisation criterion: they were tasked to decide whether a line segment of a certain length fell into one or another category and were given a feedback whether their response was correct. Thus, they learned the category boundary, but they did not have enough practice to learn it with certainty. Again, during the test, they had to make the same decision. Obviously, uncertainty would be higher, the closer the test line is to the category boundary. Uncertainty was thus defined as a function of accuracy from a preceding training session and thus inferred from people's behaviour (with the caveats mentioned above).

4.3. Perspectives

Uncertainty about sensory information probably provides one of the major examples to date of how uncertainty has been incorporated into translational (human–animal) models of affective disorder. An important example of this is the presentation of temporally unpredictable neutral sound stimuli to healthy human subjects and mouse subjects and monitoring of responses relative to those of subjects presented with the same stimuli on a predictable schedule. In humans, functional magnetic resonance imaging (fMRI) revealed that such temporal unpredictability caused sustained neural activity in amygdala and increased attention towards emotional faces. In mice, exposure to temporal unpredictability increased expression of the immediate-early gene c-*fos* and prevented rapid habituation of single neuron activity in the basolateral amygdala; at the behavioural level, it was anxiogenic in terms of increasing avoidance of illuminated, exposed areas (26). Since this encompassed a direct contrast of uncertainty with certainty, one challenge following from these results will be to quantify and experimentally manipulate uncertainty in this paradigm and investigate whether the observed behavioural and neuronal responses can be attributed to uncertainty.

In order to provide such quantification, we have discussed three paradigmatic experiments that define uncertainty either from the physical characteristics of the stimulus, or from subjects' behaviour, that is, their confidence estimate or their decision accuracy. These are the most frequent definitions in the literature;

we argue, however, that they fall short of a precise quantification. Developing a model-based approach to quantify perceptual uncertainty is beyond the scope of this chapter, but would involve estimating the probability that a given stimulus can be classified correctly and quantifying the entropy of the ensuing probability distribution (which easily generalises to multinomial classification tasks). Using accuracy as a measure for uncertainty comes closest to such an approach, although there will be a non-linear relationship between accuracy and entropy. In addition to this, uncertainty on early processing stages induces uncertainty on later stages; in order to take that influence into account as a possible confound, Bayesian models of brain function can be used; these make a formal description of how later processing stages are implemented and influenced by earlier processing steps.

5. State Uncertainty

5.1. Concepts

Uncertainty about the state one is in is a common problem in many situations where certain states require certain (known) actions or action plans. An example is sensorimotor control (e.g. (3)) where the theoretical focus is, however, more on integrating different processing stages than on isolating them. Another typical situation where state uncertainty arises is in context discrimination learning where different contexts might be associated with different outcome (e.g. reward or punishment) contingencies. State uncertainty can be conceptualised as entropy over the probabilities of a set of discrete possible states or the precision of a posterior probability distribution in Bayesian models if we assume a continuous state space.

5.2. Examples: Human Experiments

Entropy about discrete states of the environment has been investigated in a paradigmatic human fMRI experiment by Yoshida and Ishii (27). In this study, participants played a computer game where they had to find a target position in a maze. Their visual perspective was from within the labyrinth, i.e. they could only see walls and floors, so that they did not know for sure the current position in the maze. In order to solve the task, the position in the maze had to be inferred from previous experience and current information about the maze. The current position can be regarded as a particular state since specific action sequences with known rules and outcomes ensue from each position. Because subjects could move freely in the maze, the possible states at each point in time were not directly under experimental control but were inferred using a computational model of the task. In this hidden Markov model, the agent made current state estimates, the entropy of which served as quantification of state uncertainty. Yoshida and Ishii then went on to analyse brain regions where the blood oxygen level dependent (BOLD) signal covaried with state entropy.

Bestmann et al. (28) provided another example of context uncertainty in a modified Posner task. Here, there are two possible actions, and one of two target symbols signals (deterministically) which action is correct. Before the target signal is shown, a cue of the same type is provided, which most of the time is congruent (i.e. valid) with the target signal that is shown later. Because subjects are tasked to respond as quickly as possible, they can use the cue to prepare their action, but need to take into account that the cue might be incongruent. The transition from cue state to target state entails uncertainty, depending on how often the cue is congruent with the target. Bestmann et al. varied the probability of the cue–target congruence, and thus the uncertainty of the upcoming state, and its associated action. They then investigated how motor preparation of the upcoming action depended on the uncertainty of the state transition.

5.3. Perspectives

If one was interested in a more direct experimental control of state uncertainty, paradigms would be useful where several possible states at each point in time can be manipulated. This could be achieved, for example, by repeatedly endowing people with state information that needs to be interpreted and can lead to probabilistic or deterministic action sequences. This state information can be presented in such a way that the conditional probabilities of being in a current state, given this information, are under experimental control and so is entropy. Such paradigms are applicable not only to human research but also for animals, for example, by relying upon several previously conditioned contexts that require certain actions or imply particular action (or stimulus)–outcome rules.

6. Rule Uncertainty

6.1. Concepts

Uncertainty about outcome rules is a common concept in both reinforcement learning and economics. Rules are also important in social interaction and have been conceptualised, e.g. in game theory, however, with much less focus on uncertainty on these rules. The microeconomic perspective on rule uncertainty evolved around the observation that humans prefer gambles with explicitly known probabilities, a situation termed *risk*, over those involving uncertain probabilities (i.e. rules), a situation termed *ambiguity*.

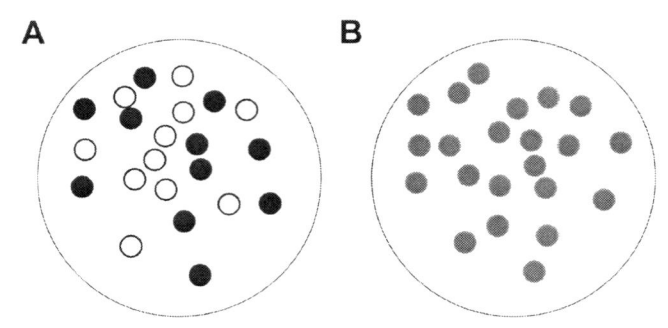

Fig. 8.3. Ellsberg's (29) description of ambiguity. **a** This urn contains 50 black and 50 white balls. **b** This urn contains 100 balls that could be either black or white; but the proportion of black and white balls is unknown to the observer who can bet on black or white in either urn **a** or urn **b**. Most people bet on a colour in urn **a**, thus avoiding the ambiguous urn **b**; although in the absence of prior knowledge, the probabilities of winning are equal for both urns.

Imagine a gamble on the two urns shown in **Fig. 8.3** (29). Urn A contains 50 black and 50 white balls – we win £10 if black is drawn. Urn B contains 100 balls each of which was previously and randomly determined to be either black or white, but we do not know the distribution of black and white balls in this urn. Again, we win £10 if black is drawn. What are our chances of winning? This is easy for urn A: it is 0.5 (i.e. 50%), because the number of black and white balls is the same. For urn B, we need to take into account each possible distribution of balls in the urn and the chances of winning from it. Because the balls were randomly determined to be black or white, there are 101 possible ball distributions. The chances of winning from each of the distributions are 0 (for 100 white balls), 0.01 (for 99 white and 1 black ball), 0.02 (for 98 white and 2 black balls), 0.03, and so on, up to 1 (for 100 black balls). If we multiply the probabilities of winning from each ball distribution (first-order probabilities) with the chances that this ball distribution is realised (second-order probabilities), and sum up these 101 values, we end up with the overall chances of winning from urn B (i.e. the expected first-order probabilities). Unsurprisingly, these turn out to be 0.5, just as in urn A.

Thus, the chances of winning are mathematically identical for both urns, but most people would prefer to bet on urn A rather than on urn B. This preference is called *ambiguity avoidance* and is a robust empirical finding even when people are told that this does not lead to (mathematically) optimal outcomes and even when they have to pay some extra money (a "premium") to avoid urn B (29–36).

Why would most individuals avoid urn B? Obviously, urn B contains some uncertainty about outcome contingencies (i.e. rules), while urn A does not. But is this really the reason why it is avoided? In fact, the experiment in **Fig. 8.3** (and most similar experiments) involves factors other than uncertainty. Some of these factors, under some circumstances, can explain ambiguity aversion: (1) There is a small chance of very unfavourable first-order probabilities, e.g. an urn with 100 white balls. Although this is very unlikely, it seems to be disproportionately weighted. If one restricts the range of possible outcome probabilities, ambiguity aversion is diminished. This factor could thus explain at least a part of ambiguity aversion without having to formally take uncertainty into account (34, 37). (2) By making a choice between urn A and urn B, people reveal their knowledge and belief about gambles and probabilities. Most people are probably more familiar with gambles similar to urn A and can be more certain that other people will have similar experiences. It turns out that when people are asked to make this choice publicly in a group, ambiguity aversion is much bigger than when they make the same choice, but write it down on a piece of paper that is only later to be read by the experimenter (32). Also, when people gamble on getting one of two movies, where the experimenter asks which of the two they prefer, they avoid gambles of type B. However, if the experimenter does not know which movie they prefer, there is no ambiguity avoidance (38). In this case, the experimenter cannot judge people's choices because he does not know what they want to obtain from the gamble. This is purely a social factor that can explain ambiguity aversion but has nothing to do with uncertainty. (3) Brain responses to ambiguity have been shown to depend on the fact that *something is hidden from the observer rather than completely unknowable* (10) and do not seem to scale with uncertainty. The fact that something is hidden might not only invoke specific behavioural responses (e.g. information seeking), but might also induce suspicions (e.g. the experimenter might be cheating and urn B is biased). Again, these factors are unrelated to uncertainty.

A number of reasons can explain people's reaction to ambiguous gambles without the need for taking uncertainty into account. One could formally test the impact of uncertainty by quantifying uncertainty as entropy over possible rules, and comparing responses to this quantity with responses to ambiguity, but this has not been done so far. At the moment, it seems that although uncertainty is a crucial factor in ambiguity, its experimental manipulation has not been developed sufficiently.

Another perspective on rule uncertainty comes from the theory of reinforcement learning. Reinforcement learning involves learning associations or rules. When we start learning, we are necessarily uncertain about the rules, because we have not yet learned them. Rule uncertainty has thus been described as a motivator for learning. In a Bayesian framework, such uncertainty can be quantified as posterior uncertainty of rule predictions, a definition that has been applied in theoretical work (39, 40). A related problem is when rules change over time in a detectable manner, and we have some notion of how much the rules are changing.

This has been termed *volatility* (41). It does, however, not equal rule uncertainty: even when volatility is zero (i.e. we know that the rule is constant) we can have maximum uncertainty about this constant rule. On the other hand, with maximum volatility, we can be fully aware of the rule at a given point in time and have no uncertainty.

In summary, although rule uncertainty is a very common problem, its quantitative experimental manipulation remains an issue to be solved.

6.2. Perspectives

An interesting application of rule uncertainty concerns models of affective disorder: rodents can be conditioned on simultaneous stimulus–response (outcome) schedules where one discriminatory stimulus signals reward following an operant response and a second, distinctly different discriminatory stimulus signals avoidance of punishment by not exhibiting the operant response. Introduction of intermediate (ambiguous) stimuli introduces uncertainty and allows for the probing of whether an animal anticipates a positive event. Rats that had experienced chronic mild stress were less likely to perform an operant response to an ambiguous stimulus suggesting that they had developed a cognitive bias to expect negative events, a state marker of human depression (42). From the perspective of uncertainty research, it would be interesting here to examine whether such a cognitive bias is dependent on the amount of uncertainty entailed in the ambiguous stimulus.

Perhaps the most promising approach to this problem in the economic context builds on the aforementioned urn problems, where rule uncertainty can be quantified as the entropy of the distribution over possible outcome contingencies. This is the distribution over possible realisations for urn B. In reinforcement learning, promising theoretical work has built on a Bayesian perspective (39, 40) which awaits its realisation in experimental research.

7. Uncertainty About Outcomes

7.1. Concepts

Outcome uncertainty is a feature of probabilistic structures and has fascinated scientists for almost 300 years (43). Outcome uncertainty denotes the perennial fact that we never precisely know what is going to happen in the future. That is, even if we know that a coin yields heads half of the time, we are not able to predict the result of the next coin flip with certainty. However, this uncertainty can be different – we can be quite sure, but we can also be very unsure. Theoretical accounts of this problem mostly come from economics and reinforcement learning. The former focuses on explicitly known rules such as in gambles and lotteries, and on the fact that people do not seem to like outcome uncertainty. The latter builds on the idea that in learning about variable outcomes, we must recognise outcomes that differ so much from our expectations that we infer a change in the rule or the context. So we need an estimate of the expected deviation or expected uncertainty.

In economics, two mainstream views aim at describing individual behaviour rather than reflecting how this behaviour could be implemented in the brain. Both acknowledge that when offered two lotteries of the same expected outcome, but different variability of the outcomes, most people consistently choose the less variable one (and some consistently choose the more variable one). But the most influential and older stream of theories (most prominently expected utility theory (44), subjective utility theory (45) and prospect theory (46)) does not have a formal notion of uncertainty or variability. So how can it account for the fact that people do not like uncertainty?

The cornerstone of this class of theories is that the usefulness of some good does not linearly relate to its value (**Fig. 8.4**); £100 is useful for us, and this usefulness can be quantified as *utility* and measured in *utils*. Imagine £100 equals 100 *utils*; £1,000 is also useful. But it is not 10 times as useful and has less than 1,000 *utils* for us (in the example in **Fig. 8.4**, it converts to about 500 *utils*). Why would that be the case? As a drastic (and evolutionary plausible) example, the difference between starving and one piece of bread is much more important than the difference between one

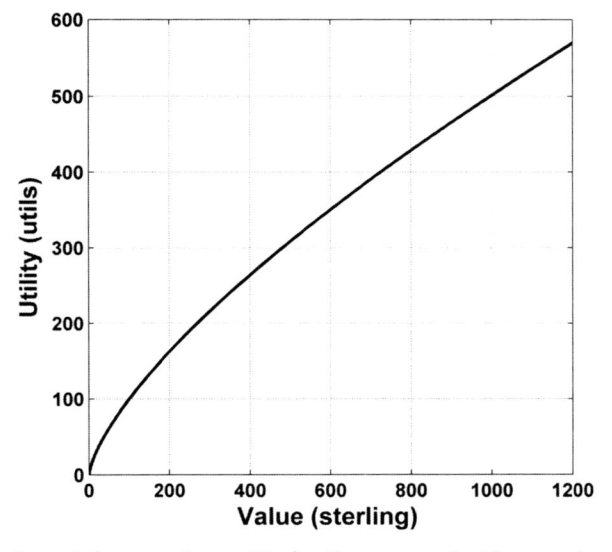

Fig. 8.4. Example for a non-linear utility function as prescribed by several mainstream economic theories: with increasing value, the increment in utility becomes smaller.

piece of bread and the next piece. Equally, for somebody who has no money, what he can buy from the first £100 is probably more important than what he can buy from the next £100 and so on. The same is even more obvious for animals, which, in natural environments, are under a constant threat of not surviving such that obtaining some resources (i.e. food, water) is much more important than acquiring a lot of it. The same holds for laboratory experiments on valuation of resources, where animals are often kept hungry and thirsty in order to observe these effects. But how can this non-linear *utility function* explain why somebody prefers a low variability gamble over a high variability gamble?

Imagine a choice between two lotteries, both of which have three outcomes with equal probabilities. The first lottery could give £400, £500 or £600, such that the expected monetary *value* is £500. The second lottery could give £100, £200 or £1,200 and has the same expected *value* of £500. How about the usefulness of that money? Using the value–utility relation shown in **Fig. 8.4**, the first lottery gives us 264, 308 or 350 *utils*, which comes up to an expected (average) utility of 307 *utils*. The second lottery gives us 100, 162 or 568 *utils* and the expected utility is 277 *utils*. That is, because £1,200 is extremely devalued by the utility function, the overall return from the second lottery is much lower than from the first one, when measured in utility. And this result can be generalised: the more the outcomes of a lottery (or an action, etc.) spread, the lower is the expected utility from that lottery (or action, etc.), given the utility function shown above. This theory can therefore explain aversion of variable outcomes without ever quantifying this uncertainty. We spent some time describing the idea behind this theory because it is prevailing in economics and also in its neuroscience sibling, neuroeconomics, and employed in most experimental work that does not explicitly deal with uncertainty (and also in some work that does deal with uncertainty). It is therefore a reference point for most other ideas and theories and a benchmark to test experimental findings against. It does, however, not help us much to find an actual measure of uncertainty.

A second strand of economic theories from finance (14) has a much simpler form and posits that lotteries (and other forms of actions with uncertain outcomes) can be represented by their mean (i.e. expected value) and variance. This leads up to a neuroeconomic view where variance is explicitly represented in the brain, thus forming a measure for uncertainty, and influencing choice behaviour.

Another perspective arises from reinforcement learning theory. Here, the problem is to learn about rules and to detect changes in these rules. In a natural environment, this could mean that a foraging environment which was once full of food options might be depleted at some point. The organism needs to detect that the fact that he does not find any food for days is not caused by bad luck any more, but by a change in the environment. So how does reinforcement learning work?

Many reinforcement learning models (such as Rescorla–Wagner (47), Pearce–Hall (48), temporal difference learning (49) and others) posit that we make a prediction about an expected (average) outcome. Violations of these predictions are signalled as prediction errors and cause us to change our predictions, i.e. to learn. If we are in a new environment, we might predict a 50% change of getting a reward per time unit. That is, in each time unit, we will predict an expected reward of 0.5 units. If we get a reward in a time unit, the prediction error is 0.5. In this case, we will update our prediction by a fraction, for example, by 10%, of the prediction error. That is, our next prediction is 0.6. If the actual chances of getting reward are 100% per time unit, we will quickly update our predictions to this outcome rule and after a few instances make a prediction of one reward per time unit. On the contrary, if the true chance of getting a reward is only 10% of the time, we will have negative prediction errors and reduce our predicted value.

So far, we only referred to expected (i.e. average) outcomes. What role does uncertainty play here? Image an environment where the true chances of getting a reward are 50%. That means, our prediction is 0.5, and on each occasion, after getting zero or one unit of reward, we get a prediction error of +0.5 or −0.5. Despite the constant prediction errors, there is no need to update our predictions any more. We could even say that we *expect* a prediction error, because we know there is some *expected uncertainty* (39) associated with this environment. Over time, the prediction will stay (rather) constant, and the prediction error has no impact any more. Imagine, however, that after a while we encounter 10 units of reward at one time. Our prediction error is now 9.5. Did we still expect this? Clearly not; this is an *unexpected uncertainty* (39) and it might mean that the environment, and the outcome rule, has changed. In order to detect this, we need a measure for the *expected uncertainty* in the first place.

So, how is expected uncertainty represented? Again, it has been proposed from economists that outcome variance (or standard deviation) is the crucial dimension and that estimates of the variance are updated in a way that is similar to models for learning expected outcome. The estimate of outcome variance is then used to scale the prediction error such that it does not signify an absolute but rather a relative deviation from expectations (50). That is, a regular prediction error of +0.5 or −0.5 will be small, if compared to the expected variance, but the prediction error of 9.5 will be very high on this relative scale.

What if outcomes cannot clearly be valued? One can imagine a learning experiment where the goal is to make correct responses without any associated reward. Assigning a value to a correct response is completely arbitrary. Here, one might want to quantify how *informative* outcomes are in the context of this learning task. This is achieved by computing Shannon entropy of outcomes (13). As pointed out in **Section 3**, one can calculate the entropy of discrete outcomes, thus ignoring outcome size. In experiments with binary outcomes and constant monetary values, variance (over values) and entropy (ignoring values) are very similar. If one is interested to find out which statistic the brain uses, one must experimentally vary outcome size.

7.2. Examples

7.2.1. Animal Experiments

Fiorillo et al. (51) were interested in responses of dopaminergic midbrain neurons to outcome uncertainty. Such neurons are known to encode a phasic prediction error signal during reinforcement learning; here the focus was on sustained (non-phasic) activity during the 2 s of reward anticipation that correlated with its uncertainty. In a first experiment, the probability of a reward with magnitude was varied (0, 0.25, 0.5, 0.75, 1.0). Uncertainty is highest with medium probabilities and lowest when the outcome is known, i.e. probabilities of 0 or 1.0. Sustained activity of dopaminergic neurons was highest in the 0.5 condition. In a second experiment, the monkey always received a reward, but there were two reward quantities with probability of 0.5. The more these reward possibilities differ from the mean reward, the higher the outcome uncertainty. This was varied in three levels. Again, sustained neuron firing correlated with outcome uncertainty. Because entropy is constant here, entropy is not driving these responses. Otherwise, the results would hold for different quantifications of uncertainty: in such a simple experiment, variance, standard deviation and coefficient of variation are so similar that neuronal responses to each of these quantities cannot be disambiguated. On the other hand, it is difficult to train animals (or humans) on a sufficient number of cues to formally disentangle these quantities.

7.2.2. Human Experiments

In humans, the bulk of experiments either take an economic perspective, using explicitly stated outcome probabilities, or a reinforcement learning perspective where the probabilities are learned. It is not clear whether explicit and learned probability representations are equivalent such that we treat these two perspectives separately here.

In a typical *economic* experiment about variance in explicitly signalled potential outcomes, Dreher et al. (52) presented virtual slot machines to their participants, clearly indicating probability and value of each possible return from this machine (i.e. 0.25/$20, 0.5/$20, 0.5/$10, 1.0/$0). Participants performed an incidental task that was not related to their winnings, such that they passively waited for their reward. During this delay period, BOLD responses were measured. Preuschoff et al. (53) provided a very similar task that was framed in terms of a card gamble. From a known stack of ten cards, one was randomly drawn. After a delay period, a second card was drawn. Participants' winnings depended on whether the first or the second card had a higher value. Thus, after the first card was drawn, the participants knew how many cards in the stack would have a higher value than the first card. Hence, there was a certain probability of the second card being higher, associated with a specific (un)certainty of winning. Both experiments made use of binary gambles where the options were to win something or to win nothing (comparable to the first experiment of Fiorillo et al. (51), described in **Section** 7.2.1).

To our knowledge, no experiments so far have investigated outcome uncertainty from the *reinforcement learning* perspective without implying uncertainty about outcome rules. Studies so far have tested responses to outcome uncertainty during learning or after short learning periods. The study of relatively isolated outcome uncertainty would involve overtraining subjects on cue–outcome associations to suppress uncertainty on these associations.

7.3. Perspectives

Outcome uncertainty is an important concept both for reinforcement learning and economics. Human neuroimaging research has so far mainly investigated this issue from the economic perspective, using explicitly signalled binary outcomes. The use of multiple outcomes with the same probability, but different magnitudes, could complement this line of research. The reinforcement learning perspective on the other hand formed the basis for the only animal study in this field. Human research on outcome uncertainty in reinforcement learning is sorely needed: recent studies have started to address this issue, but without sufficient overtraining of outcome rules, these imply rule uncertainty as well.

An important issue for human neuroimaging studies arises from the monkey study by Fiorillo et al. (51) that has suggested ramping activity in dopaminergic neurons representing uncertainty. Such ramping activity increases towards the ultimate outcome. Functional MRI has a low temporal resolution and cannot distinguish between signals shortly before and after an outcome. Because higher uncertainty involves a higher average (absolute) prediction error at the time point of the outcome, uncertainty and this absolute prediction error are confounded. Enhanced experimental designs (e.g. with hidden outcomes in the test phase) or electro- or magnetoencephalographic studies are needed in order to disentangle the effect of these two confounded independent variables.

8. Summary

In this chapter, we have provided an introduction into experimental issues that arise in the study of uncertainty. We provided a number of concepts that allow the researcher to study uncertainty on some quantity in isolation. This is a challenging and exciting new field in neuroscience, where a great deal of theoretical and empirical work is still to be done. We hope that our theoretical considerations motivate interested experimental researchers into this important branch of neurobiology.

9. Notes

1. We leave out the monetary units which have no relevance here.
2. Note that some formulations of entropy do not rely on discrete and nominal events, such as differential entropy and relative entropy, and can therefore compete with variance when values need to be taken into account. The economics literature, which is rather influential here, does, however, mainly use variance and its derivates.

Acknowledgements

The authors would like to thank Jean Daunizeau for his helpful comments on an earlier version of this manuscript.

References

1. Loh M, Rolls ET, Deco G (2007) A dynamical systems hypothesis of schizophrenia. PLoS Comput Biol 3:228
2. Hazlett-Stevens H, Borkovec TD (2004) Interpretive cues and ambiguity in generalized anxiety disorder. Behav Res Ther 42:881–892
3. Kording KP, Wolpert DM (2006) Bayesian decision theory in sensorimotor control. Trends Cog Sci 10:319–326
4. Summerfield C, Egner T (2009) Expectation (and attention) in visual cognition. Trends Cog Sci 13:403–409
5. Friston K (2009) The free-energy principle: a rough guide to the brain? Trends Cog Sci 13:293–301
6. Baker AG, Mercier P, Gabel J, Baker PA (1981) Contextual conditioning and the US preexposure effect in conditioned fear. J Exp Psychol: Anim Behav Process 7:109–128
7. Gray JA, McNaughton N (2000) *The neuropsychology of anxiety: an enquiry into the functions of the septohippocampal system.* Oxford University Press, Oxford
8. Pouget A, Dayan P, Zemel RS (2003) Inference and computation with population codes. Annu Rev Neurosci 26:381–410
9. Huettel SA, Stowe CJ, Gordon EM, Warner BT, Platt ML (2006) Neural signatures of economic preferences for risk and ambiguity. Neuron 49:765–775

10. Bach DR, Seymour B, Dolan RJ (2009) Neural activity associated with the passive prediction of ambiguity and risk for aversive events. J Neurosci 29:1648–1656
11. O'Neill M, Kobayashi S (2009) Risky business: disambiguating ambiguity-related responses in the brain. J Neurophysiol 102:645–647
12. Vickery TJ, Jiang YV (2009) Inferior parietal lobule supports decision making under uncertainty in humans. Cereb Cortex 19:916–925
13. Shannon CE (1948) A mathematical theory of communication. Bell Syst Tech J 27(379–423):623–656
14. Markowitz H (1952) Portfolio selection. J Finance 7:77–91
15. Weber EU, Shafir S, Blais AR (2004) Predicting risk sensitivity in humans and lower animals: risk as variance or coefficient of variation. Psychol Rev 111:430–445
16. Cisek P, Puskas GA, El-Murr S (2009) Decisions in changing conditions: the urgency-gating model. J Neurosci 29:11560–11571
17. Bogacz R (2007) Optimal decision-making theories: linking neurobiology with behaviour. Trends in Cog Sci 11:118–125
18. Heekeren HR, Marrett S, Ungerleider LG (2008) The neural systems that mediate human perceptual decision making. Nat Rev Neurosci 9:467–479
19. Petrusic WM, Baranski JV (2003) Judging confidence influences decision processing in comparative judgments. Psychon Bull Rev 10:177–183
20. Binder JR, Liebenthal E, Possing ET, Medler DA, Ward BD (2004) Neural correlates of sensory and decision processes in auditory object identification. Nat Neurosci 7:295–301
21. Philiastides MG, Sajda P (2007) EEG-informed fMRI reveals spatiotemporal characteristics of perceptual decision making. J Neurosci 27:13082–13091
22. Philiastides MG, Ratcliff R, Sajda P (2006) Neural representation of task difficulty and decision making during perceptual categorization: a timing diagram. J Neurosci 26:8965–8975
23. Kepecs A, Uchida N, Zariwala HA, Mainen ZF (2008) Neural correlates, computation and behavioural impact of decision confidence. Nature 455:227–231
24. Kiani R, Shadlen MN (2009) Representation of confidence associated with a decision by neurons in the parietal cortex. Science 324:759–764
25. Grinband J, Hirsch J, Ferrera VP (2006) A neural representation of categorization uncertainty in the human brain. Neuron 49:757–763
26. Herry C, Bach DR, Esposito F, Di Salle F, Perrig WJ, Scheffler K, Luthi A, Seifritz E (2007) Processing of temporal unpredictability in human and animal amygdala. J Neurosci 27:5958–5966
27. Yoshida W, Ishii S (2006) Resolution of uncertainty in prefrontal cortex. Neuron 50:781–789
28. Bestmann S, Harrison LM, Blankenburg F, Mars RB, Haggard P, Friston KJ, Rothwell JC (2008) Influence of uncertainty and surprise on human corticospinal excitability during preparation for action. Curr Biol 18:775–780
29. Ellsberg D (1961) Risk, ambiguity, and the savage axioms. Q J Econ 75:643–669
30. Becker SW, Brownson FO (1964) What price ambiguity? Or the role of ambiguity in decision making. J Pol Econ 72:62–73
31. Yates JF, Zukowski LG (1976) Characterization of ambiguity in decision-making. Behav Sci 21:19–25
32. Curley SP, Yates F, Abrams RA (1986) Psychological sources of ambiguity avoidance. Organl Behav Hum Decis Process 38:230–256
33. MacCrimmon KR, Larson S (1979) Utility theory: axioms versus 'paradoxes'. In: Allais M, Hagen O (eds) Expected utility hypotheses and the Allais paradox. D. Reidel, Dordrecht, pp 333–410
34. Keren G, Gerritsen LEM (1999) On the robustness and possible accounts of ambiguity aversion. Acta Psychol 103:149–172
35. Slovic P, Tversky A (1974) Who accepts savage's axiom? Behav Sci 19:368–373
36. Pulford BD, Colman AM (2008) Size doesn't really matter. Ambiguity aversion in Ellsberg urns with few balls. Expl Psychol 55:31–37
37. Larson JR (1980) Exploring the external validity of a subjectively weighted utility model of decision making. Org Behav Hum Perform 26:293–304
38. Trautmann ST, Vieider FM, Wakker PP (2008) Causes of ambiguity aversion: known versus unknown preferences. J Risk Uncertain 36:225–243
39. Yu AJ, Dayan P (2005) Uncertainty, neuromodulation, and attention. Neuron 46:681–692
40. Daw ND, Niv Y, Dayan P (2005) Uncertainty-based competition between prefrontal and dorsolateral striatal systems for behavioral control. Nat Neurosci 8:1704–1711

41. Behrens TE, Woolrich MW, Walton ME, Rushworth MF (2007) Learning the value of information in an uncertain world. Nat Neurosci 10:1214–1221
42. Harding EJ, Paul ES, Mendl M (2004) Animal behaviour: cognitive bias and affective state. Nature 427:312
43. Bernoulli D (1738) Specimen theoriae novae de mensura sortis. Comm Acad Sci Impers Petropoli 5:175–192
44. von Neumann J, Morgenstern O (1944) *Theory of games and economic behavior*. Princeton University Press, Princeton, NJ
45. Savage LJ (1954) *The foundations of statistics*. Wiley, New York, NY
46. Kahneman D, Tversky A (1979) Prospect theory – analysis of decision under risk. Econometrica 47:263–291
47. Rescorla RA, Wagner AR (1972) A theory of classical conditioning: variations in the effectiveness of reinforcement and non-reinforcement. In: Black AH, Prokasy WF (eds) Classical conditioning II: current research and theory. Appleton-Century-Crofts, New York, pp 64–99
48. Pearce JM, Hall G (1980) A model for Pavlovian learning – variations in the effectiveness of conditioned but not of unconditioned stimuli. Psychol Rev 87:532–552
49. Sutton R (1988) Learning to predict by the methods of temporal differences. Mach Learn 3:9–44
50. Preuschoff K, Bossaerts P (2007) Adding prediction risk to the theory of reward learning. Ann NY Acad Sci 1104:135–146
51. Fiorillo CD, Tobler PN, Schultz W (2003) Discrete coding of reward probability and uncertainty by dopamine neurons. Science 299:1898–1902
52. Dreher JC, Kohn P, Berman KF (2006) Neural coding of distinct statistical properties of reward information in humans. Cereb Cortex 16:561–573
53. Preuschoff K, Bossaerts P, Quartz SR (2006) Neural differentiation of expected reward and risk in human subcortical structures. Neuron 51:381–390

Chapter 9

Circadian Variation in the Physiology and Behavior of Humans and Nonhuman Primates

Henryk F. Urbanski

Abstract

The rhesus macaque represents a pragmatic animal model for elucidating mechanisms underlying normal and pathological human behaviors. Many of the same techniques that are used in clinical studies can be readily applied to the nonhuman primate studies. These including the use of Actiwatch recorders for monitoring of 24-h activity–rest cycles and the use of a remote blood sample collection system for assessment of changes in circadian hormone profiles. In addition, comprehensive rhesus macaque gene microarrays (Affymetrix) are now commercially available, and these can be used for profiling gene expression changes under various physiological and pathological conditions. Our recent application of these methodologies to rhesus macaque studies emphasizes that many physiological and behavioral events, and the expression of associated genes, have a distinct 24-h expression pattern. Consequently, it is important to take these circadian rhythms into account when designing experiments and interpreting the results.

1. Introduction

1.1. The Rhesus Macaque as a Pragmatic Translational Animal Model

Humans and rhesus macaques (*Macaca mulatta*) are both long-lived primates, and they show many similarities in their anatomy, physiology, and genetics (1). Consequently, these nonhuman primates are regarded as pragmatic animal models for studying mechanisms that underlie normal and pathological human physiology and behavior. Their use as translational animal models has many advantages. For example, rhesus macaques can be maintained under carefully controlled environmental conditions (e.g., photoperiod, temperature, diet, and medication). In addition, animals of a specific age, size, sex, and genetic characteristic can be selected, thereby eliminating extraneous variables and self-selection bias that are typically associated with human clinical trials. Moreover, because the timing of the necropsies can be carefully controlled in rhesus macaques, high-quality postmortem tissues and RNA samples can be collected for biochemical analysis and gene expression profiling.

The present chapter describes two methodologies that we use routinely in rhesus macaques studies to help with the interpretation of physiological and behavioral data. These include continuous monitoring of 24-h activity–rest cycles and remote collection of serial blood samples for the determination of 24-h circulating hormone profiles. Both of these methodologies have analogous applications in clinical research and highlight the important role that circadian rhythms play in human physiology and behavior. To gain further insights into the underlying mechanisms we also routinely perform comprehensive gene expression profiling, using GeneChip microarrays (Affymetrix, Inc., Santa Clara, CA, USA), and this will be discussed in the context of effective experimental designs that take circadian variations into consideration. For details of the latter procedure and data analysis the reader is referred to two recently published articles (2, 3).

1.2. Circadian Rhythms

Most animals, including humans, live in an environment that is characterized by daily and seasonal changes in lighting and temperature. Consequently, many aspects of their physiology and behavior show circadian and circannual adaptations. On the one hand, circadian rhythms are intrinsic to a wide range of body functions, including the sleep–wake cycle, metabolism, immune response, and reproduction (4), and a common human manifestation of desynchronized circadian rhythms is jet-lag. On the other hand, many mammals also show seasonal rhythms in metabolism, reproduction, and immune function (5, 6). Interestingly, in humans seasonal variations have been reported for blood pressure, immune response, birth rate, and sleep duration, as well as for behavioral traits associated with seasonal affective disorders, bulimia nervosa, anorexia, and suicide (7). Many of these physiological and behavioral alterations can be linked to underlying changes in the secretion of hormones, such as cortisol, dehydroepiandrosterone sulfate, testosterone, leptin, and melatonin, all of which have distinct 24-h release patterns in the rhesus macaque (8–11).

The present chapter focuses on two methodologies that can be readily applied to both human and nonhuman primate studies, in order to gain insights into an individual's circadian physiology. For example, performance testing, collection of diagnostic specimens, and the administration of medications, all have an optimal time of day. Hence, information about an individual's 24-h activity–rest cycles and associated 24-h hormone profiles provides an important physiological context for interpretation of test results, and also for the design of effective drug and hormone-replacement paradigms.

2. Materials and Methods

2.1. Monitoring Activity–Rest Cycles and Remote Blood Sampling

A convenient, noninvasive method of continuously monitoring motor activity in humans involves the use of Actiwatch recorders (Philips-Respironics, Bend, OR, USA; part number U198-0301-00). These unobtrusive watch-size recording devices are typically worn on the wrist, and continuously collect activity–rest data, providing information about 24-h motor activity rhythms as well as insights into the quality of sleep. These devices comprise an accelerometer and a 32-Hz microprocessor, which selects peak activity intensity during each second to sum up activity counts per sample. Using a 1-min epoch setting Actiwatches can continuously record data for 45 days, which can then be readily downloaded to a personal computer using a dedicated data reader (Philips-Respironics; part number 198-0150-00). Actiware 5.0 software can then be used to determine the total level of activity in an individual, and importantly, the mean level of activity during specific times of the day (e.g., day time or night time). Equally important, the software can analyze bouts of activities that occur during the night and provide insights into sleep quality based on parameters such as sleep latency, sleep fragmentation index, and number of wake bouts. Note, Actical recorders are also commercially available (Philips-Respironics; part number 198-0210-03). They are very similar to Actiwatch recorders except they average the activity intensity (instead of taking the peak activity intensity) during each second to sum up activity counts per sample. They are used with Actical 2.1 software and are primarily used in studies that focus on energy expenditure studies rather than sleep.

Rhesus macaques, like humans, are diurnal and also show consolidated sleep patterns. Consequently, we have made effective use of the same Actiwatch devices to study 24-h activity–rest cycles in rhesus macaques (12–15). Attaching the Actiwatch to the animal's wrist is not a viable option and so instead we usually place it inside an aluminum protective case (Philips-Respironics; part number 198-0232-00 M); this case is compatible with the nylon and aluminum primate collars available from Primate Products, Inc. (Immokalee, FL, USA) and can be unobtrusively worn around the animal's neck. For downloading the stored activity data, the animal is briefly sedated using ketamine (10 mg/kg body weight, i.m.) and the collar/recorder removed. Alternatively, if the animal is trained to enter a small transfer cage, which restricts head movement, then the Actiwatch can be removed without resorting to the use of sedation.

An alternative strategy to using the protective case and collar, which we have also found to be very effective, is to place the Actiwatch directly inside a custom-made small pocket at the rear of a nylon mesh primate vest (Lomir Biomedical, Inc., Malone, NY, USA). These protective vests are worn by our animals when they have been fitted with an indwelling subclavian or jugular vein catheter and connected to a swivel-based remote infusion/sampling system (Lomir Biomedical, Inc.). This minimally invasive long-term blood sampling system (**Fig. 9.1**), when combined with the Actiwatch recorder, enables 24-h hormone rhythms to be monitored with reference to the light–dark cycle as well as with reference to the animal's endogenous activity–rest cycle (**Fig. 9.2**). As already indicated, many hormones show a pronounced 24-h release pattern. Consequently, many serial blood samples are needed to clearly disclose changes

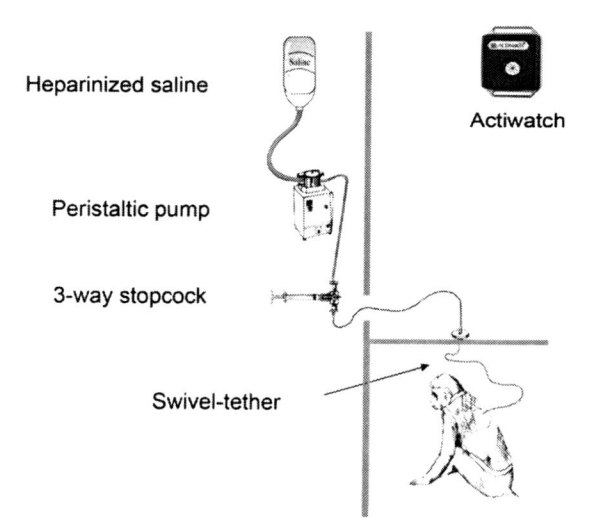

Fig. 9.1. Schematic illustration of the set up that enables serial blood samples to be collected remotely from conscious, undisturbed monkeys. The animals are surgically fitted with an indwelling subclavian vein catheter, which is channeled subcutaneously to the middle of the back. There, the catheter is connected to sampling tubing which itself is protected by a flexible stainless steel tether and, in turn, is connected to a swivel assembly at the top of the cage. The tubing then passes through a small port in the wall and into a laboratory, where it is connected via a three-way stopcock to a blood sampling syringe and also to a peristaltic pump; the latter is used to maintain catheter patency by continuously infusing a heparinized saline solution. Using this set up, serial blood samples can be collected throughout the day and night, even while the animals are asleep. The vests that the animals wear are to protect the catheter; they are routinely inspected every week and washed every 2 weeks. In addition, the vests can house a watch-sized Actiwatch recorder (*inset, upper right*) for continuous monitoring of activity–rest cycles.

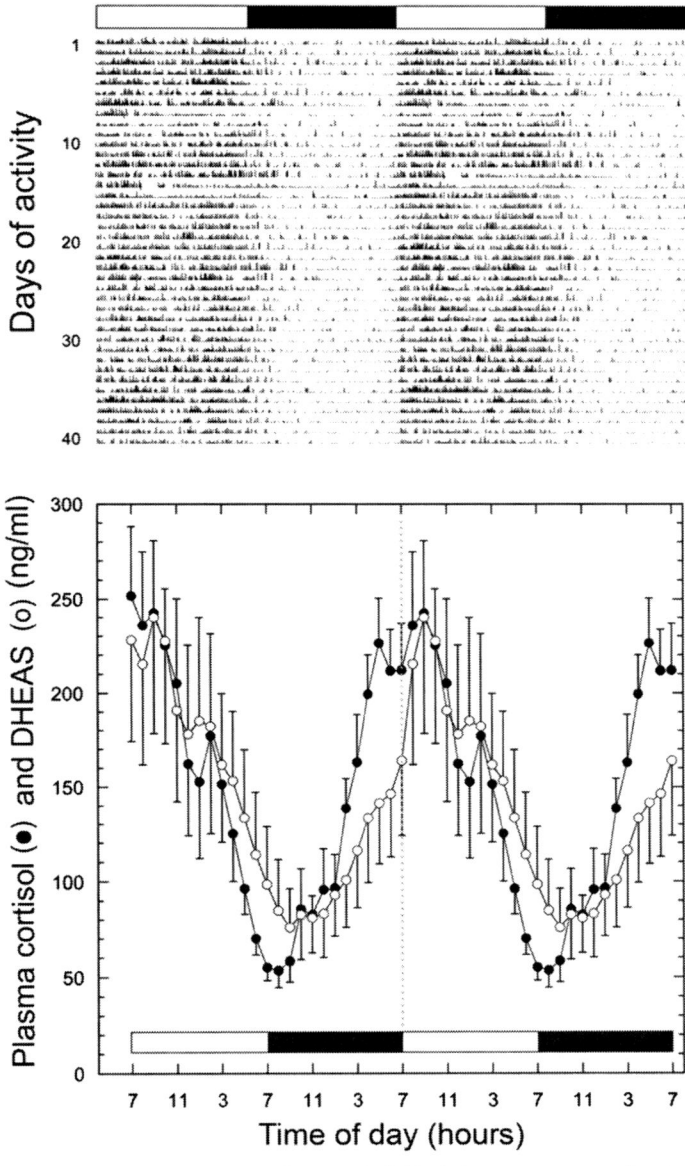

Fig. 9.2. *Upper panel*: Actogram from an individual rhesus macaque, showing diurnal activity and indication of consolidated nocturnal sleep, as in humans. Note that the data are double-plotted to facilitate viewing of the circadian rhythm, and the periods of light and dark are represented by *white* and *black horizontal bars*, respectively. *Lower panel*: Mean plasma cortisol and DHEAS profiles from young adult rhesus macaques; the serial blood samples were collected every hour, using the remote blood sampling set up depicted in **Fig. 9.1**. The hormonal data are double-plotted to facilitate visualization of the 24-h rhythms, which show peaks in the morning when the animals begin their daily activity. [Figure adapted from (12), with permission, Copyright 2006, *The Endocrine Society*.]

in the hormone secretion pattern, and often these samples need to be collected when an individual is asleep. The situation is even more complicated when attempting to disclose plasma profiles in hormones, such as luteinizing hormone (LH), which is released episodically or in a pulsatile manner. For example, the first major endocrine event associated with the onset of puberty in primates is an evening increase in the amplitude of LH pulses. Further subtle differences occur across the menstrual cycle of adults (16), in which the follicular phase is associated with high-frequency low-amplitude pulses whereas the luteal phase is associated with low-frequency high-amplitude pulses (**Fig. 9.3**). By remotely collecting small (<0.3 ml) blood samples every 10 min for 24 h, it is possible to establish detailed pulsatile LH release profiles for each individual, without perturbing its sleep–wake cycle or behavior.

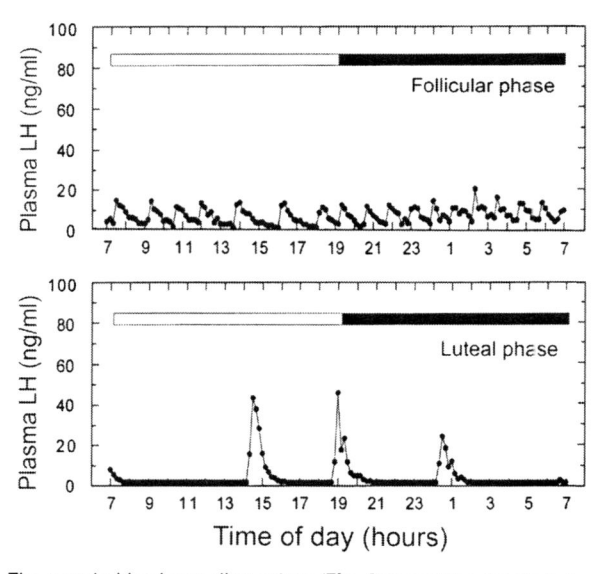

Fig. 9.3. The remote blood sampling set up (**Fig. 9.1**) enables detailed changes in the episodic release profiles of hormone such as LH to be disclosed. The *upper panel* shows a well-defined 24-h pulsatile plasma LH profile from a female rhesus macaque, sampled during the follicular phase of the menstrual cycle. The *lower panel* shows a 24-h pulsatile plasma LH profile from the same animal, sampled during the luteal phase of the menstrual cycle. These data emphasize that multiple serial blood samplings are essential for the disclosure of subtle changes in the release of some hormones.

2.2. Notes on Activity Recording

The actograms generated by the Actiware 5.0 software are typically double-plotted (**Fig. 9.2**). This means that each line of data represents two consecutive days worth of activity; the second day's data are duplicated at the start of the next line, etc. This double-plotted form of activity depiction helps with the visualization of the activity rhythm, especially when it shows daily drift (e.g., when it is uncoupled from external environmental cues and is free-running). When humans or monkeys are maintained under fixed photoperiods, such as 12 L:12D (i.e., 12 h of light and 12 h of darkness per day), analysis of the actograms reveals the intensity of total daily activity, as well as activity during the day and activity during the night. Such data are valuable in establishing whether an individual is showing normal activity behavior. For example, using a night-vision camera we have established that rhesus macaques typically wake up a few minutes before the lights come on in the morning at 7:00 h; that is, they anticipate when dawn will occur. They then show a major peak of activity in the late morning, just after their morning feed at ~8:00 h. Finally, after the lights turn off in the evening (19:00 h), the animals continue to move around for a few minutes but generally close their eyes soon afterward and cease to show major episodes of motor activity. An animal that is sick typically does not show such a biphasic diurnal activity profile; it may show an attenuated level of activity during the daytime or excessive activity at night. Consequently, an additional use for the Actiwatch is to gain insights into the general health of an animal. A way to further refine the nocturnal activity is to perform a sleep analysis. However, it should be emphasized that this component of the Actiware 5.0 software has been validated for human studies only, and so strictly speaking, when applying it to the analysis of nonhuman primate activity one cannot refer to it as "sleep analysis" but simply as "nocturnal activity analysis." Nevertheless, the algorithm has been designed to provide information about "sleep fragmentation" and "sleep latency," which are used to establish whether a human subject is experiencing perturbed sleep, or in the case of nonhuman primates, whether the animal is showing unusual activity during its normal nocturnal rest period. Note that the epoch setting of the Actiwatch needs to be set to 1-min bins, or short, in order for this function to work. A practical application of this analysis is that if an individual is not sleeping well because of some pharmacological intervention or illness, then performance in a cognitive task may be impaired – indirectly because of perturbed sleep–wake or activity–rest cycles (14). Note also that the absolute level of activity detected by the Actiwatches may be influenced by whether an animal is caged single or pair-caged. As part of a psychological enrichment strategy, rhesus macaques at ONPRC are typically group housed or pair-caged, which promotes beneficial grooming behavior. On the other hand, when animals are caged individually, as when part of a remote blood sampling study, pair caging is not recommended because the protective catheter tubing from the two animals can become entangled. So instead, we cage the animals individually but install a semi-open partition between adjacent cages; this enables neighboring animals to perform mutual grooming without invading each other's cage floor space. Animals kept under different housing/caging conditions may show difference in their absolute activity levels, and this needs to be taken into account when comparing data from animals that have been housed differently.

2.3. Notes on Remote Blood Sampling

2.3.1. Potential Biohazard

When collecting body fluids, such as blood, from nonhuman primates the investigators need to treat them as if they are potentially pathogenic and take appropriate protective measures. Typically this involves a face mask, eye protection, gloves, and gown. Moreover, the blood tubes should be capped during centrifugation, to avoid creating potentially infectious aerosols.

2.3.2. Overview of Remote Blood Sampling

In our sampling set up (**Fig. 9.1**), the dead space in the line (i.e., the volume between the sampling stopcock and the tip of the vascular catheter) is typically <2 ml. Consequently, before collecting a blood sample, we first draw out this dead volume of heparinized saline from the line and discard it. We then withdraw a saline–blood mixture, until only dark blood is visible in the line. Finally, we collect the blood sample and immediately deposit it into either a heparin-coated or EDTA-coated ice-cold borosilicate glass sample tube. Note, some hormone assays are adversely affected by specific anticoagulants; therefore, the choice of anticoagulant will depend on the hormone assay systems that will subsequently be employed. After the sample has been collected, an infusion of heparinized saline is used to flush the line and to restore vascular fluid volume.

Note, some investigators include an additional step in the blood sampling procedure, in which red blood cells from the sampled blood are re-suspended in heparinized saline and re-infused into the animal. We prefer to avoid this step as it increases the possibility of introducing a blood clot or infection. Provided the volume of sampled blood from an adult rhesus macaque does not exceed 50 ml in a 24-h period and the animal is given time to build up its hematocrit before the next sampling session, significant anemia is avoided.

2.3.3. Protocol for Remote Blood Sampling

Note, although heparinized saline (4 IU/ml of 0.9% saline solution) is continuously infused (~1 ml/h) into the animal to keep the vascular catheter patent, this concentration of heparin is too low to prevent coagulation of a drawn blood sample. Therefore, blood should not be allowed to stagnate in the sampling line for more than a minute or so. With practice and preparation (e.g., pre-filling flushing syringes with heparinized saline) the entire procedure can usually be completed within 2 min. The following protocol refers to the set up depicted in **Fig. 9.1** and describes the key operations of the remote blood sampling procedure:

1. *Removal of dead space saline*: Turn the stopcock lever toward the pump port (to close the supply line) and draw all of the clear saline from the line (animal's side) into a 3-ml syringe; the volume removed should be about 2 ml. Turn the stopcock lever to a position midway between the pump and the syringe ports (to close both ports). Remove the syringe, discard its contents into a receptacle containing bleach, and re-attach it to the stopcock.

2. *Removal of diluted blood*: Turn the stopcock lever back toward the pump port (to close the supply line) and draw the saline–blood mixture into the same syringe until only dark blood fills the line. Turn the stopcock lever to the previous midway position (to close both ports), remove the syringe, and place it on a sterile pad, taking care not to touch the exposed luer end.

3. *Collection of blood sample*: Attach a new sampling syringe (1 or 3 ml) to the stopcock. Turn the stopcock lever toward the pump (to close the supply line) and draw the desired volume of blood into the sampling syringe. Turn the stopcock lever to the previous midway position (to close both ports), remove the syringe, and immediately transfer the blood sample into an ice-cold sample tube (containing anti-coagulant). Seal the top of the tube with parafilm and maintain it in ice until centrifugation and removal of the plasma. Discard the sample syringe into a "Sharps" container. Note, occasionally, a blood sample is difficult to draw because the animal has adopted a posture that is crimping the vascular catheter. In such circumstances it is best to pull out the syringe plunger very slowly and in small steps, rather than using a continuous draw; this reduces the amount of negative pressure, which sometimes causes the vein to collapse.

4. *Replacement of diluted blood*: Re-attach the 3-ml syringe containing the saline–blood mixture to the stopcock and turn the lever toward the pump (thereby closing the pump port and opening the sampling port). Re-inject the mixture back into the line and turn the stopcock lever back to midway position (to close both ports).

5. *Flushing the line*: Remove the used syringe and discard into a "Sharps" container. Replace it with a new 3-ml syringe pre-filled with heparinized saline. Turn the stopcock toward the pump port (to close the supply line) and inject ∼2.5 ml of heparinized saline to flush the saline–blood mixture back toward the animal.

6. *Restoration of vascular fluid volume*: Repeat the previous step and inject another ∼2.5 ml of saline back into the

animal (to further clear the line and to restore vascular fluid volume).

7. *Return to continuous slow infusion of heparinized saline*: Turn the stopcock lever back to its original position (pointing toward the syringe port), thereby allowing the peristaltic pump to continue its slow infusion of heparinized saline.

8. *Preparation for collection of next blood sample*: After collection of each sample, any spilled blood should be cleaned from the outside of the stopcock; for this we recommend using a cotton wool swab soaked in a 1% solution of chlorhexidine diacetate (e.g., Nolvasan). Also, while waiting for collection of the next sample, it is a good idea to fill the next set of flushing syringes with heparinized saline, so that they are ready to be used immediately after the next blood sample is collected.

2.3.4. Long-Term Catheter Maintenance

Although vascular catheters are commonly used for acute sampling procedures, we have had great success keeping them implanted chronically for more than a year. Success with long-term catheter viability includes appropriate set up of the apparatus and the subsequent rigorous aseptic maintenance of the sampling line. Our choice of material for the catheter is Silastic tubing (0.030 inches I.D. × 0.065 inches O.D.). This material is very pliable and does not stiffen significantly with age, although it is more susceptible to kinking than PV-6 tubing (0.034 inches I.D. × 0.060 inches O.D.). The Silastic catheter is channeled subcutaneously from the subclavian vein to midscapular region of the back, where it exits the animal and is connected to the PV-6 sampling tubing using a home-made 0.5 inch-long stainless steel 19-ga connector. The PV-6 tubing is connected to the stopcock using a 19-ga luer stub connector. Before use, the blood sampling tubing and stopcock set up is assembled and sterilized in ethylene oxide, inside a standard sterilization peel pouch. Every 2 weeks, the animals are sedated with ketamine. During this time, their protective nylon mesh jackets are changed, and their cages chemically sanitized. We also use this time to flush the disconnected sampling line with ethanol, to prevent the potential build up of bacterial bio-films; the line is ultimately flushed with heparinized saline before being reconnected to the vascular catheter. Although continuous infusion of heparinized saline solution into the animal can be achieved using a variety of different infusion pumps, we prefer to use peristaltic pumps (e.g., Gilson, Inc., Middleton, WI, USA) because they are robust and compact and readily connect to bags of physiological saline (into which heparin has been added) for long-term infusion.

3. Specific Applications

3.1. Age-Related Changes in 24-h Activity–Rest Cycles

It is common for nighttime sleep quality and daytime alertness to deteriorate in the elderly (17–19). Typical manifestations include a decreased amount of sleep, an increased number and duration of intra-sleep arousals, increased time in stage 1 sleep, and decreases in slow wave sleep stages (20). Additionally, daytime sleepiness has been strongly correlated to nighttime sleep fragmentation in the elderly (21). The rhesus macaque has long been considered an ideal model for neurophysiological studies of sleep because of its well-defined nocturnal sleep organization, similarity with human physiology and sleep, and ease of handling and housing (20, 22). We recently examined the common effects of age on activity–rest cycles and sleep–wake parameters in the rhesus macaque (13), and examples from a young (10 years old) and old (26 years old) males are shown in **Fig. 9.4**. The overall intensity of daytime activity is reduced in the old rhesus macaque, along with the magnitude of the activity–rest rhythm, as shown by the light–dark activity ratio. Additionally, the inferred quality of sleep in the old animal is reduced due to an increase in arousal episodes at night resulting in >threefold greater increase in the sleep fragmentation index. These results and those of others (23, 24) suggest that fragmented and dampened activity–rest rhythms are common in old rhesus macaques. Consequently, when designing studies to evaluate cognitive performance in old rhesus macaques, it might be prudent to first establish if any of the animals show significant impairment of the activity–rest cycles, which may indirectly affect their cognitive performance (14).

3.2. Age-Related Changes in 24-h Plasma Hormone Profiles

There are several hormones that can serve as biomarkers of aging because their overall levels, as well as their rhythms, are thought to be disrupted or dampened with advancing age. One such biomarker is the adrenal steroid, dehydroepiandrosterone (DHEA) and its sulfated ester (DHEAS). In both humans and nonhuman primates, overall circulating concentrations of DHEAS show a significant postmaturational decline with age (9, 11, 25–29). Whether DHEA and DHEAS exert their physiological actions directly is unclear as specific receptors to these steroids have yet to be identified. On the other hand, there is increasing evidence that some of the action may be mediated by the conversion of DHEA and DHEAS to other bioactive steroids (e.g., testosterone and estradiol) within specific organs and tissues, including regions of the brain associated with behavior and cognitive function (30). Conversely, circulating concentrations of the stress-related glucocorticoid, cortisol, which is also produced by the adrenal glands, do not decline with age but rather appear

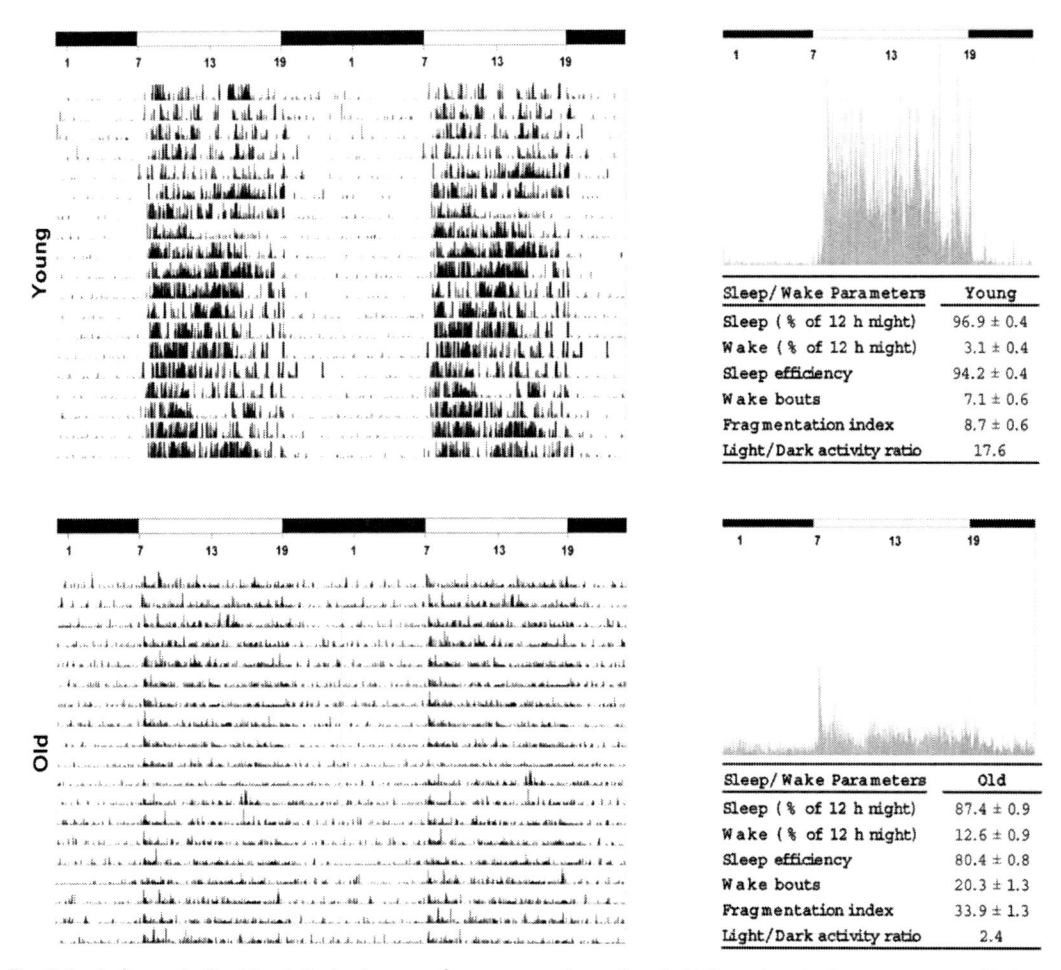

Sleep/Wake Parameters	Young
Sleep (% of 12 h night)	96.9 ± 0.4
Wake (% of 12 h night)	3.1 ± 0.4
Sleep efficiency	94.2 ± 0.4
Wake bouts	7.1 ± 0.6
Fragmentation index	8.7 ± 0.6
Light/Dark activity ratio	17.6

Sleep/Wake Parameters	Old
Sleep (% of 12 h night)	87.4 ± 0.9
Wake (% of 12 h night)	12.6 ± 0.9
Sleep efficiency	80.4 ± 0.8
Wake bouts	20.3 ± 1.3
Fragmentation index	33.9 ± 1.3
Light/Dark activity ratio	2.4

Fig. 9.4. *Left panels*: Double-plotted actograms from a young (*upper*) and old (*lower*) male rhesus macaque. Each row represents 2 days of activity recording, progressing from top to bottom. Daily periods of light and dark are depicted by the *white* and *black horizontal bars*, respectively. *Right panels*: Actiware analysis of averaged daily activity. Note the low level of diurnal activity in the old versus young animal but a relatively high level of nocturnal activity; this abnormal activity pattern manifests itself as an inferred increase in sleep fragmentation and a generally poor sleep quality index. [Figure taken from (13), with permission. Copyright 2007, Elsevier B.V.]

to be elevated (31–33). Moreover, elevated circulating cortisol has been suggested to play a role in neurodegeneration (32–34). In a recent study, we examined the 24-h patterns of circulating cortisol and DHEAS in young and old male rhesus macaques, using previously described assays (11). **Figure 9.5** depicts 24-h plasma profiles of cortisol and DHEAS (*left panel*) and the corresponding analyses (*center and right panels*). As expected, we observed robust diurnal variations in the 24-h profiles of cortisol and DHEAS in young males and a marked attenuation of the DHEAS rhythm in the old males. Conversely, plasma cortisol levels were significantly higher in the old males and the peak of the

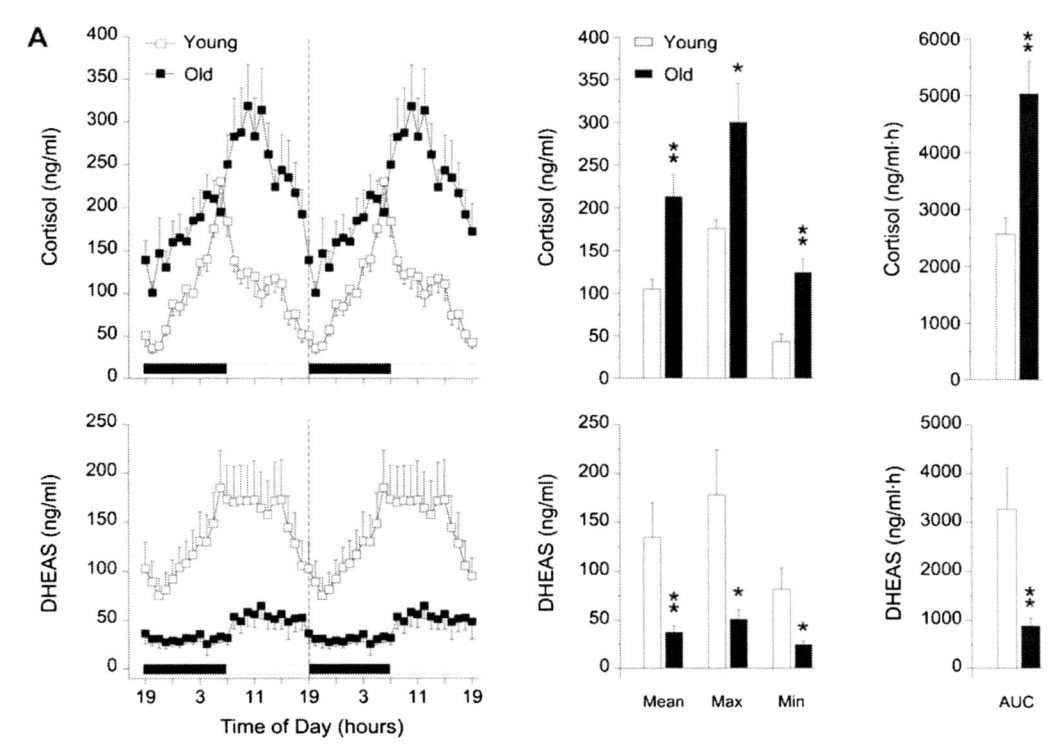

Fig. 9.5. Effect of age on circulating 24-h hormone patterns in male rhesus macaques. *Left panels:* Mean 24-h plasma cortisol and DHEAS profiles from young (~10 years old, $n = 5$) and old (~26 years old, $n = 6$) males. Although the blood samples were collected over 24-h, from 19:00 to 19:00 h, the data have been double-plotted (indicated by a *vertical dashed line*) to aid in the visualization of the night and day variations in hormone concentrations. The *horizontal black* and *white bars* on the abscissas correspond to the 12L:12D lighting schedule. *Center panels:* Analyses of age-related differences in mean, maximum, and minimum hormone values. *Right panels:* Analyses of age-related differences in the mean 24-h area under the curve (AUC) of cortisol and DHEAS concentrations. Values are expressed as mean ± SEM. *$P < 0.05$, **$P < 0.01$. [Figure taken from (10), with permission. Copyright 2006, The Society for Endocrinology.]

cortisol rhythm appeared to be slightly phase-delayed compared to the young group. Elevated DHEA and DHEAS levels have been shown to reduce male aggression and to boost cognition and learning in rodents; it is unclear, however, if similar beneficial effects can be obtained in humans and nonhuman primates. Other studies have suggested that DHEA and DHEAS can act as potent antiglucocorticoids, and so the age-associated decline in DHEAS/cortisol ratio may increase the brain's susceptibility to cortisol-induced neurotoxicity (30).

3.3. Seasonal Changes in 24-h Activity–Rest Cycles

In humans, seasonal variations have been reported for many physiological and behavioral traits, including changes in activity–rest patterns (7). Using Actiwatch recorders, we recently showed that the activity–rest cycle of rhesus macaques also changes significantly depending on the photoperiod (15). The animals showed considerably more nocturnal activity when maintained under short winter day lengths (comprising 8 h of light per day; 8L:16D), compared to when they were maintained under long summer day lengths (**Fig. 9.6**), even though they showed no difference in average total activity. Furthermore, exposure to the 8L:16D photoperiod was associated with earlier activity onset

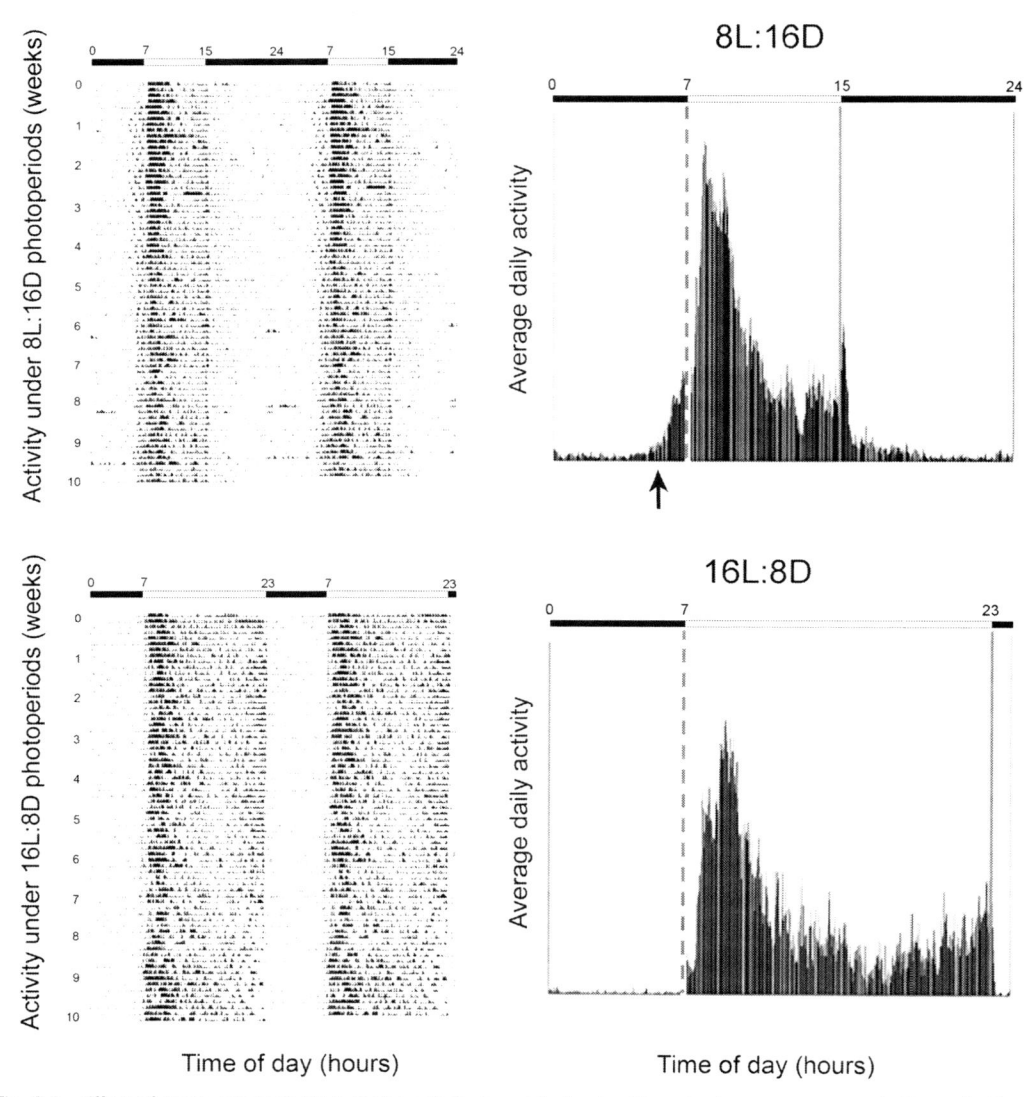

Fig. 9.6. Effect of photoperiod on 24-h motor activity in ovariectomized female rhesus macaques. *Left panels*: Representative actograms from an individual animal that was exposed for 10 weeks to short winter photoperiods (8L:16D) and subsequently to long summer photoperiods (16L:8D); the activity *data* are double-plotted to aid visualization of circadian changes. *Right panels*: Representative mean activity profiles during the corresponding 10-week periods. The *arrow* indicates the advancement of activity onset, and the *vertical dashed line* indicates the beginning of the diurnal phase. The *horizontal white* and *black bars* indicate day and night, respectively. [Figure adapted from (15), with permission. Copyright 2009, The Society for Endocrinology.]

each day, which occurred while the lights were still off, as well as a significant advancement of the peak of the daily activity rhythm.

In natural environments, short photoperiods are generally associated with the onset of unfavorably low temperatures and a scarce food supply. In this context, the advancement of the activity rhythm and other physiological functions in the winter would help to optimize the use of a shorter light phase. Another way of interpreting these data is that the animals centralize their daily activity around the middle of the day, regardless of photoperiod. This means that in short winter photoperiods their daily activity onset occurs several hours before dawn. Note that in our study we kept dawn fixed for each of the photoperiods, at 7:00 h. Thus, under 8L:16D the animals woke up at ~5:00 h, 6 h before the middle of the day (i.e., at 11:00 h). Under 16L:8D, however, they woke up at ~7:00 h, which was 8 h before the middle of the day (i.e., at 15:00 h). When viewed from this alternative perspective (i.e., relative to midday rather than dawn), our monkey activity data agree with previously published human data (35) showing that humans wake up earlier in summer than in winter.

3.4. Seasonal Changes in 24-h Plasma Hormone Profiles

In addition to continuous activity monitoring mentioned above, we also monitored 24-h plasma cortisol and DHEAS concentrations in rhesus macaques after 10 weeks of exposure to either short or long photoperiods (15). No significant differences were detected in either the mesor or amplitude of the plasma cortisol rhythm (**Fig. 9.7**). However, the cortisol peak was attained around the time when lights came on (i.e., at 07:00 h), when the animals were maintained under the 16L:8D photoperiod but showed a significant phase advancement when they were maintained under 8L:16D; notably, plasma cortisol levels reached a maximum while the animals were still in the dark phase of their daily photoperiodic cycle. No significant differences were detected in either the mesor or amplitude of the plasma DHEAS rhythm (**Fig. 9.7**), although both parameters showed a tendency to decrease under the 16L:8D photoperiod. Under the 16L:8D photoperiod the peak of the DHEAS rhythm occurred approximately 3–4 h after the lights came on in the morning, whereas under 8L:16D it occurred significantly earlier, within an hour of the beginning of the light phase.

Interestingly, the phase advancement of the cortisol rhythm under short days (**Fig. 9.7**) parallels that of the activity rhythm (**Fig. 9.6**), and so a causal relationship is likely to exist between the two rhythms. In view of its physiological functions, an earlier peak of cortisol might help an individual to achieve a state of arousal earlier in the day, thus facilitating an earlier awakening relative to dawn and so maximizing the animal's ability to forage under short winter photoperiods.

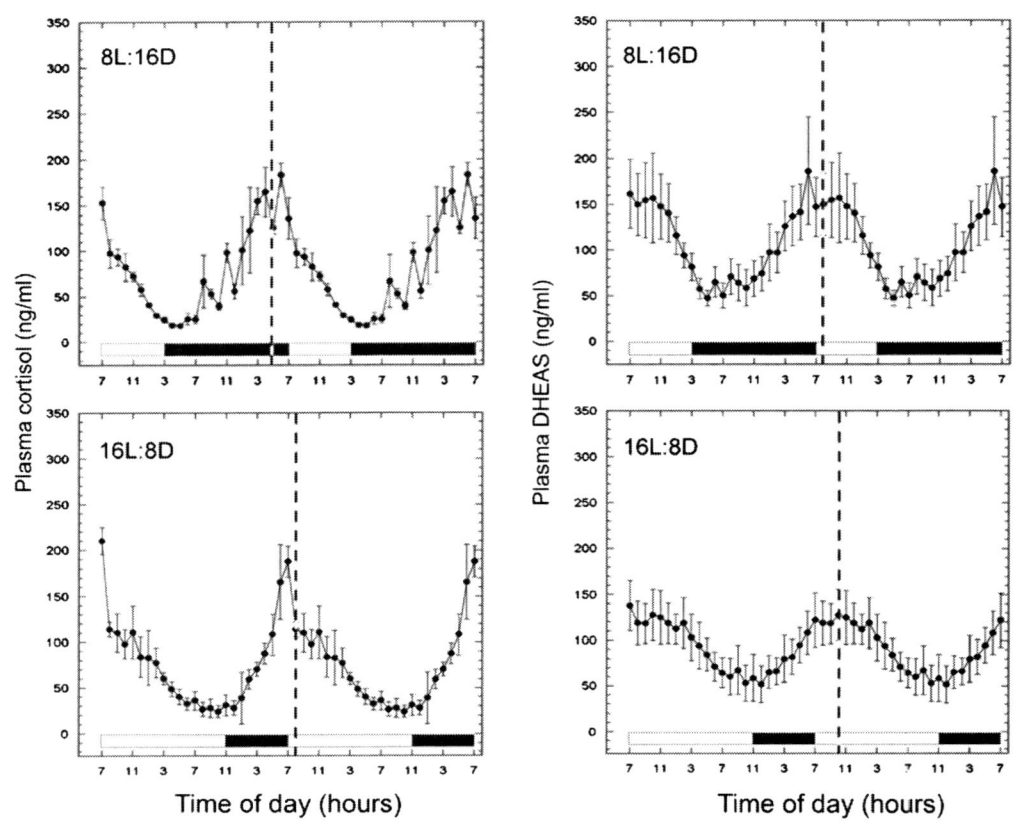

Fig. 9.7. Effect of photoperiod on the 24-h circulating cortisol and DHEAS rhythms in female rhesus macaques (*left* and *right panels*, respectively). Plasma samples were collected from the same animals after 10 weeks of exposure to either short winter photoperiods (8L:16D) or long summer photoperiods (16L:8D). Values are expressed as means ± SEM ($n = 3$); the data are double-plotted to aid visualization of circadian changes. *Vertical dashed lines within each panel* indicate the hormonal peaks, and the *horizontal white* and *black bars* represent day and night, respectively. Note the phase advancement of the two hormone rhythms under short days, which resembles the phase advancement in the onset of daily activity (**Fig. 9.6**). [Figure adapted from (15), with permission. Copyright 2009, The Society for Endocrinology.]

4. Impact of Biological Rhythms on Experimental Design

Section 3 emphasizes how circadian and seasonal rhythms can influence mechanisms that underlie many physiological and behavioral functions in vivo. Equally important, these rhythms can profoundly affect the outcome of studies designed to elucidate gene expression changes in vitro. For example, specific gene microarrays for both humans and rhesus macaques are now available commercially (e.g., Affymetrix), and these can be used for cost-effective gene profiling studies; details of the methodology used for analyzing the data from rhesus macaque gene microarrays are described elsewhere (2, 3). Our recent rhesus macaque gene microarray studies emphasize that caution needs to be exercised when designing such studies, because the expression of many genes, like the release of many hormones, is profoundly affected by circadian and seasonal rhythms (12, 15). It has been estimated that at least 10% of the expressed genes are likely to show a 24-h expression pattern, and so the time of day when RNA samples are collected can influence detection of significant changes in gene expression levels.

For example, the gene expression profiles of different experimental groups may appear to be similar if the RNA samples are collected at a time of day when the genes of interest are at the nadir of their circadian expression. In such circumstances, the optimal time of day for collecting RNA samples should be empirically determined for each gene of interest, and this information incorporated into the experimental design. This is particularly important when comparing gene expression profiles from a nocturnal rodent with that of a human or rhesus macaque, as data points obtained during an investigator's normal working hours would correspond to the rodent's subjective night but to the primate's subjective day, and so the results from the two animal models might not be directly comparable. Similarly, when photoperiodic species such as the rhesus macaque are housed outdoors, some of their genes show differential expression according to the time of year, and this needs to be taken into account when performing physiological and behavioral testing. By carefully controlling for circadian and seasonal variation one can optimize the physiological relevance of differential gene expression results and gain more meaningful insights into the mechanism that underlie normal and pathological human physiology and behavior.

Acknowledgments

This work was supported by National Institutes of Health grants: AG29612, HD29186, and RR00163.

References

1. Gibbs RA, Rogers J, Katze MG, Bumgarner R, Weinstock GM, Mardis ER et al (2007) Evolutionary and biomedical insights from the rhesus macaque genome. Science 316:222–234
2. Urbanski HF, Noriega NC, Lemos DR, Kohama SG (2009) Gene expression profiling in the rhesus macaque: experimental design considerations. Methods 49:26–31
3. Noriega NC, Kohama SG, Urbanski HF (2009) Gene expression profiling in the rhesus macaque: methodology annotation and data interpretation. Methods 49: 42–49

4. Hastings M, O'Neill JS, Maywood ES (2007) Circadian clocks: regulators of endocrine and metabolic rhythms. J Endocr 195:187–198
5. Bilbo SD, Dhabhar FS, Viswanathan K, Saul A, Yellon SM, Nelson RJ (2002) Short day lengths augment stress-induced leukocyte trafficking and stress-induced enhancement of skin immune function. Proc Natl Acad Sci 99:4067–4072
6. Nakao N, Ono H, Yamamura T, Anraku T, Takagi T, Higashi K (2008) Thyrotrophin in the pars tuberalis triggers photoperiodic response. Nature 20:317–322
7. Bronson FH (2004) Are humans seasonally photoperiodic? J Biol Rhyth 19:180–192
8. Garyfallou VT, Brown DI, Downs JL, James JL, Urbanski HF (2005) Effect of aging on circulating testosterone levels and on the expression of genes associated with testosterone biosynthesis. Endocr Soc Abstr 87:OR4
9. Urbanski HF, Downs JL, Garyfallou VT, Mattison JA, Lane MA, Roth GS, Ingram DK (2004) Effect of caloric restriction on the 24-hour plasma DHEAS and cortisol profiles of young and old male rhesus macaques. Ann NY Acad Sci 1019: 443–447
10. Downs JL, Urbanski HF (2006) Aging-related sex dependent loss of the circulating leptin 24-hour rhythm in the rhesus monkey. J Endocr 190:117–127
11. Downs JL, Mattison JA, Ingram DK, Urbanski HF (2008) Effect of age and caloric restriction on circadian adrenal steroid rhythms in rhesus macaques. Neurobiol Aging 29:1412–1422
12. Lemos DR, Downs JL, Urbanski HF (2006) Twenty-four hour rhythmic gene expression in the rhesus macaque adrenal gland. Molec Endocrinol 20:1164–1176
13. Downs JA, Dunn MR, Borok E, Shanabrough M, Horvath TL, Kohama SG, Urbanski HF (2007) Orexin neuronal changes in the locus coeruleus of the aging rhesus macaque. Neurobiol Aging 28: 1286–1295
14. Haley G, Landauer N, Renner L, Hooper Km, Urbanski HF, Kohama SG, Neuringer M, Raber J (2009) A maze to assess spatial learning and memory in aging rhesus monkeys. Exp Neurol 217:55–62
15. Lemos DR, Downs JL, Raitiere MN, Urbanski HF (2009) Photoperiodic modulation of activity–rest cycles, adrenal steroid rhythms and adrenal gland gene expression in ovariectomized rhesus macaques. J Endocr 201: 1–12
16. Downs JL, Urbanski HF (2006) Neuroendocrine changes in the aging reproductive axis of female rhesus macaques (*Macaca mulatta*). Biol Reprod 75:539–546
17. Bliwise DL (2000) Normal aging. In: Kryger MH, Roth T, Dement WC (eds) Principles and practice of sleep medicine, 3rd edn. WB Saunders, Philadelphia, PA, pp 26–42
18. Dijk DJ, Duffy JF, Riel E, Shanahan TL, Czeisler CA (1999) Ageing of the circadian and homeostatic regulation of human sleep during forced desynchrony of rest, melatonin and temperature rhythms. J Physiol (Lond) 516:611–627
19. Miles L, Dement W (1980) Sleep and aging. Sleep 3:119–220
20. Tobler I (1989) Napping and polyphasic sleep in mammals. In: DingesDF, Broughton RJ (eds) Sleep and alertness: chronobiological, behavioral, and medical aspects of napping. Raven Press, New York, NY, pp 9–30
21. Carskadon MA, Brown ED, Dement WC (1982) Sleep fragmentation in the elderly: relationship to daytime sleep tendency. Neurobiol Aging 3:321–327
22. Balzamo E, Santucci V, Seri B, Vuillon-Cacciuttolo G, Bert J (1977) Nonhuman primates: laboratory animals of choice for neurophysiologic studies of sleep. Lab Anim Sci 27:879–886
23. Emborg ME, Ma SY, Mufson EJ, Levey AI, Taylor MD, Brown WD, Holden JE, Kordower JH (1998) Age-related declines in nigral neuronal function correlate with motor impairments in rhesus monkeys. J Comp Neurol 401:253–265
24. Walton A, Branham A, Gash DM, Grondin R (2006) Automated video analysis of age-related motor deficits in monkeys using EthoVision. Neurobiol Aging 27: 1477–1483
25. Orentreich N, Brind JL, Rizer RL, Vogelman JH (1984) Age changes and sex differences in serum dehydroepiandrosterone sulfate concentrations throughout adulthood. J Clin Endocrinol Metab 59:551–555
26. Orentreich N, Brind JL, Vogelman JH, Andres R, Baldwin H (1992) Long-term longitudinal measurements of plasma dehydroepiandrosterone sulfate in normal men. J Clin Endocrinol Metab 75:1002–1004
27. Lane MA, Ingram DK, Ball SS, Roth GS (1997) Dehydroepiandrosterone sulfate: a biomarker of primate aging slowed by caloric restriction. J Clin Endocrinol Metab 82:2093–2096
28. Labrie F, Bélanger A, Luu-The V, Labrie C, Simard J, Cusan L, Gomez J-L, Candas B (1998) DHEA and the intracrine formation

of androgens and estrogens in peripheral target tissues: its role during aging. Steroids 63:322–328
29. Mattison JA, Lane MA, Roth GS, Ingram DK (2003) Calorie restriction in rhesus monkeys. Exp Gerontol 38:35–46
30. Sorwell K, Urbanski HF (2010) Dehydroepiandrosterone and age-related cognitive decline. *Age* 32:61–67
31. Van Cauter E, Leproult R, Kupfer DJ (1996) Effects of gender and age on the levels and circadian rhythmicity of plasma cortisol. J Clin Endocrinol Metab 81:2468–2473
32. Magri F, Locatelli M, Balza G, Molla G, Cuzzoni G, Fioravanti M, Solerte SB, Ferrari E (1997) Changes in endocrine circadian rhythms as markers of physiological and pathological brain aging. Chronobiol Int 14:385–396
33. Ferrari E, Arcaini A, Gornati R, Pelanconi L, Cravello L, Fioravanti M, Solerte SB, Magri F (2000) Pineal and pituitary-adrenocortical function in physiological aging and in senile dementia. Exp Gerontol 35:1239–1250
34. Porter NM, Landfield PW (1998) Stress hormones and brain aging: adding injury to insult? Nat Neurosci 1:3–4
35. Honma K, Honma S, Kosaka M, Fukuda N (1992) Seasonal changes of human circadian rhythms in Antarctica. Am J Physiol Regul Integr Comp Phsysiol 262:885–891

Chapter 10

Traumatic Brain Injury in Animal Models and Humans

Hita Adwanikar, Linda Noble-Haeusslein, and Harvey S. Levin

Abstract

Clinical/behavioral measures have traditionally been used to assess neurologic outcomes after human traumatic brain injury (TBI) as well as in experimental models of TBI. In this chapter, we address the metrics to assess injury/recovery in human TBI and consider the determinants of outcome. Further, we describe the commonly used experimental rodent models of TBI and the behavioral assays employed in these models for three major categories of neurologic assessments: sensorimotor (aggregate and individual tests), cognitive (memory and avoidance paradigms), and affective (novelty, social interaction, and anxiety) behaviors. Finally, we discuss the issues underlying use of behavioral assays in successful translation of candidate therapeutics from experimental models to human TBI.

1. Introduction

Traumatic brain injury (TBI) initiates a cascade of events that collectively contribute to secondary pathogenesis and in some cases reparative processes and behavioral impairments (1). It is the latter that is the focus of this review. Here we address the behavioral sequelae following human TBI, the determinants of outcome, as well as the outcome measures used to assess function. We further consider behavioral paradigms in experimental models of TBI, as these measures have been traditionally used to screen candidate therapeutics for the brain-injured patient.

2. Human TBI: Overview

In the following sections, we briefly review the epidemiology and pathophysiology of human TBI. Following this introduction, we present a summary concerning the determinants of outcome and describe neurobehavioral measures for various domains of outcome. Caveats regarding translational aspects of patient-oriented TBI research are also mentioned.

Epidemiology and pathophysiology: Human TBI occurs with an incidence that is related to age and gender reaching a peak of about 400/100,000 in adolescents and young adults (2). The incidence of TBI is generally higher in males than in females, but this gender disparity is reduced at both extremes of age (2). Closed head traumas resulting from blunt trauma, falls, sports-related injuries, and acceleration/deceleration forces associated with motor vehicle crashes account for most cases of TBI in civilians. Falls predominate in adults over 60 years old and are common in young children, whereas vehicular injury and sports-related injuries are more common in adolescents and young adults.

Briefly, the pathophysiology of TBI associated with closed head trauma is heterogeneous and can be divided between diffuse brain injury and focal or multifocal lesions (3, 4). Diffuse brain injury includes shearing and stretching of axons, reflecting the biomechanics of injury, and secondary insult due to excitotoxic cascades over the hours and possibly days after injury. The designation "diffuse axonal injury" implies widespread injury to white matter, but axonal injury can also be focal and is not necessarily equally distributed throughout the brain. With evidence that many cognitive functions depend on widely distributed networks that are often prefrontally guided (5), disruption of white matter connections is thought to be an important determinant of cognitive deficits. Inflammation, brain swelling, and oxygen toxicity also contribute to secondary injury (3). Extraparenchymal lesions include expanding epidural and subdural hematomas which can result in increased intracranial pressure pending surgical evacuation. Parenchymal focal lesions include cortical contusions and intracerebral hematomas which generally do not require surgery but could contribute to neuropsychological deficit at least during the initial months following injury. TBI in the moderate-to-severe range often has both diffuse injury and focal lesion components, thus contributing heterogeneity to the pathophysiology (6). In addition, acute complications such as hypoxia and hypotension can exacerbate the brain injury and worsen the outcome (7).

3. Severity of TBI

Impairment of consciousness: Glasgow Coma Scale (8): The Glasgow Coma Scale (GCS) is the most widely used measure of acute TBI severity especially for closed head trauma. The GCS evaluates impairment of consciousness based on the best eye opening, motor response, and verbal response that the examiner can elicit. The total GCS score, which is the sum of these three component scores and ranges from 3 to 15, is an index of overall severity. By convention, mild TBI is defined by a GCS score of 13–15 where a score of 13–14 denotes disorientation and confusion and a score of 15 indicates normal consciousness. Moderate TBI corresponds to a GCS score of 9–12 indicative of impaired consciousness but not coma. Severe TBI is defined by a GCS score ≤8, indicating coma as defined by no eye opening, inability to follow commands, and no comprehensible speech. However, it has been increasingly recognized that the post-resuscitation GCS score obtained after arrival in an emergency center may be insufficient to classify injury severity for research purposes (6). Pathology of injury seen on brain imaging (7), neurosurgical findings in patients with evacuation of mass lesions, and early neurologic deficit such as impaired pupillary response (9) are robust indicators of injury severity that must be considered along with the trajectory of GCS scores over the first 24–72 h. Rapid recovery of consciousness reflected by a steep trajectory of increased GCS scores during the first 24–72 h post-injury is prognostic of a better global outcome than a more protracted course of slowly improving or highly variable GCS scores even though the initial post-resuscitation GCS score may be identical in these two hypothetical patients. Other caveats include associated injuries that interfere with performing the GCS (e.g., ocular injury which complicates assessment of eye opening), obscuration of the early GCS by intoxication due to alcohol or drugs, and pharmacologic sedation, which confounds measuring conscious level with the GCS.

Post-traumatic amnesia (10): In post-acute studies, hospital records from the acute injury phase may be unavailable. Post-traumatic amnesia (PTA), the length of time after injury for which the patient has no lasting memory of events when queried at a later date, provides an index of TBI severity (10). A caveat is that retrospective estimation of PTA duration depends on the patient's recall and cooperation. Consequently, retrospective estimation of PTA duration is best approached by a careful interview and categorization of the duration, i.e., <30 min, >30 min but <24 h, >24 h but <7 days, and >7 days. Brief tests designed to assess orientation, appreciation of the circumstances of injury, attention, and memory for ongoing events have been developed to monitor PTA directly in patients during their initial hospitalization or rehabilitation (11, 12). Retrograde amnesia (RA) for events before the injury can also occur, especially for the period immediately prior to trauma. As with PTA, the length of RA tends to recede over the days and weeks following injury. In contrast to animal models which can include pre-injury training on a task to provide a baseline for measuring post-injury RA, it is difficult to establish a baseline in patients. One approach has been to test memory for widely accessible information that the individual is likely to have been exposed, including public events and popular media (13).

Acute computed tomography (CT): Complimentary to the impairment of consciousness and duration of PTA, the acute brain pathology seen on computed tomography (CT) has also been used to classify severity of TBI (14). Variables of interest include the size of mass lesion, extent of midline shift toward the opposite hemisphere caused by a unilateral lesion, and the presence of diffuse brain swelling as reflected by obscuration of the ventricles and cisterns.

Penetrating missile wounds of the brain: Penetrating brain injuries due to gunshot wounds and other missiles are far less common than closed head trauma in civilians. In contrast to closed head trauma, these injuries are relatively focal and related to the track of the foreign body. Consequently, the neurobehavioral sequelae are more strongly related to the localization of the neuroanatomic site of injury and are generally less dependent on diffuse brain insult (15–17). Consequently, a penetrating injury to the left hemisphere may produce aphasic disorder and right hemiparesis despite relatively mild impairment of consciousness. Posttraumatic epilepsy, which is especially prevalent following penetrating missile injuries of the brain, exacerbates cognitive sequelae (18).

4. Determinants of Outcome

Table 10.1 presents the patient (host) and injury variables that have been shown to predict outcome of TBI associated with closed head trauma. Although there is variation across studies in the predictors analyzed and the samples of patients have differed in severity of injury, follow-up interval, and subject variables, the general findings are summarized in Table 10.1. It is seen that severity of TBI is a major determinant of outcome as is age at injury. Mortality and disability are highest at the extremes of the age distribution.

Table 10.1
Determinants of outcome of human TBI[a] (4, 19)

Category	Determinants of outcome
Patient (host) variables	• *Age* • *Intelligence* • Comorbidities: substance abuse, neurologic disorders, psychiatric disorders • Genotype • Resilience
Injury variables	• *Severity and duration of impaired consciousness* • *Pupillary response to light* • *Mass lesion and location (especially penetrating missile)* • *Intracranial pressure* • Complications: hypotension, hypoxia, post-traumatic epilepsy • Other injuries such as limb fractures
Environment	• Medications and associated neurotoxicity • Rehabilitation and neuroplasticity • Social support and environmental enrichment

[a]Determinants that are most strongly supported by prospective, longitudinal outcome studies are presented in italics.

5. Assessment of Outcome of TBI

An interagency Common Data Elements Traumatic Brain Injury Outcomes Workgroup (19) identified domains of outcome, including recovery of consciousness, global level of function, neuropsychological impairment, psychological (emotional) status and post-concussion symptoms, performance of everyday activities, social role participation, and perceived health-related quality of life. Within each domain, the Workgroup distinguished core measures that have been well validated for general application to the adult TBI population and shown to be sensitive to change over time. Specialty measures recommended for use with specific populations or to address specific research questions were identified for each domain as were future measures that are under investigation. **Table 10.2** provides a summary of the outcome measures recommended by the TBI Outcomes Workgroup. Under neuropsychological impairment, the subdomains including episodic memory, attention/processing speed, and executive function encompass the most frequent impairments reported in patients following TBI due to closed head trauma. Although quantitative neuropsychological tests are widely used to measure the subdomains noted in **Table 10.2**, problems in executive function and memory may also impact performance of daily activities such as planning, monitoring performance, and flexibility in switching tasks or dividing one's attention between tasks.

Table 10.2
Outcome measures recommended by interagency TBI Outcomes Workgroup (19)

Outcome	Core measure(s)	Specialty measures
Global outcome	• Glasgow Outcome Scale-Extended (GOSE) (23, 85)	• Mayo-Portland adaptability inventory (MPAI-4) (86, 87) • Disability Rating Scale (DRS) (88) • Short Form-36 medical outcome study (SF-36v2) (89, 90)
Recovery of consciousness		• JFK Coma Recovery Scale-Revised (CRS-R) (91)
Neuropsychological impairment		
Episodic memory	• Rey Auditory Verbal Learning Test (RAVLT) (92)	• Brief Visuospatial Memory Test-Revised (BVMT-R) (93)
Attention/processing speed	• Processing Speed Index from the WAIS-III (94)	• Digit Span subtest of the Wechsler Adult Intelligence Scale (WAIS-III) (94)
Executive function	• Trail Making Test (TMT) (95)	• Letter–Number Sequencing subtest of the Wechsler Adult Intelligence Scale (WAIS-III) (94, 96) • Controlled Oral Word Association Test (COWAT) (97, 98) • Color-Word Interference Test (99)
Motor speed	• Grooved pegboard test (96)	
		• †*NIH toolbox cognitive battery*
Psychological status	• Brief Symptom Inventory-18 Item (BSI-18) (100)	• Minnesota Multiphasic Personality Inventory-2-Restructured Form (MMPI-2-RF) (101) • Alcohol Use Disorders Identification Test (AUDIT) (102–104) • Substance Use Questions from the TBI Model Systems Dataset (105, 106) • Alcohol, Smoking, and Substance Use Involvement Screening Test (ASSIST) (107, 108) • PTSD Checklist – Civilian/Military/Stressor Specific (PCL – C/M/S) (109, 110) • The Family Assessment Device (FAD) (111, 112)
		• †*NIH toolbox emotional battery*
TBI-related symptoms	• Rivermead Post-concussive Symptom Questionnaire (RPQ) (113, 114)	• Neurobehavioral Symptom Inventory (NS) (115)
Behavioral function		• The Frontal Systems Behavior Scale (FrSBe) (116–118)
Cognitive activity limitations	• Functional Independence Measure-Cognition Sub-scale (Cog-FIM) (119–121)	

(continued)

Table 10.2 (continued)

Outcome	Core measure(s)	Specialty measures
Physical function	• Functional Independence Measure-Motor Subscale (FIM) (122, 123)	
		• [†] *NIH Toolbox Motor and Sensory Batteries* • [†] *Neurological Outcome Scale for Traumatic Brain Injury (NOS-TBI).*
Social role participation	• Craig Handicap and Assessment Reporting Technique (CHART-SF) (124, 125)	
		• [†] *Participation Assessment with Recombined Tools (PART)*
Perceived generic and disease-specific health-related quality of life	• Satisfaction With Life Scale (SWLS) (126, 127)	
		• [†] *Quality of Life after Brain Injury (QOLIBRI)* (128)
Health-economic measures		• EuroQOL (129, 130)
Patient reported outcomes (future multidimensional tools)		
		• [†] *Patient Reported Outcome Measurement Information System (PROMIS)* (131) • *Neuro-QOL* (132, 133) • *TBI-QOL* (134, 135)

[†] Future measures are presented in italics.

Psychological (emotional) status and behavior: This domain of outcome is also sensitive to prefrontal–temporal injury including effects of TBI on the orbitofrontal–amygdala system. Major depression, disinhibition, and irritability are frequent behavioral sequelae of TBI associated with closed head trauma. These behavioral sequelae may also result from penetrating missile wounds of the brain, depending on the neuroanatomic site of injury. Depression and anxiety may persist despite good cognitive recovery following mild TBI (20). In addition to self-report measures, scales completed by a caregiver and rating forms completed by an examiner based on a structured interview are informative. Behavioral sequelae been implicated in residual problems in social functioning, including loss of friends and loneliness. Families have reported that the behavioral effects of severe TBI constitute the greatest burden on caregivers (21), but also see Temkin et al. (22).

Challenges to researchers investigating outcome: The diversity and number of relevant outcome measures are often a challenge for examiners to implement. First, patients tend to fatigue easily during the early post-injury phase thus limiting their capacity to complete a lengthy examination. Catastrophic TBI may also result in cognitive deficits that limit the patient's capability of comprehending and completing more difficult neuropsychological tests, thus resulting in selection bias.

Assessment of global outcome and functional recovery depends on integration of sources of information, raising a question of interexaminer reliability and sufficient training of the examiners. Multicultural issues must also be addressed, including the assessment of individuals whose primary language is not English. Analysis of multiple outcome measures raises questions about spurious results, an issue which can be potentially mitigated by developing composite measures. Models of outcome that emphasize participation in activities within the family, community, and at work have drawn a distinction between impairment identified by a neuropsychological test and disability which reflects curtailment of participation.

Heterogeneity in outcomes: In general, cognitive and behavioral sequelae contribute more to disability following TBI than motor or other specific neurologic deficits (23). Cognitive deficits at least partially resolve during the first 6–12 months after moderate-to-severe TBI, whereas cognitive recovery is generally complete within 1–3 months following mild TBI. However, impaired episodic memory on tests of recalling word lists or spatial location of items persists following moderate-to-severe TBI in about 20–25% of adults whose other cognitive functions recover relatively well during the first year post-injury (24). This dissociation may be attributed to hippocampal damage, but recent studies using diffusion tensor imaging (DTI) have shown that compromise of the microstructure of white matter connections including prefrontal and temporal white matter and corpus callosum is related to the cognitive sequelae of TBI (25, 26).

However, there is heterogeneity in outcomes and the most salient deficits among patients sustaining moderate-to-severe TBI. A patient whose TBI is complicated by extensive focal injury in the posterior region of the right hemisphere may be disabled by visuospatial neglect despite otherwise good recovery of cognition whereas other patients' functional recovery may be limited by language deficits though these severe, specific deficits persist in a relatively small subgroup of recovering TBI patients whose injury mechanism was closed head trauma.

Heterogeneity seems to be one of the fundamental barriers to successful therapeutic trials in TBI. A clear disconnect exists between preclinical research and the lack of treatment effects in clinical trials (27). In the search for neuroprotective therapeutics, animal models play an important role in elucidating the molecular and cellular mechanisms underlying secondary pathogenesis. While individual models of TBI may lack the heterogeneity seen in human TBI, the diversity of models provides a means for validating robust neuroprotection that is not constrained to a single model. In the following sections, we review the most common rodent models of TBI with special emphasis on those behavioral metrics that are most closely aligned to human TBI.

6. Animal Models of TBI: Overview

A number of tests have been developed to profile emerging deficits and/or recovery after experimental TBI. Behavioral assays, conducted at varying times post-injury, provide an essential background for understanding the underlying relationships between pathogenesis, wound healing, and clinical measures of recovery. Broadly, these tests fall into three categories: sensorimotor, cognitive, and affective behaviors. In the following sections, we review these measures of function in the context of experimental models of TBI in the rodent. We limit this review to rodent models, the species that have been most studied in experimental models of TBI, and to the most common models of TBI; namely, fluid percussion injury, controlled cortical impact, impact acceleration injury, and closed head injury (**Table 10.3**). Readers are referred to recent reviews that describe these models in detail (28, 29–31). While the unique characteristics of these models may impart behavioral "signatures," they also produce some similar findings across different models. In these sections, we highlight both the differences and similarities in behavioral profiles across the different models of TBI. In general, comparisons described in this review are against sham-injured animals.

Early determinants of outcome (**Table 10.4**): With this focus on rodent models, we acknowledge differences that exist between species and strains (32, 33). Such differences may influence how the brain responds to injury and the profile of behavioral recovery. Thus, we present findings in the context of both the type of injury, and when available, the species/strain studied.

Beyond differences related to species/strain, a number of early determinants of behavioral outcomes have been identified (**Table 10.4**), paralleling the determinants of outcome following human TBI. These include type of injury, location and severity of

Table 10.3
Characteristic pathologic features of the most commonly studied animal models of TBI. For reviews, refer to (28–30, 34)

Experimental model	Characteristic pathologic features
Fluid percussion injury	• Focal cortical contusion • Intraparenchymal hemorrhage • Marked diffuse axonal injury • Widespread cortical and subcortical damage
Controlled cortical impact	• Focal cortical contusion • Intraparenchymal hemorrhage • Some diffuse axonal injury • Widespread cortical and subcortical damage
Impact acceleration injury	• Absence of focal lesions • Mild subarachnoid hemorrhage • Marked diffuse axonal injury • Cortical damage directly beneath impact site and in hippocampus
Closed head injury	• Focal cortical contusion • Intraparenchymal hemorrhage • Some diffuse axonal injury • Increased possibility of skull fracture

Table 10.4
Major determinants of outcome following experimental TBI. For details, refer to (29, 35, 39, 40, 42, 43, 47, 52, 54)

Determinants	Comments
Location of injury	• Midline brain injuries have a greater brain stem component than lateral injuries. Midline injury leads to greater asymmetrical dysfunction than anterior or posterior injuries
Severity of initial insult	• Mild, moderate, or severe injury yields graded effects on the behavioral sequelae
Gender	• Sensorimotor deficits after TBI are greater in males compared to females
Species and strain	• Rats and mice have different responses to injury, with mice showing less severe injury • C57BL/6 are more sensitive to injury on the beam walking task, compared to FVB/N or 129/SvEMS. The latter strains are also unable to learn the Morris water maze or Barnes circular maze tasks
Age at time of injury	• The developing brain is sensitive to age at time of injury • Both the immature and the aging brain show enhanced response to injury

the initial insult, gender, and age at time of injury. The pathophysiology of TBI occurs over time and in a pattern consistent with the biomechanics of the initial impact. Acceleration/deceleration, focal, and diffuse models of injury have differing but well-defined cascades of temporal progression and lead to specific patterns of cellular and neurophysiological damage (34). Moreover, the biomechanics of injury are differentially impacted by location and severity of injury, leading to varying outcome (35). While animal models replicate different aspects of human TBI, the choice of experimental model depends on the underlying objectives of a study (28, 30) and an understanding of the relationship between the model and the emerging behavioral phenotype.

Gender is also a determinant of the behavioral phenotype. Experimental models of TBI have historically focused on males. However, over the past decade it has become increasingly clear that gender is a determinant of performance in behavioral tasks (36, 37) including those that assess working, reference, and spatial memory (38, 39). The underlying differences between genders may at least in part reflect diverging pathophysiology and behavior that show a differential sensitivity to a given therapeutic. For example, sensorimotor deficits, measured by foot faults in the grid walk test, are significantly greater in male C57Bl/6 mice compared to females, 1 week following a controlled cortical impact, while spatial learning deficits are similar between genders (40). While human studies have noted a gender difference in the epidemiology of TBI (2), the differences in injury response and behavioral phenotype due to gender are not well understood.

Age at the time of injury also influences outcome. For example, trauma to the developing brain modifies subsequent maturation and interferes with the acquisition of age-appropriate skills (41). The developing, immature brain is uniquely vulnerable to injury. Moreover, the enhanced post-injury plasticity attributed to the developing brain limits normal developmental neuroplasticity, especially at selective time periods after injury (42). At the opposite end of the temporal spectrum, TBI to the aging rodent brain results in a greater loss of function compared to the injured adult brain. For example, when adult and aging rodents are subjected to controlled cortical impact, the latter shows more pronounced early edema, neurodegeneration, and functional deficits (43). These findings suggest that both early and prolonged secondary pathogenesis in the aging brain contribute to poorer behavioral outcomes.

7. Experimental TBI: Behavioral Assays

Various behavioral tests have been used to assay specific components of neurological dysfunction following TBI. Here we summarize these findings according to sensorimotor, cognitive, and affective behaviors.

Table 10.5
Examples of sensorimotor behavioral tests according to species/strain and type of TBI. For reviews, refer to (47, 51)

Behavioral measures	Species/strain	Type of injury
Aggregate tests		
Composite neuroscore	Rats/Sprague Dawley	Fluid percussion injury
	Mice/C57Bl/6	Controlled cortical impact
Neurologic severity score	Rats/Sprague Dawley	Closed head injury
	Mice/Sabra	Impact acceleration injury
SNAP assessment	Mice/C57Bl/6	Controlled cortical impact
Individual tests		
Rotarod	Rats/Sprague Dawley	Controlled cortical impact
	Rats/Sprague Dawley	Fluid percussion injury
	Rats/Sprague Dawley	Impact acceleration injury
	Mice/C57Bl/6, FVB/N, 129/SvEMS	Controlled cortical impact
Grid walk	Mice/C57Bl/6	Controlled cortical impact
Spontaneous forelimb use	Mice/C57Bl/6	Controlled cortical impact
Wire hang	Mice/CF-1	Closed head injury
Beam balance	Rats/Sprague Dawley	Fluid percussion injury
	Rats/Sprague Dawley	Controlled cortical impact
	Mice/C57Bl/6	Controlled cortical impact
Beam walk	Rats/Sprague Dawley	Fluid percussion injury
	Rats/Sprague Dawley	Controlled cortical impact
Adhesive tape removal	Rats/Sprague Dawley	Fluid percussion injury

Sensorimotor behavior (**Table 10.5**): Motor function is integrated and mediated by a complex network of nuclei of the vestibulomotor pathway (44). Motor deficits, arising from TBI, may in part reflect damage to this pathway. A number of tests measure the ability of the brain-injured animal to locomote and maintain balance. Few of these tasks, however, purely test motor behavior. Each has a component of sensory feedback and proprioception, and therefore, most tasks are considered to be sensorimotor in nature.

One approach to profiling motor/sensory function is the use of aggregate tests that incorporate different measures of function. In the context of TBI, aggregate tests include the composite neuroscore, the neurological severity score (NSS), and the Simple Neuroassessment of Asymmetric imPairment (SNAP).

The composite neuroscore includes tests of flexion of forelimb and hindlimb, walking ability in a line or in a circle, abnormal movements, proprioception, as well as reflexes. This test was first described in rats and then modified for mice (45). Animals are scored on a 4 (normal) to 0 (severely impaired) scale using each of the following indices: (a) forelimb flexion, (b) resistance to lateral pulsion, (c) circling behavior in spontaneous ambulation, (d) ability to stand on an inclined plane, and (e) open-field activity. A total composite functional neuroscore of 0–20 is determined by combining the scores for the various tests so that 20 = normal, 15 = slightly impaired, 10 = moderately impaired, 5 = severely impaired, and 0 = non-functional. Fluid percussion injury in Sprague Dawley rats produces a chronic neurological deficit lasting up to 4 weeks that reflects to the severity of the initial injury (46). A modified composite neuroscore, ranging from 0 (non-functional) to 12 (normal), has been described for mice (C57Bl/6) and shown to predict marked behavioral impairments (scores of 9) when animals are subjected to a controlled cortical impact (45).

The NSS is another example of an aggregate measure (**Table 10.6**). Animals are scored on an all or none scale on multiple tests, with the overall NSS score being the sum of these tests. Individual tests include limb flexion, head movement, ability to walk straight, abnormal movements such as immobility, tremor, and myodystony, sensory tests, and tests of reflexes (47). Sprague Dawley rats with a closed head injury and evaluated on a scale of 1–24 show a decrease in the score over 30 days (48). The NSS likewise reveals deficits in murine (Sabra mice) models of TBI including impact acceleration injury that persist for at least 30 days post-injury (49).

The aggregate scoring test known as the SNAP test has recently been developed to evaluate the relationship between magnitude of injury and functional recovery in a murine model of controlled cortical impact (50). This test incorporates measures of vision, proprioception, motor strength, and posture, and the SNAP score reflects a summation of these assessments. The SNAP is based on a scale from 0 to 25, with a neurologically intact animal expected to have a score of 0. C57Bl/6 mice, when subjected to a moderate controlled cortical impact, show scores of about 8.4 at 3 days post-injury and persistent deficits (SNAP score = 6.1) at 12–14 months post-injury.

While aggregate tests provide an overall behavioral profile, individual tests offer insight into sensorimotor functions that may be affected by the injury. These include assessments of performance on a rotarod, ability to locomote across a wire grid, spontaneous forelimb use, beam balance, beam walk, and adhesive removal test. The rotarod test, in particular, is a very sensitive and efficient index for assessing sensorimotor impairments produced by TBI (51). This test assays the ability of the animal to maintain balance on a rotating rod by measuring latency to end of a trial

Table 10.6
Measurement of neurological severity score. For details, refer to (48). One point is given for failure to perform a task, thus, a higher score indicates a more severe injury. The animals are evaluated at 1 h and again at predetermined time points between 1 and 30 days. The difference between the new score and the 1 h score, ΔNSS, defines the extent of recovery of the tested animal

Neurological severity score	Evaluation (time after injury)	
	1 h	>1 h
Inability to exit from a circle (50 cm in diameter) when left in its center		
For 30 min after injury	1	–
For 60 min after injury	1	–
> 60 min after injury	1	–
Loss of righting reflex		
For 20 min after injury	1	–
For 40 min after injury	1	–
> 60 min after injury	1	1
Hemiplegia – inability to resist forced changes in position	1	1
Flexion of hindlimb when raised by the tail	1	1
Inability to walk straight when placed on the floor	1	1
Reflexes		
Pinna reflex	1	1
Corneal reflex	1	1
Startle reflex	1	1
Clinical grade		
Loss of seeking behavior	1	1
Prostration	1	1
Limb reflexes		
Loss of placing reflexes		
Forelimbs left/right	1	1
Hindlimbs left	1	1
Hindlimbs right	1	1
Functional test		
Failure in beam balancing task (1.5 cm wide)		
For 20 s	1	1
For 40 s	1	1
For >60 s	1	1
Failure in beam walking task		
8.5 cm wide	1	1
5 cm wide	1	1
2.5 cm wide	1	1
Maximum points	24	20

(falling off the rotarod or gripping the device for two consecutive revolutions). Test paradigms include different rotation speeds and accelerating rotation. Mice (C57Bl/6, FVB/N, 129/SvEMS), subjected to a controlled cortical impact, show deficits in performance up to 7 days following a moderate insult (52) whereas Sprague Dawley rats, subjected to a similar injury, show deficits in the rotarod test for several weeks (53).

The grid walk test allows for the independent assessment of all four limbs. Animals locomote on an elevated grid surface, and the percentage of stepping errors (foot faults) is calculated. The asymmetry of foot faults is measured by the difference between ipsilateral and contralateral foot faults. Injury results in location-dependent contralateral forelimb deficits. Increased contralateral foot faults are evident up to 4 weeks following controlled cortical impact in C57Bl/6 mice (54).

The spontaneous forelimb-use task allows the exclusive evaluation of forelimb function. It assays function of the forelimbs in vertical wall exploration (54). When placed in a Plexiglass cylinder, the animal rears to a standing position on the cylinder wall, supporting its weight with either one or both of its front limbs. The forelimb preference during vertical exploratory movements has been quantitatively evaluated by measuring the time spent supporting its weight on each foot. C57Bl/6 mice, sustaining a controlled cortical impact, show asymmetry in their forelimb use, with the ipsilateral paw being favored. This task is also sensitive to the location of the injury. Animals with midline TBI show greater asymmetry than animals with anterior or posterior injuries. These deficits may persist up to 5 months after injury (54).

The wire hang task is typically assayed in murine models of TBI. The latency of the animal to hang on a taut string is measured up to a maximum of 60 s. Decreased wire hang scores are reported acutely after closed head injury or impact acceleration injury in CF-1 mice (47, 55).

The beam balance or beam walk tasks assess the more complex components of vestibulomotor function and coordination. The beam balance task has been studied in brain-injured rats and mice. The animal is placed on a narrow wooden beam and its ability to maintain equilibrium is scored from 1 to 5 based on predetermined criteria. Based upon early studies, C57Bl/6 mice, subjected to controlled cortical impact, are impaired in their performance on the beam balance by 48 h post-injury but recover over 3 weeks (56). Sprague Dawley rats show beam balance deficits early after moderate fluid percussion or controlled cortical impact, which are then resolved by about 10 days post-injury (57, 58). In the beam walk task, animals are trained to escape a bright light and loud white noise by traversing an elevated narrow wooden beam to enter a darkened goal box at the opposite end of the beam. Performance is measured by the animal's latency to traverse the beam. Sprague Dawley rats show increased latency in traversing the beam in the first 5 days following lateral fluid percussion injury (57, 59) or controlled cortical impact (58).

The magnitude of somatosensory asymmetry is measured using the adhesive tape removal test, which evaluates the ability to purely integrate sensory information into motor responses. Circular adhesive stimuli are attached to the distal-radial portion of both forelimbs and the order and latency in which the adhesives are removed are noted. Following severe lateral fluid percussion injury in Sprague Dawley rats, there is a notable asymmetry in removal of adhesive up to 1 month following injury (60).

Cognitive behavior (**Table 10.7**): Experimental models of TBI produce hippocampal pathology as well as impairments in the ability to perform hippocampal-dependent behavioral tasks (47, 49, 58, 61–63). The Morris water maze (MWM) is one of the most commonly used assays of cognitive function in experimental models of TBI (64). It involves training the animal to swim in

Table 10.7
Examples of cognitive behavioral tests according to species/strain and type of TBI. For reviews, refer to (47, 64)

Behavioral measures	Species/strain	Type of injury
Morris water maze paradigms		
Acquisition and retention	Rats/Sprague Dawley	Controlled cortical impact
Learning	Rats/Sprague Dawley	Fluid percussion injury
	Rats/Sprague Dawley	Controlled cortical impact
	Mice/C57Bl/6, FVB/N, 29/SvEMS	Controlled cortical impact
Serial testing	Rats/Sprague Dawley	Fluid percussion injury
Working memory	Rats/Sprague Dawley	Fluid percussion injury
Memory	Rats/Sprague Dawley	Fluid percussion injury
	Mice/C57Bl/6	Controlled cortical impact
Memory and relearning	Mice/Sabra	Closed head injury
Other cognitive tests		
Delayed non-matching to place task	Rats/Sprague Dawley	Fluid percussion injury
	Rats/Sprague Dawley	Controlled cortical impact
Radial arm maze	Rats/Sprague Dawley	Fluid percussion injury
	Rats/Sprague Dawley	Controlled cortical impact
Circular Barnes maze	Mice/C57Bl/6	Controlled cortical impact
Passive avoidance	Rats/Wistar	Fluid percussion injury
Conditioned fear response	Mice/C57Bl/6	Fluid percussion injury
Place avoidance paradigm	Rats/Sprague Dawley	Controlled cortical impact

a circular tank to locate an escape platform, whose location can normally be identified using spatial memory. Spatial cues in the room containing the water maze allow the animal to navigate to the platform. In the most common method of using this test, the latency of the animal to reach the escape platform after training is measured. In probe trials, the platform is removed, and time spent by the animal looking for the platform in the correct quadrant is quantified. Place navigation in the water maze is a standard measure of cognitive ability and spatial memory following a disturbance of the nervous system.

The MWM test is very versatile because distinct complex protocols are used to investigate specific questions. Training before injury and evaluating following the injury addresses the animal's ability to remember previously learned tasks and is considered a measure of retrograde amnesia (63). Similarly, different training and testing paradigms are executed to test acquisition and retention of memories (65), reference and working memory (66), or extinction and relearning (67).

Memory dysfunction has been documented in different models of TBI, such as controlled cortical impact and impact acceleration injury. The degree of dysfunction following lateral controlled cortical impact is related to severity of injury and is concurrent with hippocampal cell loss on the ipsilateral side (63), specifically in the hilus of the dentate gyrus (68). Similarly, several studies (58, 65) describe MWM deficits in Sprague Dawley rats subjected to a controlled cortical impact. Brain-injured rats show longer escape latencies, with deficits in search time and relative target visits. Even mildly injured animals that do not demonstrate obvious tissue pathology show deficits in performance in both controlled cortical impact and fluid percussion models of injury (58, 69). Sprague Dawley rats, subjected to unilateral parietal controlled cortical impact, have been evaluated for spatial learning ability between 10 and 20 days following injury (70). These animals show deficits in latency to reach the platform, as well as in their ability to reach the learning criteria, despite no major differences in swim speed. These animals not only reach fewer criteria, but also require more days of training. Such findings reveal deficits in both the acquisition and retention paradigms in the MWM in brain-injured rats. The longer-term effects of injury on MWM behavior have been investigated following controlled cortical impact. Place learning has been tested at 2, 4 weeks, 3, 6, and 12 months post-injury, with the escape platform in a different maze quadrant for each time point (71). This paradigm addresses learning ability and spatial memory of the animals at each time point. The escape latency of Sprague Dawley rats, subjected to a controlled cortical impact, is increased at 3 and 12 months after injury. These results suggest that the maze deficits at 12 months are due to an inability of the injured animals to benefit from repeated MWM exposure, which could represent a deficit in procedural memory independent of changes in the hippocampus. Thus, a combination of different analyses in the MWM may be required to get a fuller picture of the abilities of the animal.

Most current analyses that rely on the MWM focus on latency and swimming distance. However, current tracking systems are also capable of measuring the average distance to platform, number and frequency of turns, number of goal crosses, and percentage time spent in goal quadrant. Such parameters may reveal deficits in search strategy that may be impaired following TBI (47).

To further understand the mechanisms of memory impairment, a water T-maze version of the delayed non-matching to place (DNMP) task has been studied in experimental models of TBI. In this task, the animal is habituated to swimming along both arms of a water T-maze to locate a platform at each end. Habituation training is followed by acquisition training, where the animal swims to one of the arms in a "forced choice" sample phase, during which the other arm is blocked off. In a subsequent "free-choice" phase 7 s later, the animal is allowed to choose between to two arms and is rewarded for choosing the arm opposite to the one it went to in the sample phase. During the delay testing, the delay between the sample phase and the choice phase is progressively increased. Moderate fluid percussion injury in Sprague Dawley rats produces a delay-dependent memory impairment, even when the initial degree of learning is controlled for, suggesting that the animals are impaired in memory retention rather than learning (72).

The radial arm maze and the circular Barnes maze also provide information about navigation. The radial arm maze consists of eight arms extending outward from a central platform. Cups of food are located at the end of each arm, some of which are baited with food. The test assays the ability of the animal to remember the baited and unbaited arms and correctly traverse the radial maze. Working and reference memory errors are recorded when an animal returns to an arm it had previously chosen during a particular session or when an animal enters an arm that was never baited, respectively. Sprague Dawley rats, subjected to fluid percussion injury, show an increase in the number of working memory errors in the first 2 weeks following injury, with no effect on reference memory (73).

The Barnes maze is used to assess spatial reference memory by training an animal to locate a hidden escape tunnel located directly under one of the holes at the perimeter of a brightly lit circular platform. The latency of the animals to escape into the tunnel is recorded. C567Bl/6 mice, subjected to a controlled cortical impact, show much higher latency to find the escape tunnel (up to 1 month following injury) and typically take much longer to develop an efficient searching strategy, as implied by the use of serial search rather than spatial cues (74). This suggests a deficit in learning of search strategy following injury.

The passive avoidance test measures the latency of the animal to remain in a safe box to avoid a foot shock. The animal is initially trained by placing it in a light compartment and allowed to enter a dark box. Immediately after entry, a scrambled foot shock (5 V, 50 Hz for 2 s) is delivered through the grid floor, which the animal escapes by stepping back in to the safe compartment. In the retention test, the animal is again placed in the light compartment and the latency to enter the dark side is measured. Brain-injured Wistar rats show shorter latencies (up to 14 days) after lateral fluid percussion injury (75).

Cognitive function has been measured using a more complex avoidance paradigm. This paradigm involves a hierarchy of tests with increasing cognitive demand; open-field exploration; passive place avoidance; and followed by two tests of active place avoidance. Phase 1 examines habituation of the animal as well as innate exploratory behavior in an open-field arena. Phase 2 tests whether injury alters conditioned behavior, specifically, the ability to inhibit exploration of the stationary arena to avoid entering a stationary shock zone. A foot shock is administered after the animal enters the shock zone, followed by shocks every 1.5 s until the animal vacates the shock zone, by passing a constant current (< 0.3 mA, 60 Hz, 500 ms) through the cable to a low-impedance (1 kΩ) shock electrode implanted in the paw. The high impedance contact between the paws of the rat and the grounded arena surface (400 kΩ) caused a major voltage drop across the paws. The number of entrances into the shock zone measures lack of avoidance. Phase 3 is similar to phase 2, except that the shock zone remains stationary while the arena rotates. This tests whether injury alters the ability of the animal to actively avoid the shock zone. Time before the first entrance assays memory of the shock zone location, and the number of shocks per entrance assays the ability and motivation to escape shock. Phase 4 involves a similar arena as phase 3, with the shock zone shifted 180° from its original location. It tests whether the avoidance memory being learned in phase 4 conflicts with the avoidance memory from phase 3. When tested in Sprague Dawley rats after mild or moderate controlled cortical impact, this paradigm shows no effect on passive avoidance or open-field exploration at 7 days following injury (76), suggesting comparable sensory, motor skills, and contextual memory. However, moderately injured animals are impaired in their ability to avoid a stationary shock zone in a rotating arena. In the fourth phase of testing, even mildly injured animals show impairments in their active avoidance abilities. Thus, this set of tests can be used to discriminate injury severity and related cognitive deficits (76).

Table 10.8
Examples of affective behavioral measures according to species/strain and type of TBI. For details refer to (77–80)

Behavioral measures	Species/strain	Type of injury
Gustatory neophobia	Rats/Sprague Dawley	Controlled cortical impact
Y maze	Rats/Sprague Dawley	Controlled cortical impact
Free-choice novelty	Rats/Sprague Dawley	Controlled cortical impact
Open field	Rats/Sprague Dawley	Controlled cortical impact
	Rats/Wistar	Impact acceleration injury
Elevated plus maze	Rats/Wistar	Impact acceleration injury
Social interaction	Rats/Wistar	Impact acceleration injury
Hyperemotionality	Rats/Wistar	Impact acceleration injury
Marble burying	Rats/Wistar	Impact acceleration injury
Object recognition	Mice/BALB/c	Closed head injury

Affective behavior (**Table 10.8**): There is an emerging interest in evaluating the more generalized affective behaviors, such as anxiety and depression, which have been frequently observed following human TBI. Such assays have provided inroads to understanding the neuropsychiatric disorders associated with human TBI.

Gustatory neophobia was one of the earliest assays of an affective disorder in experimental TBI. The animal is habituated in a testing chamber. After 24 h of food deprivation, the animal is re-introduced to the chamber containing measured amounts of hidden foods, both familiar (rat chow) and unfamiliar (raisins, potatoes, chocolate chip cookies). After testing, the weight of uneaten food is measured. Sprague Dawley rats with lateral fluid percussion injury lose the typical gustatory neophobia response and the preference for familiar foods at 7 days post-injury. Injured rats sample all food types equally, although they retain the ability to find hidden food, suggesting an absence of olfactory impairment (77).

Additionally, a decrease in some novelty-related behaviors has been identified in the Y maze, free-choice novelty exploration, and the object recognition test. In the Y-maze test, each of the three arms of the Y maze contains a unique texture. One of the arms is closed off and the animal is allowed to habituate to the other arms for 10 min. Following the habituation period, the animal is allowed to explore all arms for 5 min, and the number of entries and total time spent in each arm are recorded. Sprague Dawley rats, subjected to controlled cortical impact, spend less time in a novel arm of the Y maze within the first several weeks post-injury (78).

In the free-choice novelty test, the animal chooses between a familiar chamber and a similar chamber containing several objects of different textures, sizes, and shapes with which the animal could interact. After habituation in the familiar (empty) environment, the animal is allowed to explore the novel environment for 15 min. The amount of time in the novel environment and actions performed while doing so are recorded as a measure of exploratory behavior. Sprague Dawley rats that sustained a controlled cortical impact are impaired on various measures of the free-choice novelty test. Injured animals display fewer active behaviors in the novel environment and spend less time in the novel environment interacting with the novel objects (78).

In a similar object recognition test adapted for mice, the animal explores two objects in a familiar box for 5 min. Four hours later, it is introduced to the same cage where one of the objects is replaced with a new one. The cumulative time spent by the mouse exploring each of the objects is recorded. The ability of injured BALB/c mice to discriminate between the familiar and the new objects is impaired at 3 days following closed head injury, with considerable improvements by 1 month (79).

Anxiety-like and depressive behaviors have also been evaluated in brain-injured animals using a battery of tests that collectively model the disabling neuropsychological deficits associated with TBI. This is exemplified by a study that examined a modified open-field behavior test, hyperemotionality behavior in response to stimuli, social behavior, socio-sexual behavior, elevated plus maze, and marble burying behavior test, following impact acceleration injury in Wistar rats (80). The open-field test consists of an arena with a white floor divided into squares. The ambulation score (total number of squares crossed), rearing episodes, and defecation, during a 5-min observation period are measured. Injured rats show an increase in anxiety-related behaviors, such as increased ambulation and rearing in the modified open field in the first 3–4 weeks post-injury. Animals are also scored in an aggregate test of hyperemotionality behaviors. This is based on a graded score from 0 (no reaction) to 4 (extreme response) to various stimuli as follows: (1) startle response to a stream of air, (2) struggle response to handling, and (3) fight response to tail pinch. Results are expressed as the sum of the individual scores, with controls scoring about 5. Injury leads to an increase in the total hyperemotionality scores to about 10.

Social behaviors between two animals are assessed by measuring the time spent in social interaction (grooming, mounting, and crawling under), and passive interaction (number of times the animals cross each other). Injury leads to a decreased interaction time in the social behavior tests. The socio-sexual interactions between a male test subject and a receptive female were measured using the following parameters of socio-sexual behavior: latency of genital probing, number of episodes of genital probing, thrusting latency, number of thrusting episodes, and pursuit. Injured male rats exhibit an increase in the latency of socio-sexual behaviors and a decrease in the number of episodes and pursuit time (80).

The elevated plus maze consists of two open arms and two enclosed arms joined by a central platform. The number of entries and time spent in each of the arms are noted in a 5-min observation period. Injured rats exhibit an increase in percentage of open to total arm entries, which suggests a decreased defensive behavior in an unfamiliar environment (80).

Finally, anxious rats respond to unfamiliar objects by exhibiting a burying behavior. This is tested by placing the animal in a cage with the floor covered in 5 cm of bedding material; 25 marbles are arranged in the cage, and the number of marbles buried in a 30-min period is measured. The increased burying behavior described in the brain-injured rat reveals an injury associated neophobia and compulsiveness (80).

8. Summary

The heterogeneity of human TBI makes it difficult to develop a single experimental model that accurately mimics the human condition. Despite this challenge, each of the models highlights certain features that are seen in human TBI. While assays of motor/sensory functions are common in most experimental models and are indeed seen in human TBI, cognitive sequelae contribute more to disability. Although global outcome has been the primary measure in large outcome studies and clinical trials, translational studies may focus on specific outcome measures that more closely parallel animal models. **Figure 10.1** summarizes the categories of behavioral tests used in experimental and human TBI. As such, the Morris water maze has become one of the "gold standards" to assess memory impairments based upon navigational cues (64). Research concerning hippocampal damage and episodic memory might focus on measures of episodic or working memory (81–83) that could be translated to both animal models and humans. However, measures of episodic memory that test recall of a word list over trials (**Table 10.2**) or recall of spatial locations of stimuli (**Table 10.2**) have been used extensively in clinical outcome studies. It is particularly exciting that a virtual maze has recently been developed for humans thus allowing possible comparisons to findings in experimental models (84).

There are also emerging new tools, driven by human findings, to address the psychologic status in animal models, including

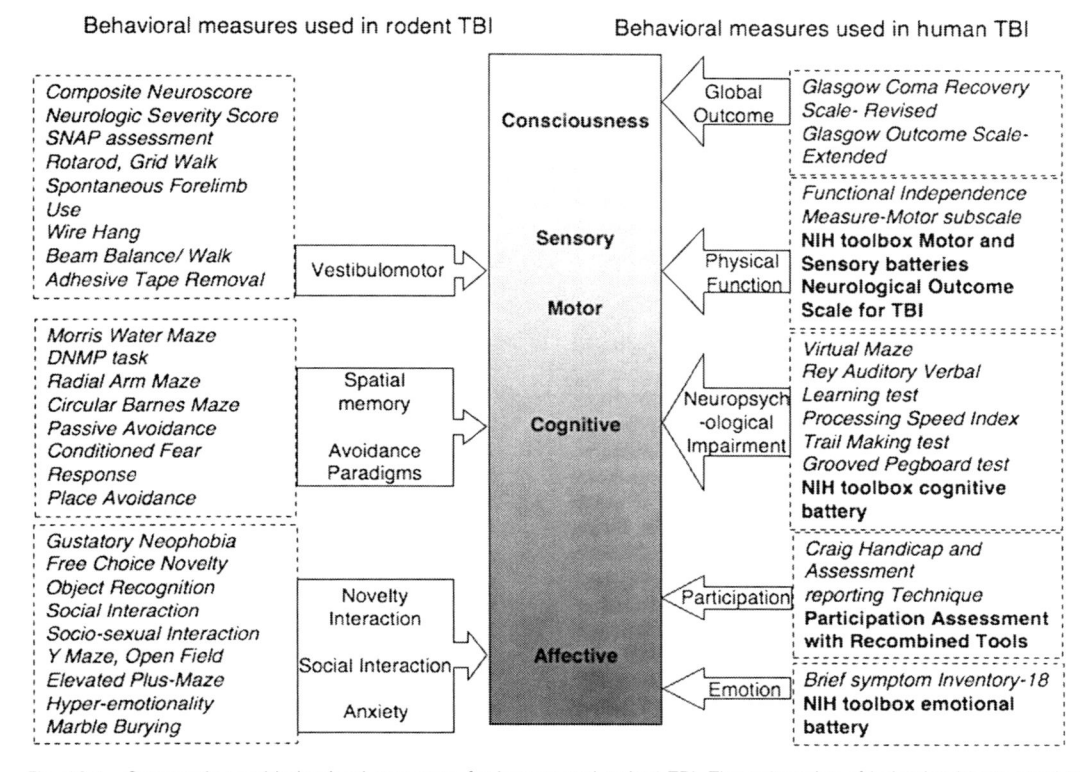

Fig. 10.1. Commonly used behavioral measures for human and rodent TBI. The categories of behavioral tests used in experimental models of rodent TBI, along with the individual tests, are listed on the *left*. The recommended core and future measures for human TBI are listed on the *right*. Bold indicates recommended future measures.

those paradigms that assay a hierarchy of tests with increasing cognitive demand, measures of anxiety-like and depressive behaviors, and aberrant social interactions. While understanding the pathobiology of TBI is key to developing a pipeline of candidate therapeutics, successful translation to human TBI mandates reproducible behavioral improvements across models of TBI with sensitivity to location and magnitude of the insult, gender, and age at time of injury.

Acknowledgments

This research was supported by grant B4596 by the Department of Veterans Affairs, Veterans Health Administration, Rehabilitation Research and Development Service and the Houston VA HSR&D Center of Excellence (HFP90-020) and by grant numbers R21NS065937, RO1NS050159, NS-21889, NS-056202, and P01NS056202 from the National Institute of Neurological Disorders and Stroke. The content is solely the responsibility of the authors and does not necessarily represent the official views of the National Institute of Neurological Disorders and Stroke or the National Institutes of Health. We gratefully acknowledge the assistance of Stacey K. Martin for her assistance in document preparation.

References

1. Rutherford G, Cernak BJ, Dikman I, Grady S, Hesdorffer S, Kraus D, Levin J, Noble H, Potolicchio L, Rauch S, Stiers S, Tamminga W, Temkin C, Weisskopf N (2009) Long-term consequences of traumatic brain injury. In: Gulf War and Health. Vol. 7. Long-term Consequences of Traumatic Brain Injury. Committee on Gulf War and Health: Brain Injury in Veterans and Long-term Health Outcomes, Board on Population Health and Public Health Practise. Institute of Medicine of the National Academics. The National Academics Press, Washington DC 2009
2. Langlois JA, Rutland-Brown W, Thomas KE (2006) Traumatic brain injury in the United States: emergency department visits, hospitalizations, and deaths. National Center for Injury Prevention and Control, Centers for Disease Control, Atlanta, GA
3. Povlishock JT, Katz DI (2005) Update of neuropathology and neurological recovery after traumatic brain injury. J Head Trauma Rehabil 20:76–94
4. IOM (Institute of Medicine) (2009) Gulf war and health, volume 7: long-term consequences of traumatic brain injury. The National Academic Press, Washington, DC
5. Miller EK, Cohen JD (2001) An integrative theory of prefrontal cortex function. Annu Rev Neurosci 24:167–202
6. Saatman KE, Duhaime AC, Bullock R, Maas AI, Valadka A, Manley GT (2008) Classification of traumatic brain injury for targeted therapies. J Neurotrauma 25:719–738
7. Eisenberg HM, Gary HE Jr, Aldrich EF, Saydjari C, Turner B, Foulkes MA, Jane JA, Marmarou A, Marshall LF, Young HF (1990) Initial CT findings in 753 patients with severe head injury. A report from the NIH Traumatic Coma Data Bank. J Neurosurg 73:688–698
8. Teasdale G, Jennett B (1974) Assessment of coma and impaired consciousness. A practical scale. Lancet 2:81–84
9. Levin HS, Gary HE Jr, Eisenberg HM, Ruff RM, Barth JT, Kreutzer J, High WM Jr, Portman S, Foulkes MA, Jane JA (1990) Neurobehavioral outcome 1 year after severe head injury. Experience of the Traumatic Coma Data Bank. J Neurosurg 73: 699–709
10. Russell WR (1932) Cerebral involvement in head injury. Brain 55:549–603
11. Levin HS, O'Donnell VM, Grossman RG (1979) The Galveston orientation and amnesia test. A practical scale to assess cognition after head injury. J Nerv Ment Dis 167: 675–684
12. Shores EA, Lammel A, Hullick C, Sheedy J, Flynn M, Levick W, Batchelor J (2008) The diagnostic accuracy of the Revised Westmead PTA Scale as an adjunct to the Glasgow Coma Scale in the early identification of cognitive impairment in patients with mild traumatic brain injury. J Neurol Neurosurg Psychiatry 79:1100–1106
13. Levin HS, High WM, Meyers CA, Von Laufen A, Hayden ME, Eisenberg HM (1985) Impairment of remote memory after closed head injury. J Neurol Neurosurg Psychiatry 48:556–563
14. Marshall LF, Marshall SB, Klauber MR, van Berkum Clark M, Eisenberg HM, Jane JA, Luerssen TG, Marmarou A, Foulkes MA (1991) A new classification of head injury based on computerized tomography. J Neurosurg 75:S14
15. Raymont V, Greathouse A, Reding K, Lipsky R, Salazar A, Grafman J (2008) Demographic, structural and genetic predictors of late cognitive decline after penetrating head injury. Brain 131:543–558
16. Grafman J, Schwab K, Warden D, Pridgen A, Brown HR, Salazar AM (1996) Frontal lobe injuries, violence, and aggression: a report of the Vietnam head injury study. Neurology 46:1231–1238
17. Grafman J, Salazar AM, Weingartner H, Amin D (1986) Face memory and discrimination: an analysis of the persistent effects of penetrating brain wounds. Int J Neurosci 29:125–139
18. Grafman J (2007) Vietnam head injury study phase III: a 30-year post-injury follow-up

study. The Henry M. Jackson Foundation for the Advancement of Military Medicine, Rockville, MD
19. Wilde EA, Whiteneck GG, Bogner J, Bushnik T, Cifu DX, Dikmen S, French L, Giacino JT, Hart T, Malec J, Millis SR, Novack TA, Sherer M, Tulsky DS, Vanderploeg RD, von Steinbuechel N(in press) Recommendations for the use of common outcome measures in traumatic brain injury research. Arch Phys Med Rehabil
20. Levin HS, McCauley SR, Josic CP, Boake C, Brown SA, Goodman HS, Merritt SG, Brundage SI (2005) Predicting depression following mild traumatic brain injury. Arch Gen Psychiatry 62:523–528
21. Oddy M, Humphrey M, Uttley D (1978) Subjective impairment and social recovery after closed head injury. J Neurol Neurosurg Psychiatry 41:611–616
22. Temkin NR, Corrigan JD, Dikmen SS, Machamer J (2009) Social functioning after traumatic brain injury. J Head Trauma Rehabil 24:460–467
23. Jennett B, Snoek J, Bond MR, Brooks N (1981) Disability after severe head injury: observations on the use of the Glasgow Outcome Scale. J Neurol Neurosurg Psychiatry 44:285–293
24. Levin HS, Goldstein FC, High WM Jr, Eisenberg HM (1988) Disproportionately severe memory deficit in relation to normal intellectual functioning after closed head injury. J Neurol Neurosurg Psychiatry 51:1294–1301
25. Kraus MF, Susmaras T, Caughlin BP, Walker CJ, Sweeney JA, Little DM (2007) White matter integrity and cognition in chronic traumatic brain injury: a diffusion tensor imaging study. Brain 130:2508–2519
26. Sidaros A, Engberg AW, Sidaros K, Liptrot MG, Herning M, Petersen P, Paulson OB, Jernigan TL, Rostrup E (2008) Diffusion tensor imaging during recovery from severe traumatic brain injury and relation to clinical outcome: a longitudinal study. Brain 131:559–572
27. Aarabi B, Simard JM (2009) Traumatic brain injury. Curr Opin Crit Care 15:548–553
28. Cernak I (2005) Animal models of head trauma. NeuroRx 2:410–422
29. Thompson HJ, Lifshitz J, Marklund N, Grady MS, Graham DI, Hovda DA, McIntosh TK (2005) Lateral fluid percussion brain injury: a 15-year review and evaluation. J Neurotrauma 22:42–75
30. Morales DM, Marklund N, Lebold D, Thompson HJ, Pitkanen A, Maxwell WL, Longhi L, Laurer H, Maegele M, Neugebauer E, Graham DI, Stocchetti N, McIntosh TK (2005) Experimental models of traumatic brain injury: do we really need to build a better mousetrap? Neuroscience 136:971–989
31. Flierl MA, Stahel PF, Beauchamp KM, Morgan SJ, Smith WR, Shohami E (2009) Mouse closed head injury model induced by a weight-drop device. Nat Protoc 4:1328–1337
32. Kacew S, Dixit R, Ruben Z (1998) Diet and rat strain as factors in nervous system function and influence of confounders. Biomed Environ Sci 11:203–217
33. Voikar V, Koks S, Vasar E, Rauvala H (2001) Strain and gender differences in the behavior of mouse lines commonly used in transgenic studies. Physiol Behav 72:271–281
34. Gaetz M (2004) The neurophysiology of brain injury. Clin Neurophysiol 115:4–18
35. Yu S, Kaneko Y, Bae E, Stahl CE, Wang Y, van Loveren H, Sanberg PR, Borlongan CV (2009) Severity of controlled cortical impact traumatic brain injury in rats and mice dictates degree of behavioral deficits. Brain Res 1287:157–163
36. Bimonte HA, Hyde LA, Hoplight BJ, Denenberg VH (2000) In two species, females exhibit superior working memory and inferior reference memory on the water radial-arm maze. Physiol Behav 70:311–317
37. Bimonte HA, Denenberg VH (2000) Sex differences in vicarious trial-and-error behavior during radial arm maze learning. Physiol Behav 68:495–499
38. Roof RL, Stein DG (1999) Gender differences in Morris water maze performance depend on task parameters. Physiol Behav 68:81–86
39. Wagner AK, Kline AE, Ren D, Willard LA, Wenger MK, Zafonte RD, Dixon CE (2007) Gender associations with chronic methylphenidate treatment and behavioral performance following experimental traumatic brain injury. Behav Brain Res 181:200–209
40. Xiong Y, Mahmood A, Lu D, Qu C, Goussev A, Schallert T, Chopp M (2007) Role of gender in outcome after traumatic brain injury and therapeutic effect of erythropoietin in mice. Brain Res 1185:301–312
41. Prins ML, Hovda DA (2003) Developing experimental models to address traumatic brain injury in children. J Neurotrauma 20:123–137
42. Giza CC, Kolb B, Harris NG, Asarnow RF, Prins ML (2009) Hitting a moving target: basic mechanisms of recovery from acquired developmental brain injury. Dev Neurorehabil 12:255–268

43. Onyszchuk G, He YY, Berman NE, Brooks WM (2008) Detrimental effects of aging on outcome from traumatic brain injury: a behavioral, magnetic resonance imaging, and histological study in mice. J Neurotrauma 25:153–171
44. Hamm TM (1990) Recurrent inhibition to and from motoneurons innervating the flexor digitorum and flexor hallucis longus muscles of the cat. J Neurophysiol 63:395–403
45. Raghupathi R, Fernandez SC, Murai H, Trusko SP, Scott RW, Nishioka WK, McIntosh TK (1998) BCL-2 overexpression attenuates cortical cell loss after traumatic brain injury in transgenic mice. J Cereb Blood Flow Metab 18:1259–1269
46. McIntosh TK, Vink R, Noble L, Yamakami I, Fernyak S, Soares H, Faden AL (1989) Traumatic brain injury in the rat: characterization of a lateral fluid-percussion model. Neuroscience 28:233–244
47. Fujimoto ST, Longhi L, Saatman KE, Conte V, Stocchetti N, McIntosh TK (2004) Motor and cognitive function evaluation following experimental traumatic brain injury. Neurosci Biobehav Rev 28:365–378
48. Shohami E, Novikov M, Bass R (1995) Long-term effect of HU-211, a novel non-competitive NMDA antagonist, on motor and memory functions after closed head injury in the rat. Brain Res 674:55–62
49. Chen Y, Constantini S, Trembovler V, Weinstock M, Shohami E (1996) An experimental model of closed head injury in mice: pathophysiology, histopathology, and cognitive deficits. J Neurotrauma 13:557–568
50. Shelton SB, Pettigrew DB, Hermann AD, Zhou W, Sullivan PM, Crutcher KA, Strauss KI (2008) A simple, efficient tool for assessment of mice after unilateral cortex injury. J Neurosci Methods 168:431–442
51. Hamm RJ, Pike BR, O'Dell DM, Lyeth BG, Jenkins LW (1994) The rotarod test: an evaluation of its effectiveness in assessing motor deficits following traumatic brain injury. J Neurotrauma 11:187–196
52. Fox GB, LeVasseur RA, Faden AI (1999) Behavioral responses of C57BL/6, FVB/N, and 129/SvEMS mouse strains to traumatic brain injury: implications for gene targeting approaches to neurotrauma. J Neurotrauma 16:377–389
53. Lindner MD, Plone MA, Cain CK, Frydel B, Francis JM, Emerich DF, Sutton RL (1998) Dissociable long-term cognitive deficits after frontal versus sensorimotor cortical contusions. J Neurotrauma 15:199–216
54. Baskin YK, Dietrich WD, Green EJ (2003) Two effective behavioral tasks for evaluating sensorimotor dysfunction following traumatic brain injury in mice. J Neurosci Methods 129:87–93
55. Panter SS, Braughler JM, Hall ED (1992) Dextran-coupled deferoxamine improves outcome in a murine model of head injury. J Neurotrauma 9:47–53
56. Scherbel U, Raghupathi R, Nakamura M, Saatman KE, Trojanowski JQ, Neugebauer E, Marino MW, McIntosh TK (1999) Differential acute and chronic responses of tumor necrosis factor-deficient mice to experimental brain injury. Proc Natl Acad Sci USA 96:8721–8726
57. Dixon CE, Lyeth BG, Povlishock JT, Findling RL, Hamm RJ, Marmarou A, Young HF, Hayes RL (1987) A fluid percussion model of experimental brain injury in the rat. J Neurosurg 67:110–119
58. Hamm RJ, Dixon CE, Gbadebo DM, Singha AK, Jenkins LW, Lyeth BG, Hayes RL (1992) Cognitive deficits following traumatic brain injury produced by controlled cortical impact. J Neurotrauma 9:11–20
59. Lyeth BG, Gong QZ, Shields S, Muizelaar JP, Berman RF (2001) Group I metabotropic glutamate antagonist reduces acute neuronal degeneration and behavioral deficits after traumatic brain injury in rats. Exp Neurol 169:191–199
60. Riess P, Bareyre FM, Saatman KE, Cheney JA, Lifshitz J, Raghupathi R, Grady MS, Neugebauer E, McIntosh TK (2001) Effects of chronic, post-injury Cyclosporin A administration on motor and sensorimotor function following severe, experimental traumatic brain injury. Restor Neurol Neurosci 18:1–8
61. Colicos MA, Dixon CE, Dash PK (1996) Delayed, selective neuronal death following experimental cortical impact injury in rats: possible role in memory deficits. Brain Res 739:111–119
62. Colicos MA, Dash PK (1996) Apoptotic morphology of dentate gyrus granule cells following experimental cortical impact injury in rats: possible role in spatial memory deficits. Brain Res 739:120–131
63. Smith DH, Okiyama K, Thomas MJ, Claussen B, McIntosh TK (1991) Evaluation of memory dysfunction following experimental brain injury using the Morris water maze. J Neurotrauma 8:259–269
64. D'Hooge R, De Deyn PP (2001) Applications of the Morris water maze in the study of learning and memory. Brain Res Brain Res Rev 36:60–90

65. Scheff SW, Baldwin SA, Brown RW, Kraemer PJ (1997) Morris water maze deficits in rats following traumatic brain injury: lateral controlled cortical impact. J Neurotrauma 14:615–627
66. Hamm RJ, Temple MD, Pike BR, O'Dell DM, Buck DL, Lyeth BG (1996) Working memory deficits following traumatic brain injury in the rat. J Neurotrauma 13:317–323
67. Chen Y, Shohami E, Constantini S, Weinstock M (1998) Rivastigmine, a brain-selective acetylcholinesterase inhibitor, ameliorates cognitive and motor deficits induced by closed-head injury in the mouse. J Neurotrauma 15:231–237
68. Hicks RR, Smith DH, Lowenstein DH, Saint Marie R, McIntosh TK (1993) Mild experimental brain injury in the rat induces cognitive deficits associated with regional neuronal loss in the hippocampus. J Neurotrauma 10:405–414
69. Gurkoff GG, Giza CC, Hovda DA (2006) Lateral fluid percussion injury in the developing rat causes an acute, mild behavioral dysfunction in the absence of significant cell death. Brain Res 1077:24–36
70. Griesbach GS, Sutton RL, Hovda DA, Ying Z, Gomez-Pinilla F (2009) Controlled contusion injury alters molecular systems associated with cognitive performance. J Neurosci Res 87:795–805
71. Dixon CE, Kochanek PM, Yan HQ, Schiding JK, Griffith RG, Baum E, Marion DW, DeKosky ST (1999) One-year study of spatial memory performance, brain morphology, and cholinergic markers after moderate controlled cortical impact in rats. J Neurotrauma 16:109–122
72. Whiting MD, Hamm RJ (2006) Traumatic brain injury produces delay-dependent memory impairment in rats. J Neurotrauma 23:1529–1534
73. Lyeth BG, Jenkins LW, Hamm RJ, Dixon CE, Phillips LL, Clifton GL, Young HF, Hayes RL (1990) Prolonged memory impairment in the absence of hippocampal cell death following traumatic brain injury in the rat. Brain Res 526:249–258
74. Fox GB, Fan L, LeVasseur RA, Faden AI (1998) Effect of traumatic brain injury on mouse spatial and nonspatial learning in the Barnes circular maze. J Neurotrauma 15:1037–1046
75. Yamaguchi T, Ozawa Y, Suzuki M, Yamamoto M, Nakamura T, Yamaura A (1996) Indeloxazine hydrochloride improves impairment of passive avoidance performance after fluid percussion brain injury in rats. Neuropharmacology 35:329–336
76. Abdel Baki SG, Kao HY, Kelemen E, Fenton AA, Bergold PJ (2009) A hierarchy of neurobehavioral tasks discriminates between mild and moderate brain injury in rats. Brain Res 1280:98–106
77. Hamm RJ, Pike BR, Phillips LL, O'Dell DM, Temple MD, Lyeth BG (1995) Impaired gustatory neophobia following traumatic brain injury in rats. J Neurotrauma 12:307–314
78. Wagner AK, Postal BA, Darrah SD, Chen X, Khan AS (2007) Deficits in novelty exploration after controlled cortical impact. J Neurotrauma 24:1308–1320
79. Levy A, Bercovich-Kinori A, Alexandrovich AG, Tsenter J, Trembovler V, Lund FE, Shohami E, Stein R, Mayo L (2009) CD38 facilitates recovery from traumatic brain injury. J Neurotrauma 26:1521–1533
80. Pandey DK, Yadav SK, Mahesh R, Rajkumar R (2009) Depression-like and anxiety-like behavioural aftermaths of impact accelerated traumatic brain injury in rats: a model of comorbid depression and anxiety? Behav Brain Res 205:436–442
81. Hanten G, Stallings-Roberson G, Song JX, Bradshaw M, Levin HS (2003) Subject ordered pointing task performance following severe traumatic brain injury in adults. Brain Inj 17:871–882
82. Delis D, Kramer L, Kaplan E (1986) The California verbal learning test. Psychological Corporation, San Antonio, TX
83. Petrides M, Milner B (1982) Deficits on subject-ordered tasks after frontal- and temporal-lobe lesions in man. Neuropsychologia 20:249–262
84. Livingstone SA, Skelton RW (2007) Virtual environment navigation tasks and the assessment of cognitive deficits in individuals with brain injury. Behav Brain Res 185:21–31
85. Wilson JT, Pettigrew LE, Teasdale GM (1998) Structured interviews for the Glasgow Outcome Scale and the extended Glasgow Outcome Scale: guidelines for their use. J Neurotrauma 15:573–585
86. Malec JF, Kragness M, Evans RW, Finlay KL, Kent A, Lezak MD (2003) Further psychometric evaluation and revision of the Mayo-Portland Adaptability Inventory in a national sample. J Head Trauma Rehabil 18:479–492
87. Malec JF (2004) Comparability of Mayo-Portland Adaptability Inventory ratings by staff, significant others and people with acquired brain injury. Brain Inj 18:563–575
88. Rappaport M, Hall KM, Hopkins K, Belleza T, Cope DN (1982) Disability rating scale for severe head trauma: coma to community. Arch Phys Med Rehabil 63:118–123

89. Ware JE Jr, Sherbourne CD (1992) The MOS 36-item short-form health survey (SF-36). I. Conceptual framework and item selection. Med Care 30:473–483
90. Ware JE, Gandek B, Aaronson NK, Acquadro C, Alonso J, Apolone G, Bech P, Brazier J, Bullinger M, Fukuhara S, Kaasa S, Keller S, Leplege A, Razavi D, Sanson-Fisher R, Sullivan M, Wagner A, Wood-Dauphinee S (1994) The SF-36 health survey: development and use in mental health research and the IQOLA project. Int J Ment Health 23:49–74
91. Giacino JT, Kalmar K, Whyte J (2004) The JFK Coma Recovery Scale-Revised: measurement characteristics and diagnostic utility. Arch Phys Med Rehabil 85:2020–2029
92. van den Burg W, Kingma A (1999) Performance of 225 Dutch school children on Rey's Auditory Verbal Learning Test (AVLT): parallel test–retest reliabilities with an interval of 3 months and normative data. Arch Clin Neuropsychol 14:545–559
93. Benedict RH (1997) Brief visuospatial memory test-revised. Psychological Assessment Resources, Inc., Odessa, FL
94. Wechsler D (1997) Wechsler Adult Intelligence Scale III. Harcourt Assessment, Inc., San Antonio, TX
95. Mitrushina, M., Boone, K. B., Razani, J., and D'Elia, L. F. (2005) *Handbook of Normative Data for Neuropsychological Assessment* (2nd Edn). New York, NY: Oxford University Press.
96. Strauss E, Sherman EMS, Spreen O (2006) A compendium of neuropsychological tests: administration, norms, and commentary, 3rd edn. Oxford University Press, New York, NY
97. Tombaugh TN, Kozak J, Rees L (1999) Normative data stratified by age and education for two measures of verbal fluency: FAS and animal naming. Arch Clin Neuropsychol 14:167–177
98. Iverson GL, Franzen MD, Lovell MR (1999) Normative comparisons for the controlled oral word association test following acute traumatic brain injury. Clin Neuropsychol 13:437–441
99. Batchelor J, Harvey AG, Bryant RA (1995) Stroop colour word test as a measure of attentional deficit following mild head injury. Clin Neuropsychol 9:180–186
100. Meachen SJ, Hanks RA, Millis SR, Rapport LJ (2008) The reliability and validity of the brief symptom inventory-18 in persons with traumatic brain injury. Arch Phys Med Rehabil 89:958–965
101. Tellegen A, Ben-Porath YS (2008) MMPI-2-RF (Minnesota Multiphasic Personality Inventory-2): technical manual. University of Minneapolis Press, Minneapolis, MN
102. Reinert DF, Allen JP (2007) The alcohol use disorders identification test: an update of research findings. Alcohol Clin Exp Res 31:185–199
103. Moussas G, Dadouti G, Douzenis A, Poulis E, Tzelembis A, Bratis D, Christodoulou C, Lykouras L (2009) The Alcohol Use Disorders Identification Test (AUDIT): reliability and validity of the Greek version. Ann Gen Psychiatry 8:11
104. Hides L, Cotton SM, Berger G, Gleeson J, O'Donnell C, Proffitt T, McGorry PD, Lubman DI (2009) The reliability and validity of the Alcohol, Smoking and Substance Involvement Screening Test (ASSIST) in first-episode psychosis. Addict Behav 34:821–825
105. Stein AD, Lederman RI, Shea S (1993) The behavioral risk factor surveillance system questionnaire: its reliability in a statewide sample. Am J Public Health 83:1768–1772
106. Miller JW, Gfroerer JC, Brewer RD, Naimi TS, Mokdad A, Giles WH (2004) Prevalence of adult binge drinking: a comparison of two national surveys. Am J Prev Med 27:197–204
107. ASSIST (2002) The Alcohol, Smoking and Substance Involvement Screening Test (ASSIST): development, reliability and feasibility. Addiction 97:1183–1194
108. Humenik, R. (2006) Validation of alcohol, smoking and substance involvement screening test (ASSIST) and pilot brief intervention [electronic resource]: a technical report of phase II findings of the WHO ASSIST Project, prepared by R. Humeniuk, and R. Ali, on behalf of the WHO ASSIST Phase II Study Group. Available at: http://www.who.int/substance_abuse/activities/assist_technicalreport_phase2_final.pdf (accessed 24 April 2007).
109. Weathers F, Litz B, Herman D, Huska J, Keane T (1993) The PTSD Checklist (PCL): reliability, validity, and diagnostic utility. Paper presented at the 9th annual convention of the International Society For Traumatic Stress Studies, San Antonio, TX
110. Blanchard EB, Jones-Alexander J, Buckley TC, Forneris CA (1996) Psychometric properties of the PTSD checklist (PCL). Behav Res Ther 34:669–673
111. Epstein NB, Baldwin LM, Bishop DS (2000) Family assessment device (FAD). In: Rush, J.A. (Ed). Handbook of psychiatric measures. American Psychiatric Association, Washington, DC

112. Kabacoff RI, Miller IW, Bishop DS, Epstein NB, Keitner GI (1990) A psychometric study of the McMaster Family Assessment Device in psychiatric, medical, and nonclinical samples. J Fam Psychol 3:431–439
113. King NS, Crawford S, Wenden FJ, Moss NE, Wade DT (1995) The Rivermead Post Concussion Symptoms Questionnaire: a measure of symptoms commonly experienced after head injury and its reliability. J Neurol 242:587–592
114. Crawford S, Wenden FJ, Wade DT (1996) The Rivermead head injury follow up questionnaire: a study of a new rating scale and other measures to evaluate outcome after head injury. J Neurol Neurosurg Psychiatry 60:510–514
115. Schwab KA, Ivins B, Cramer G, Johnson W, Sluss-Tiller M, Kiley K, Lux W, Warden D (2007) Screening for traumatic brain injury in troops returning from deployment in Afghanistan and Iraq: initial investigation of the usefulness of a short screening tool for traumatic brain injury. J Head Trauma Rehabil 22:377–389
116. Grace J, Malloy PF (2001) Frontal systems behavior scale professional manual. Psychological Assessment Resources, Inc., Lutz, FL
117. Caracuel A, Verdejo-Garcia A, Vilar-Lopez R, Perez-Garcia M, Salinas I, Cuberos G, Coin MA, Santiago-Ramajo S, Puente AE (2008) Frontal behavioral and emotional symptoms in Spanish individuals with acquired brain injury and substance use disorders. Arch Clin Neuropsychol 23:447–454
118. Stout JC, Ready RE, Grace J, Malloy PF, Paulsen JS (2003) Factor analysis of the frontal systems behavior scale (FrSBe). Assessment 10:79–85
119. Stineman MG, Shea JA, Jette A, Tassoni CJ, Ottenbacher KJ, Fiedler R, Granger CV (1996) The functional independence measure: tests of scaling assumptions, structure, and reliability across 20 diverse impairment categories. Arch Phys Med Rehabil 77:1101–1108
120. Hobart JC, Lamping DL, Freeman JA, Langdon DW, McLellan DL, Greenwood RJ, Thompson AJ (2001) Evidence-based measurement: which disability scale for neurologic rehabilitation? Neurology 57:639–644
121. Corrigan JD, Smith-Knapp K, Granger CV (1997) Validity of the functional independence measure for persons with traumatic brain injury. Arch Phys Med Rehabil 78:828–834
122. Lee JE, Stokic DS (2008) Risk factors for falls during inpatient rehabilitation. Am J Phys Med Rehabil 87:341–350
123. Heinemann AW, Linacre JM, Wright BD, Hamilton BB, Granger C (1993) Relationships between impairment and physical disability as measured by the functional independence measure. Arch Phys Med Rehabil 74:566–573
124. Whiteneck GG, Charlifue SW, Gerhart KA, Overholser JD, Richardson GN (1992) Quantifying handicap: a new measure of long-term rehabilitation outcomes. Arch Phys Med Rehabil 73:519–526
125. Mellick D, Walker N, Brooks CA, Whiteneck GG (1999) Incorporating the cognitive independence domain into chart. J Rehabil Outcomes Meas 3:12–21
126. Diener E, Emmons RA, Larsen RJ, Griffin S (1985) The satisfaction with life scale. J Pers Assess 49:71–75
127. Pavot W, Diener E (1993) Review of the satisfaction with life scale. Psychol Assess 5:164–172
128. von Steinbuechel N, Petersen C, Bullinger M (2005) Assessment of health-related quality of life in persons after traumatic brain injury – development of the Qolibri, a specific measure. Acta Neurochir Suppl 93:43–49
129. The EuroQol Group (1990) EuroQol – a new facility for the measurement of health-related quality of life. The EuroQol Group. Health Policy 16:199–208
130. van Agt HME, Essink-Bot ML, Krabbe PFM, Bonsel GJ (1994) Test–retest reliability of health state valuations collected with the EuroQol questionnaire. Soc Sci Med 39:1537–1544
131. Ader D (2007) Developing the Patient-Reported Outcomes Measurement Information System (PROMIS). Medical care 45:S
132. Perez L, Huang J, Jansky L, Nowinski C, Victorson D, Peterman A, Cella D (2007) Using focus groups to inform the Neuro-QOL measurement tool: exploring patient-centered, health-related quality of life concepts across neurological conditions. J Neurosci Nurs 39:342–353
133. Miller D, Nowinski C, Victorson D, Peterman A, Perez L (2005) The Neuro-QOL project: establishing research priorities through qualitative research and consensus development. Qual Life Res 14:2031
134. Tulsky DS, Carlozzi NE, Kisala P, Victorson D, Cella D (2009) New initiatives to develop patient reported outcomes measures for research with neurological populations. International Neuropsychological Society, Atlanta, GA
135. Tulsky DS, Kisala P, Victorson D, Carlozzi NE, Cella D (2009) Development of tailored outcomes measures for traumatic brain injury and spinal cord injury. American Academy of Neurology, Toronto, ON

Chapter 11

Behavioral Sensitization to Addictive Drugs: Clinical Relevance and Methodological Aspects

Tamara J. Phillips, Raúl Pastor, Angela C. Scibelli, Cheryl Reed, and Ernesto Tarragón

Abstract

Sensitization to the locomotor stimulant effects of abused drugs provides a behavioral measure thought to reflect underlying neural adaptations to repeated drug exposure. Neurochemical measures have provided information about the specific neural systems impacted and altered by repeated drug exposure. In pre-clinical studies, sensitized animals exhibit facilitated acquisition of drug self-administration and preference for cues associated with past drug experiences. This has suggested a role for sensitization in the development of drug abuse and in relapse. In humans, self-reports of sensitized vigor and energy levels have been described that may relate to the more direct measurements of locomotor sensitization in animals. Described in this chapter are methods used to measure psychomotor sensitization in mice, which are partially dependent upon the drug under investigation. The advantages to the use of mice in pre-clinical research are (1) that they readily sensitize to all drugs of abuse, (2) many methods have been developed for studying other aspects of their behavior that may be related to sensitization, and (3) they are an excellent species for genetic investigations aimed at determining susceptibility to behavioral sensitization and thus neuroadaptations related to drug abuse. Factors to consider when designing a study of drug-induced psychomotor sensitization include dose, number of treatments, frequency or interval between treatments and challenge, and duration of testing. First, a measure of baseline level of activity should be obtained, followed by measurement of the initial drug effect, measures of the change in initial effect with repeated administration, and a subsequent measure of baseline to see how it may have changed after repeated drug testing. Depending upon the goal of the research, a drug withdrawal period may be desirable, followed by another drug challenge to determine whether sensitization is still present. Such a withdrawal or "incubation" period has been instated to allow for the establishment of long-term central nervous system changes that may accompany sensitization in studies of mechanism. The recommended frequency of dosing is dependent upon characteristics of the drug, particularly its clearance rate. More intermittent schedules of administration are particularly important for inducing robust sensitization to classical psychostimulant drugs like cocaine and methamphetamine. The recommended duration of testing is influenced by the duration of drug effect, but data should be collected in isolated time units so that the time response curve can be examined. Finally, associative conditioning and stress-related factors can have large impacts on sensitization and should be carefully considered in all aspects of the research design, including whether drug treatment is linked to the test environment or not, density of housing, and specifics of handling.

1. Introduction

The study of sensitization provides a model for the study of the bases of long-term behavioral and neural plasticity. The term sensitization, as used in this chapter, refers to the phenomenon whereby a physiological or behavioral effect increases in magnitude as a consequence of repeated exposure to a particular stimulus. It may also be the case that, after sensitization has been induced, a smaller dose of a sensitizing drug or a smaller intensity of a stimulus induces a larger response than that seen initially. Many drugs with abuse potential have been found to induce both neurochemical and behavioral sensitization, which has been sometimes called reverse tolerance. However, it is not necessarily the case that sensitization occurs as the result of tolerance to a competing stimulus effect (e.g., tolerance to sedative effects of the drug). The behavioral augmentation may be incremental, but has been seen after a single prior exposure (1) and may be maximal even after a single pre-exposure (e.g., 2–4). Repeated drug exposure can render the subjects hypersensitive for a long period of time (5) and the underlying neural adaptations associated with sensitization have been suggested by some to play a role in the development of drug use disorders or in relapse (6–12); the persistent alterations of neural pathways that accompany sensitization may support the recurring desire to take drugs even after years of abstinence.

From a historical perspective, sensitization as a model of drug-induced neuroplasticity has its origins in the pioneering work done by Eric Kandel and colleagues beginning in the early 1970s. Dr. Kandel investigated the cellular correlates of simple forms of learning and memory, with the invertebrate sea snail *Aplysia californica*. In this animal model, sensitization was investigated by studying a progressively enhanced defensive reflex response and described as "behavioral arousal" (13–15). Experiments to determine the cellular correlates underlying sensitization showed strengthened synaptic connections, increases in neurotransmitter release, and subsequent second messenger systems (14, 16). Interestingly, many of these systems are the same as those later implicated in the drug literature, as involved in drug-induced sensitization. The understanding of the mechanisms and behavioral relevance of drug-induced neuroplasticity is also linked to Bliss and Lomo's (17) work, which first described the phenomenon of long-term potentiation (LTP), a long-lasting enhancement in signal transmission between two neurons in the hippocampus. Besides this landmark finding, they shaped the field by suggesting that this response was not simply a correlate of synaptic plasticity (18), but might actually underlie learning and memory (17). This idea further developed into an expanding theoretical framework suggesting that drug effects and their associated stimuli produce long-lasting memories which impact the development of drug abuse and relapse (19–24). In 1989, Karler and colleagues presented data showing that sensitization to the behavior-activating effects of psychostimulants did not occur when an NMDA receptor antagonist preceded each repeated treatment with cocaine or amphetamine (25). This finding was notable as NMDA receptors were known to be critical in the induction of LTP in hippocampal CA1 pyramidal cells (26–28). In addition, research findings indicating the contribution of the mesolimbic dopamine (DA) pathway to sensitization were reminiscent of the "transfer" of signaling in LTP, involving the same pathway (28, 29). Behavioral sensitization and sensitization-associated neuroplasticity have now been widely investigated in animal models to study learning and memory, bipolar and psychotic disorders, impulsivity, stress-induced alterations in drug effects, and drug abuse. This chapter will focus on methodological details for inducing and measuring drug-induced sensitization and on the evidence supporting the theoretical role of sensitization in drug addiction and relapse.

1.1. Psychomotor Sensitization

In early descriptions of psychostimulant sensitization, heightened locomotor responses were the main measure (30, 31). Repeated exposure to abused drugs can result in sensitization of several behaviors and this change in drug sensitivity has been demonstrated not only for locomotor effects, but also for some measures of arousal (e.g., stereotypic behaviors, exploration, approach, and attention) (32–36) and also for some effects seen during drug withdrawal (e.g., anxiety-like behavior) (37). Psychomotor sensitization (changes in motor behavior) is the form of behavioral sensitization that has received the most attention, with horizontal locomotor activity used most often as the dependent variable (*see* Section 3 for details). Rats and mice have been most often used in studies of drug-induced behavioral sensitization. Two distinct phases of the sensitization process have been considered: induction (also referred as development or acquisition) and expression. The dissociation between those two phases has been established at both pharmacological and neuroanatomical levels (27, 38, 39) and has important clinical implications. Interventions targeting the induction and expression of drug-induced sensitization are both potentially approachable. However, clinically intervening in the expression of sensitization by preventing activation of the altered brain mechanism is likely to be more feasible than preventing the changes in the brain from occurring in the first place. In animal research in which drug exposure can be completely controlled, a great deal of information about factors that influence these two phases has been obtained, including experimental determinants (e.g., drug dose, frequency, and duration of treatment), information about the role of specific neurotransmitter systems and receptors, genetic influences, and environmental factors (e.g., contextual stimuli, the role of stressors), many of which are discussed below.

Psychomotor sensitization presents some key features of addiction: its noteworthy persistence, influence by genotype, and facilitation by stress (34, 40, 41). In animal research, it has been shown that psychomotor sensitization can be found, when tested by drug challenge, for a month to up to a year after drug treatment cessation (42–46). There is evidence for genotype dependence of susceptibility, although this has been relatively little studied. The few studies that do exist have shown that, similar to what happens with vulnerability to drug abuse in humans, animals of some genotypes sensitize readily, whereas others are more resistant (2, 34, 47–54). In addition, single-gene mutants have been studied for differences in susceptibility to sensitization, providing some clues to mechanisms that may be involved (e.g., 38, 55–59). Finally, like certain drug effects in humans, psychomotor sensitization can be facilitated by stress axis (hypothalamic–pituitary–adrenal or HPA axis) activation (60, 61), to the extent that stress exposure can substitute for drug exposure in the development of sensitization (a phenomenon referred to as "cross-sensitization") (62–66). This suggests an important role of the HPA axis in the neuroplastic changes associated with repeated drug exposure. These important factors, combined with a relatively easy-to-implement methodology, have made psychomotor sensitization a convenient and relevant behavioral procedure for studying drug-induced neuroplasticity.

1.2. Neural Sensitization

Because the focus of this chapter is on methods, we will touch only briefly on the topic of neural changes and neurochemical processes involved in sensitization and will direct our comments largely to approaches used for these investigations. Examinations have focused on neural mechanisms associated with both the acquisition and expression of sensitization and have used neurochemical analysis methods such as tissue content measurement by high performance liquid chromatography (67), in vitro voltammetry (67, 68), and in vivo microdialysis (69, 70). In addition, pharmacological and in vivo genetic manipulations have been used to get at underlying mechanisms. Changes in transmitter receptor and transporter densities or function (56, 71) have been examined, and receptor-specific antagonists and agonists have been used in attempts to block acquisition or expression of sensitization (28, 38, 63, 66, 72–75). Drugs or genetic manipulations, such as lentiviral gene delivery, have been used to disrupt signaling molecules (76–78) or trophic factors (79). Finally, neuroimaging methods have also been used (80) and are discussed in some detail in **Section 1.5.2**.

Although some pharmacological manipulations have clearly attenuated the locomotor response used to index sensitization, interpretation of this outcome must be made with some caution. A pharmacological agent that attenuates a sensitized locomotor response could do so by blocking a mechanism activated by the sensitizing drug, by initiating a neurochemical response that competes with the mechanism activated by the sensitizing drug, or by activating a similar mechanism to that activated by the sensitizing drug. In the final example, the drugs may act additively and place the behavior at a different point along the behavioral dose–response curve, perhaps at a point where there is profound behavioral depression, rather than restoration of basal levels of activity. Drugs with their own effects on locomotor behavior are particularly subject to having effects that may be undesirable for a treatment agent, particularly when given in combination with an addictive drug (e.g., 81, 82). Their use for treatment of non-drug-induced sensitization (e.g., sensitized neural processes associated with pathological gambling or eating) may be less problematic. Use of a behavior rating scale (83), or measurement of behavioral coordination after each drug alone and the drug combination (81), can assist in the interpretation.

Another method used for getting at mechanism is single-gene mutation in mice. One strength of this method is that it removes interpretational complications associated with some of the drugs used for pharmacological approaches. Another is the ability to study mechanisms for which there are no specific pharmacological tools. For example, existing muscarinic receptor antagonists are not completely receptor subtype specific, but recent data using m5 muscarinic receptor gene mutant mice suggest a role for this receptor in amphetamine-induced sensitization (58). However, there are complicating factors associated with interpreting data from single-gene mutant mice, which have been well discussed (84–86), and the combined use of single-gene mutant mice with existing pharmacological tools can facilitate interpretation (e.g., 38, 87).

1.3. Sensitization and Theories of Addiction

Behavioral sensitization and the neural correlates associated with this process have been at the core of one of the most influential theories of addiction, the incentive sensitization theory of addiction (88, 89). Before this theory was proposed, a number of authors had made important theoretical contributions to the idea that the neurobiology underlying the behavioral stimulant effects of abused drugs might be a useful readout of their reinforcing properties. In 1987, Wise and Bozarth put forth the seminal psychomotor theory of addiction. In this view, all drugs of abuse share the ability to stimulate locomotor behavior, resulting from activation of a common mesolimbic DA brain pathway which also was proposed to underlie primary positive reinforcement (29). This leading theory was a shift away from dependence models and focused attention on locomotor activity as an animal model to investigate euphorigenic properties of drugs and ultimately addiction. It took only a small step to extend the theory to include sensitization; if the neurobiology of psychomotor activation overlaps with that supporting positive reinforcement, then it was logical to propose that sensitized psychomotor responses to drugs would reflect greater susceptibility to positively reinforcing drug effects. The evolution of this notion has paralleled the revisions and refinements of ideas regarding the role of mesolimbic DA in drug and non-drug reinforcement (90–94). Accumulating evidence indicates that mesolimbic DA responses might not directly mediate the emotional aspects of reinforcement, commonly referred to as "reward" (92, 93, 95). In line with these postulates, the incentive theory of addiction (88, 89) postulates that sensitization of accumbens DA activity to abused drugs contributes to an explanation of pathological motivation toward drugs and to the transition from controlled consumption to compulsive patterns of drug-seeking and drug-taking that are characteristic of addiction. More specifically, the incentive sensitization theory of addiction proposes that repeated exposure to addictive drugs induces enduring changes in brain systems (neuroadaptations mainly of the mesocorticolimbic DA system) that regulate the attribution of incentive salience (i.e., motivational drive, attractiveness) to stimuli. In this way, repeated drug administration can increase (sensitize) incentive to consume a drug or to experience drug-seeking behavior induced by stimuli associated with drug effects. Consequently the drug and its associated stimuli become highly desired or "wanted." Thus, changes in incentive salience as a consequence of sensitization are linked to profound changes in attentional and motivational processes that ultimately support drug-seeking and drug-consumption behaviors. According to this theory, drug-induced neuroplasticity renders these brain circuits hypersensitive even after long periods of drug abstinence, which would explain pathological, unmanageable motivation underlying relapse in addicts. In this incentive sensitization view, Pavlovian conditioned incentive motivational processes play an important role, as conditioned stimuli can trigger the activation of sensitized neural pathways, promoting relapse. The incentive sensitization theory proposes a critical dissociation between brain mechanisms underlying the hedonic properties of drugs, or "liking," and those supporting their motivational properties, or "wanting." According to this theory, the motivational attraction-like features that make unconditioned and conditioned stimuli "wanted" undergo long-lasting changes (sensitization) as a consequence of repeated drug exposure and wanting is transformed into craving (88, 89). Whereas previously there was considerable theorizing about changes in hedonic strength of drugs, with repeated administration, this theory suggests no change or even a decrease in strength. This view is compatible with allostatic theories, which view dysregulation of brain hedonic systems as a major component of drug addiction (96).

Conceptually, the hypothesis behind the role of neural sensitization in drug addiction, which is reflected in behavioral measures, incorporates a very inclusive framework. For instance, it explains how an individual with a drug use disorder could show "cross-sensitization" to other drugs or non-drug events, such as sex, food, or gambling; the neurobiology underlying reinforcer-driven motivation is not exclusive for abused drugs, which would explain comorbidity between drug addiction and other bingeing disorders (97–105). Also, it incorporates an explanation about how stress factors can impact and predispose to drug use. In addition to behavioral evidence of cross-sensitization, it has been shown that stress can produce neurochemical sensitization in pathways altered by addictive drugs (50, 64, 66, 106–108). Further, brain stress systems figure prominently in the idea that negative reinforcement drives drug-seeking and thus, relapse, in the abstaining addict (96).

Combined with pathological motivation toward drugs, it is also important to mention that parallel to sensitization-related neuroplasticity, drug-induced adaptations resulting in dysfunctional cortical mechanisms and impairment of executive control have also been shown; this has been related to habit formation and decision-making processes, which might be related to narrowing decision-making about drug use (109, 110). Together, all of these drug effects could explain many aspects of addictive behavior (111–114).

1.4. Pre-clinical Relevance of Behavioral Sensitization

As mentioned above, behavioral sensitization (mainly psychomotor sensitization) has been widely studied over the last two decades in drug research to identify neural changes and other consequences of repeated or chronic drug exposure. A PubMed (http://www.ncbi.nlm.nih.gov/sites/entrez) search (7-August-2010), using the keywords "behavioral sensitization drug abuse" pulled up 900 citations between 1990 and 2010. From a clinical point of view, a critical question has been whether sensitization, specifically incentive sensitization, has a role in the development and maintenance of key features of addictive behavior. In pre-clinical studies, it has been shown that behaviorally sensitized animals exhibit facilitated acquisition of drug self-administration and drug conditioned preference, as well as increased motivation in their efforts to obtain the drug (115–119). Further, the effects of presentation of drug-associated stimuli that were paired with a drug during the acquisition of sensitization have shown that they support incentive motivation on their own and also facilitate learning of new drug-oriented or natural reinforcer-oriented behaviors (120–124). At a neurochemical level, it has been shown that drug sensitization in rats increases specific firing patterns of neurons in mesolimbic structures that code the incentive salience of a reward-related conditioned stimulus (125). In conclusion, a solid body of data indicate that past drug exposure resulting in psychomotor sensitization correlates with increased accumbens DA function and has been seen to support increased drug-seeking and increased motivation toward drugs and drug-associated cues.

1.5. Drug Sensitization in Humans, Can It Be Measured?

1.5.1. Behavioral Sensitization

Robust drug-induced behavioral sensitization has been found in all mammalian species in which this process has been investigated: rodents (mice and rats), dogs, cats, guinea pigs, and non-human primates (31, 62, 126–132). It has also been found in invertebrate species, like *Drosophila* (133, 134). In all of the species in which DA changes in sensitized animals have been assessed (mainly rodents and non-human primates), behavioral sensitization has often been seen to develop parallel to an increased response of the mesolimbic DA system to drug. Increased levels of DA have been measured in the nucleus accumbens after challenge following repeated administration of amphetamines, cocaine, nicotine, ethanol, and morphine (69, 70, 135–140). However, there is at least one report suggesting that sensitization of the mesolimbic DA system may not be a requirement for behavioral sensitization to ethanol (141).

Behavioral and neurochemical sensitization have also been studied in humans (142–163) (*see* **Table 11.1** for a summary of studies). However, drug-induced motor activation has not been the trait of choice in studies of behavior. Rather, self-reports of activation and vigor, as well as experimenter ratings of energy and activation, attention tasks, and eye-blinking tests (*see* **Chapter 1** by Woodruff-Pak for detailed information about eye-blink conditioning in humans and animals) have been used to investigate this process in humans. Although the methods for studying drug-induced behavioral sensitization in humans have not been optimized, compared to those used in animal research (see below), over the last decade an increasing number of studies have demonstrated drug-induced sensitization in humans at both behavioral and neurochemical levels (*see* (164) for a recent review on sensitization in humans).

Table 11.1
Summary of behavioral and neurochemical sensitization studies in humans

References	Drug	Treatment	Withdrawal period[a]	Measured variables	Subjects	Behavioral or neurochemical outcome
Berger et al. (142)	COC	Exposure to COC cues with or without oral 4 mg/kg haloperidol	Unknown	Plasma HVA, ACTH, and cortisol; craving and anxiety measured in response to visual cue exposure	Male COC-dependent inpatients	Increases in anxiety, craving, ACTH, cortisol, and HVA following cue exposure; increases in anxiety, and craving attenuated by haloperidol
Boileau et al. (143)	AMP	On days 1, 3, 5, then after 2 weeks and 1 year, subjects received 0.3 mg/kg AMP or placebo p.o.; PET scans were performed on day 1, 2 weeks and 1 year; all treatments were given in the same context	2 weeks, 1 year	Striatal DA release; behavioral and psychomotor responses	Healthy males without a history of substance abuse	Greater psychomotor response (vigor and eye-blink rates) and increased DA release observed at 2 weeks and 1 year, as compared to day 1
Boileau et al. (144)	AMP	Days 1, 3, and 5 subjects received 0.3 mg/kg AMP p.o.; ~2 weeks later, placebo was administered p.o. in the same context; PET scans were performed on day 1 after AMP and at 2 weeks after placebo designed to look like AMP	2 weeks (for most subjects)	Striatal DA release; behavioral and psychomotor responses	Healthy males without a history of substance abuse	AMP increased DA release in the VS under both AMP and placebo conditions; both AMP and placebo increased subjective ratings of euphoria, etc.
Breier et al. (145)	AMP	Subjects received 0.2 mg/kg AMP i.v.; PET scans were performed during the treatment	NA	Striatal DA release (PET); behavioral and psychomotor responses	Healthy vs. schizophrenic male and females without a history of substance abuse	AMP increased striatal DA in both healthy and schizophrenic subjects, but to a greater extent in schizophrenics

Table 11.1 (continued)

References	Drug	Treatment	Withdrawal period[a]	Measured variables	Subjects	Behavioral or neurochemical outcome
Cox et al. (146)	COC	1 mg/kg intranasal COC or placebo	Unknown	Striatal DA release; behavioral and psychomotor responses	Male and female non-dependent COC users	Number of past COC/AMP experiences was positively correlated with DA increase in VS
Foltin et al. (147)	COC Base (Crack)	Smoked COC twice/day for 3 consecutive days; escalating dose group (12 mg × 1, 25 mg × 1, 50 mg × 4) vs. fixed dose group (50 mg × 6)	2 weeks	Physiological measures (heart rate, blood pressure); subjective drug ratings	Male, non-treatment seeking COC users	Escalating dose group displayed dose-dependent increases in all dependent variables; no changes in fixed group
Johanson and Uhlenhuth (148)	AMP	Alternating p.o. 5 mg AMP vs. placebo over 4 days; drug order was counterbalanced and pill color indicated AMP or placebo; on days 5–9, subjects chose either AMP or placebo; this sequence of events occurred three times in the same subjects	None	Subjective drug effects (POMS, vigor, and elation)	Healthy men and women without a history of substance abuse	AMP increased vigor, elation, arousal, and positive mood similarly across each of the three sessions; however, preference for AMP declined over the three sessions
Kegeles et al. (149)	AMP	Two i.v. administrations of 0.3 mg/kg AMP spaced 16 days apart; SPECT scan was performed during each treatment	16 days	Synaptic DA release	Healthy males without history of stimulant use	AMP elicited a similar within-subjects DA increase and subjective activation across the two sessions (i.e., no sensitization)

Table 11.1 (continued)

Kelly et al. (150)	AMP	15 day in-house study, counterbalanced for drug exposure order; two oral doses per day of 0.14 mg/kg AMP in 10% EtOH for 3 consecutive days; 3 days of twice daily 6% EtOH "placebo" treatment separated a second period of 3 more active days of drug treatment	None	Subjective effects, food intake, accuracy on some work tasks, verbal interaction, and cigarette smoking	Healthy males	AMP decreased food consumption and increased task performance, verbal interaction, and cigarette smoking similarly in both sessions; AMP increased ratings of potency, liking, stimulated anxious, and sedated vs. placebo in the first session; tolerance developed to these effects in the second drug exposure session
Leyton et al. (151)	COC	Intranasal COC 3.0 mg/kg on day 1, and then three doses of 0.6, 1.5, and 3.0 mg/kg COC on each of the following 3 consecutive days. Effects of pre-treatment with an amino acid mixture and of L-dopa/carbidopa on COC effects were also evaluated	None	Subject drug effects; subjective visual COC cue response	Male, non-treatment seeking COC users	COC dose-dependently increased euphoria ratings. L-dopa/carbidopa treatment reduced COC- and COC cue-induced craving. Frequency of past year COC use predicted the cue-induced item "want cocaine;" data for each day were not shown separately, not permitting evaluation of possible sensitization

Table 11.1 (continued)

References	Drug	Treatment	Withdrawal period[a]	Measured variables	Subjects	Behavioral or neurochemical outcome
Martinez et al.-Exp. 1 (152)	AMP	Subjects received 0.3 mg/kg AMP i.v.; PET scans were performed during the treatment	At least 14 days before study initiation	DA release in striatal subregions; vital signs; subjective effects	Male and female healthy controls and COC-dependent users	COC-dependent users had blunted DA responses in the striatum and rated AMP as less euphoric than healthy controls
Martinez et al.-Exp. 2 (152)	COC	Three sample sessions: users self-administered 0, 6, or 12 mg COC. Three choice sessions: users received a priming dose of COC and were given the choice to self-administer 0, 6, or 12 mg COC or receive $5	At least 14 days before study initiation	Subjective effects; number of times COC was chosen over money (from 0 to 5)	COC-dependent users	Users rated 12 mg COC as the most euphoric; users chose the 12 mg dose of COC over money 3.3 out of 5 times; magnitude of decrease in AMP-induced DA response in the striatum in Exp. 1 was inversely correlated with likelihood of choosing COC over money
Nagoshi et al. (153)	COC	Initial test doses were placebo and 10–60 mg/kg COC; retest doses were placebo and 40 mg/kg COC	1–3 weeks	Cardiovascular and subjective responses	Male i.v. COC users	Heightened heart rate for both the placebo and common 40 mg/kg COC dose on retest compared to initial response; similar subjective responses to the 40 mg/kg COC dose were found across time (no sensitization)

Table 11.1 (continued)

Newlin and Thomson (154)	EtOH	Three identical sessions of oral 0.5 g/kg of EtOH followed by a fourth placebo session	None	Heart rate, finger temperature, finger pulse amplitude, skin conductance, and general motor activity	Healthy male college students with or without a parental history of alcoholism	Greater motor activity across sessions in sons of alcoholics; sensitization to finger pulse amplitude; and lack of tolerance to changes in skin conductance and finger temperature in sons of alcoholics
Newlin and Thomson (155)	EtOH	Four identical sessions of oral 0.5 g/kg of EtOH followed by a fifth placebo session	None	Heart rate, pulse transit time, finger temperature, cheek temperature and body sway	Healthy male college students with or without a parental history of alcoholism	Sensitization to pulse transit time and body sway in sons of alcoholics
Rothman et al.(156)	COC	Day 1: i.v. saline or 40 mg/kg COC in two contexts; day 2: i.v. saline or 25 mg/kg COC in "test" context-only	None	Physiological and subjective responses; hormone levels	Healthy male and female subjects with a history of i.v. COC use (at least three times in the month prior to admission)	No conditioned or unconditioned sensitization to COC; acute COC increased cortisol
Sax and Strakowski (157)	AMP	Placebo and 0.25 mg/kg AMP were administered on alternating days so that each subject received three doses of AMP separated by 48-h intervals; order of treatment was randomized	None	Eye-blink rate, mood, motor activity/energy, speech level, subjective drug effects, TPQ	Healthy male and female subjects without a history of alcohol or drug abuse	Greater change in elevated mood between AMP doses 1 and 3 (sensitization) was associated with higher rating of novelty seeking on the TPQ abuse

Table 11.1 (continued)

References	Drug	Treatment	Withdrawal period[a]	Measured variables	Subjects	Behavioral or neuro-chemical outcome
Strakowski and Sax (158)	AMP	Placebo and 0.25 mg/kg AMP were given on alternating days so that each subject received three doses of AMP separated by 48-h intervals; order of treatment was randomized	None	Eye-blink rate; vigor, talking ease; subjective ratings of euphoria, vigor, and drug-liking; TPQ	Healthy males and females without a history of stimulant use or abuse	Progressively increased subjective responses and eye-blinking following repeated AMP administration (sensitization)
Strakowski et al. (159)	AMP	Placebo and 0.25 mg/kg AMP were given on alternating days so that each subject received two doses of AMP separated by a 48-h interval; order of treatment was randomized	None	Eye-blink rate and clinician-rated scales for manic symptoms	Healthy males and females without a history of stimulant use or abuse	Increased eye-blink rate, ratings of energy level, mood and talkativeness following the second AMP administration (sensitization) compared to the first
Strakowski et al. (160)	AMP	Treatments were given on days 1, 3, and 5; subjects were given placebo on all 3 days, 0.25 mg/kg AMP on all 3 days, or placebo on days 1 and 3 then 0.25 mg/kg AMP on day 5	None	Eye-blink; subjective ratings of vigor and euphoria	Healthy males and females without a history of stimulant use	Decreased drug liking in the repeated AMP, compared to single AMP administration group. Sensitized vigor rating in the group that received three administrations of AMP, compared to the group that received one, in females only
Szechtman et al. (161)	APO	12 s.c. injections of 0.0107 mg/kg APO every 2 weeks	None	Yawning, growth hormone, and drug-induced nausea	Healthy males without a history of substance abuse	Sensitization to yawning occurred, shown as shorter latency to onset, and an increase in peak activity

Table 11.1 (continued)

Volkow et al. (162)	COC	PET scan was performed in the presence of radiotracer [^{11}C]raclopride; exposure to neutral or COC cues	None	DA level changes inferred from occupancy of dopamine D2-like receptors; COC craving questionnaire; addiction severity index; COC selectivity assessment scale	Male and female COC-dependent users	Individual differences in cue-induced DA changes correlated with differences in subjective drug craving; high scores on measures of withdrawal symptoms and addiction severity were associated with larger striatal dopamine changes in response to COC cues
Wong et al. (163)	COC	PET scan was performed in the presence of radiotracer [^{11}C]raclopride; exposure to neutral or COC cues	24 h before study initiation	DA level changes inferred from occupancy of dopamine D2-like receptors; subjective visual cue responses	Male and female COC-dependent users	Increased cue-induced craving was associated with increased D2 receptor occupancy in putamen

ACTH: adrenocorticotropin hormone; AMP: d-amphetamine; APO: apomorphine; COC: cocaine; D2: D2-like dopamine receptor; DA: dopamine; DSST: digit-symbol substitution task; EtOH: ethanol (alcohol); HVA: homovanillic acid; i.v.: intravenous; NA: not applicable; p.o.: oral; POMS: profile of mood states; PET: positron emission tomography; s.c.: subcutaneous; SPECT: Single-photon emission computed tomography; TPQ: tridimensional personality questionnaire; VAS: visual analog scales; VS: ventral striatum.

[a]Information in this column refers to that given in the publication regarding the period of time individuals were drug-free prior to drug challenge; in the case of repeated drug exposures, the period given is that between the final drug administration and challenge, if a withdrawal period was used, rather than a series of closely spaced treatments. NA was entered when no information about previous use of the study drug was provided and the drug was administered only once during the study.

Before sensitization-related theories of addiction became acknowledged in the addiction research field, a common assumption was that repeated drug exposure would lead to tolerance to the majority of the behavioral drug effects. However, clinicians noted that some motor responses to psychostimulants did not follow a tolerance-like profile. With amphetamines, for instance, clinicians observed that their repeated abuse induced prolonged increases in stereotypies (repetitive movements) and psychotic-like, paranoia-related episodes (165, 166). More recently, several systematic studies of behavioral sensitization in humans have shown that the repeated intermittent administration of amphetamine can produce persistent potentiation of eye-blink responses, drug-cue-biased attention, and also increases in subjective euphoria (143, 158, 159). There are data, however, indicating that subjective ratings of euphorigenic or "pleasurable" drug effects might not be the most appropriate dependent variable to test sensitized responses (activation or vigor ratings have shown results that are more consistent with sensitization theories). Some studies indicate that drug abusers, in particular psychostimulant abusers, describe increases in drug-seeking behavior even when they show tolerance to the euphorigenic effects of the drug or have decreased drug liking (96). Also, as mentioned before, accumulating evidence suggests that the "pleasurable" or rewarding effects of drugs might not be mediated by DAergic mechanisms (those that undergo enduring drug-induced sensitization). Accumbens DA functions appear to be more closely related to motivation, activation, vigor, and salience of reinforcers and reinforcer-associated cues than to drug-induced positive emotional responses or pleasure (92, 93, 167–169). Indeed, self-reports of sensitized vigor and energy levels have been described in studies with repeated d-amphetamine administration in subjects with past drug exposure, but no history of substance dependence (143, 160); the same results were found when vigor and energy levels were rated by clinicians (158, 159). Interestingly, in some of those studies, sensitized behavioral responses to amphetamine were found in the absence of an increase in drug liking; in fact, self-reported drug liking was either not altered or decreased with repeated drug exposure in these studies (158, 160).

The majority of human studies that have found behavioral sensitization to psychostimulants were conducted in healthy subjects, without a drug abuse history. Although at face value this sounds similar to how animal research is conducted – with drug-naïve subjects – there are important differences. Because, for ethical reasons, it is rarely possible to justify the use of completely drug-naïve subjects in drug exposure research, when subjects were naïve to the particular drug being studied, they were not naïve to other drugs, some of which have been demonstrated to induce cross-sensitization to the study drug. This raises the question of what the "actual" baseline response for each individual subject may be, making determination of the change in response difficult to measure. This is not an issue in animal research in which drug exposure can be completely controlled.

Sensitization has also been studied in individuals with drug use disorders. In long-term cocaine abusers, sensitized subjective effects or physiological responses were reported to be absent (e.g., 156); however, without initial response data, the accuracy of this conclusion cannot be evaluated. Even after a drug-free period, the response of these individuals to the initial drug exposure in these studies may have been sensitized from the history of drug exposure, resulting in no change in response upon drug re-exposure (i.e., a ceiling effect). Also, it is common for these studies to use only two drug exposures to assess the presence or absence of sensitization, which may have been inadequate for demonstration of this phenomenon. Other important factors in human studies examining behavioral sensitization are dose, dosing interval, and time elapsed between the cessation of treatment and test. Animal research shows that for classical psychostimulant drugs like cocaine and amphetamines, intermittent drug administrations of moderate or high doses are optimal for achieving drug-induced behavioral sensitization (5, 127, 170). Similarly, studies in humans using amphetamine have shown that repeated administrations of higher doses (20–30 mg; p.o.) of this drug, but not lower doses (5–10 mg), are required to observe sensitized responses (143, 148, 150). Some animal research has also indicated that an "incubation" period after the cessation of drug treatment may be important for the establishment of long-term central nervous system changes that accompany sensitization (72, 171, 172). These factors are discussed in greater detail in **Section 4.2**. To the best of our knowledge, this factor has not been explored in studies involving human subjects.

1.5.2. Neural Sensitization

Behavioral sensitization in the form of stereotypic behaviors, as an apparent result of chronic pro-DAergic treatment, has been seen in Parkinson's disease patients undergoing DA replacement therapies and described in relation to the similar abnormal involuntary movements (i.e., stereotypies) seen in methamphetamine abusers (173, 174). In fact, the mechanisms of behavioral sensitization during DA replacement therapy have been suggested to be very close to the neurobiology underlying the sensitizing effects of methamphetamine. This homology has been seen not only at a behavioral level, but also when several neurochemical measures are taken into account in human studies (i.e., DA receptor stimulation, changes in transduction pathways, gene expression, and alterations in the phenotype of striatal neurons) (*see* 173, 174).

Evidence for sensitization of amphetamine-induced striatal DA release has been described recently in humans. In 2006, Boileau and colleagues (143) used the radiolabeled tracer [^{11}C]raclopride (a DA D2/D3 antagonist) in a positron emission tomography (PET) study that also included registration of anatomical magnetic resonance imaging (MRI). In this study, subjects without a history of drug abuse received three amphetamine (0.3 mg/kg) administrations, orally, given every other day with a withdrawal period of 2 weeks or 1 year. Then, a fourth amphetamine administration was given in the same context (the PET scan). A decrease in [^{11}C]raclopride binding was interpreted as an increase in dopamine release. A greater decrease in binding was seen after the fourth administration as compared to the first, which was interpreted as sensitization to the DA-releasing effect of amphetamine. This effect was found at the level of the ventral striatum, which includes the nucleus accumbens, as well as the sensorimotor putamen. Corresponding with the neurochemical outcome was sensitization of behavioral variables, including ratings of alertness and energy (143).

A number of studies in humans have failed to find behavioral or neurochemical sensitization (some of these studies are also included in **Table 11.1**). In those in which sensitization was demonstrated, healthy subjects without a history of drug abuse were used, drug administration and tests were context-dependent, and moderate-to-high doses of psychostimulants were used. It is noteworthy that unlike pre-clinical research, to the best of our knowledge, there are no data available for drugs like ethanol or opiates showing neurochemical sensitization in humans; only cocaine and amphetamines have been studied.

1.5.3. The Ideal Human Study?

Because we do not perform research using human subjects, we are not qualified to provide a detailed description of a human study of sensitization, as we do for measurement of the same in mice (*see* **Section 3**). However, we will take this opportunity to express our thoughts about what characteristics we believe to be important in the ideal human study. The first would be that individuals that are relatively drug naïve be used so that a measure of initial drug sensitivity can be obtained. Repeated drug exposure and response measurement in the same drug-associated context would also be part of the design. Another important characteristic would be that objective measures of sensitivity be used. Depending upon the drug, these could include heart rate, skin temperature, eye-blink rate, the use of actimeters for objective measurement of activity, and neural imaging. However, clearly an advantage to using human subjects over animal subjects is their ability to express how they are feeling, and descriptive questionnaires should also be part of the study design. In this case, it would be beneficial to consider questions or objective ratings that specifically target vigor and energy apart from general mood. Also important would be examination of the data for individual variability in initial drug response and change in response. As has been seen in animals, some individuals may be more susceptible to the development of sensitization than others and this susceptibility may or may not be associated with their initial sensitivity. It is important to take this into consideration when drawing conclusions about whether humans did or did not display sensitization in a particular study. Finally, because animal studies have suggested that an incubation period may reveal unique information about drug-induced sensitization, the ideal human study would also include a drug wash-out period followed by a final drug challenge. All of this said, we recognize that because drug-induced sensitization has been shown to be long-lasting and to influence drug self-administration, there are certain risks in performing the ideal study in humans that may be impossible to overcome.

2. Materials

The methods for measuring drug-induced sensitization in animals and humans are quite different and require different materials. In addition, methods for measuring behavioral versus neurochemical sensitization require completely different analytical tools. We will describe herein only the materials and methods relevant to psychomotor sensitization measurement in mice, using injected drugs.

(1) Calibrated 1-ml injection syringes with 0.4-mm, 27-ga hypodermic needles
(2) Vehicle for injection
(3) Drug in the appropriate concentration(s) for injection
(4) Automated activity monitoring equipment – we use 40 × 40 × 30 cm AccuScan monitors (AccuScan Instruments, Columbus, OH)
(5) Ventilated housing chambers for the monitoring equipment to exclude external light and noise
(6) A scale for measuring body weight
(7) Bedding-lined holding cages

3. Methods

(1) On each day, move mice in their home cages to the testing room about 1 h prior to injection or testing and leave

them undisturbed to allow them to acclimate to the test room.

(2) On each test day, prepare one bedding-lined holding cage for each locomotor chamber to be used. Weigh each mouse to be tested in the first test pass and place each in a separate holding cage within 10 min of testing.

(3) On test days 1 and 2, inject each mouse intraperitoneally (i.p.) with vehicle (saline) immediately prior to placement in the activity chamber and test for 5–20 (ethanol or other rapid, short-acting drugs) or 15–60 (methamphetamine or other drugs with longer-duration effects) min; the duration of test can be as long as desired, but data should be collected in relatively short-time intervals (1–5 min) so that the time course can be examined. Data collected on day 1 will provide baseline activity data in a novel environment. Data collected on day 2 will provide baseline activity data in a familiar environment.

(4) Return mice to their home cages after testing on each day and return mice to colony room after all mice have been tested.

(5) Examine day 2 baseline activity data to be certain that groups to be treated differently on subsequent days are well matched for activity level. We recommend that if photocell beam interruptions are used to provide the measure of activity, then a monitoring system that can translate these data into distance traveled be chosen. Arrange mice into treatment groups so that they are matched for baseline activity level, also taking into consideration the important factors in your experiment, such as litter, strain, and sex.

(6) On test day 3, inject one group of mice i.p. with vehicle (Vehicle Control) and the other group i.p. with drug (Drug Group; this is for a single-dose study). Begin testing immediately after injection as on days 1 and 2. Data collected on day 3 will provide a measure of acute stimulation, when compared to the day 2 baseline.

(7) Treatment on subsequent days will depend upon the drug to be tested:

For methamphetamine and cocaine (*see* **Table 11.2**)

(a) On days 4, 6, 8, and 10, leave animals undisturbed in their colony room.

(b) On days 5, 7, and 9, inject with vehicle or the same treatment dose as on day 3, and test as on day 3.

(c) On day 11, inject *all mice* with drug and test. Day 11 data will provide a measure of sensitization by comparing the repeated Vehicle Control group with the

Table 11.2
Summary of test protocol for cocaine- or methamphetamine-induced locomotor sensitization

Day:	1	2	3	4	5	6	7	8	9	10	11	12
Vehicle Control	Veh	Veh	Veh	None	Veh	None	Veh	None	Veh	None	Drug	Veh
Drug Group	Veh	Veh	Drug	None	Drug	None	Drug	None	Drug	None	Drug	Veh
	Test	Test	Test	–	Test	–	Test	–	Test	–	Test	Test

Veh: vehicle treatment; None: no treatment; Drug: cocaine, methamphetamine, or other psychostimulant drug treatment; Test: locomotor test; –: no locomotor testing.

repeated Drug Group. In addition, the increase in drug response on day 11 for the repeated Drug Group, compared to their response on day 3 will provide a within-group measure of sensitization. A blood sample can be obtained on this day after testing to measure blood drug concentration in mice receiving drug for the first time versus those that have received drug repeatedly.

(d) Finally, on day 12, inject *all mice* with vehicle and test. Day 12 data will provide a measure of conditioned activation – mice that have received drug repeatedly paired with the test chamber may exhibit higher levels of activity on this day than those that have received vehicle on most days in the same environment.

(e) If desired, wait for a period of 1–3 weeks and challenge the repeated Drug Group mice with drug to examine whether sensitization continues to be expressed.

Figure 11.1 gives an example of data for a cocaine sensitization study through step (d). Mice in this experiment were adult, male and female mice from a line selectively bred for low levels of voluntary methamphetamine consumption. The dose of cocaine was 10 mg/kg. They were tested for 15 min, with data collected in 5-min time intervals. Note (1) the similarity of the two groups in locomotor activity after saline on days 1 and 2, (2) the acute response to cocaine of the Cocaine Group on day 3, (3) the similarity of the acute cocaine response of the Vehicle Control on day 11 to that of the Cocaine Group on day 3, (4) the gradual increase in cocaine stimulant response of the Cocaine Group from day 3 to day 11 (this is the measure of within-group sensitization), (5) the difference in response to cocaine between the two groups on day 11 (this is the measure of between-groups sensitization), and finally (6) the slightly increased locomotor behavior after saline

Example of Cocaine-Induced Sensitization

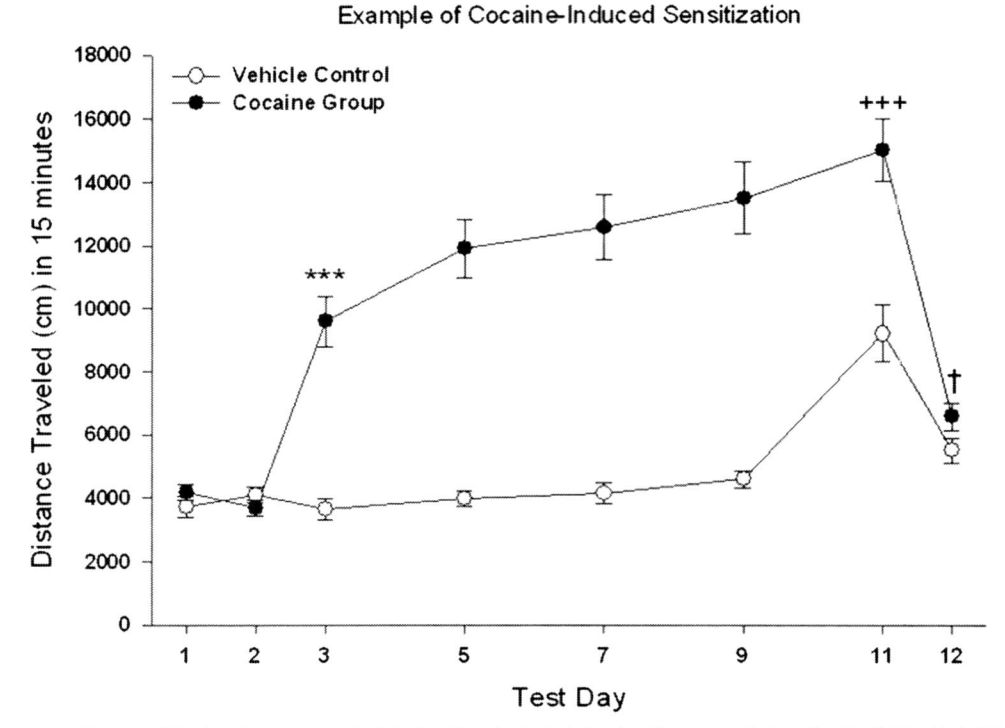

Fig. 11.1. Mean ± SEM for distance traveled during 15-min tests following the protocol described in **Table 11.2**. Vehicle Control mice received 0.9% saline on all days shown except day 11, when they received an i.p. injection of 10 mg/kg cocaine HCl. The Cocaine Group received 0.9% saline on days 1, 2, and 12, and 10 mg/kg cocaine on all other days. Mice were left undisturbed on days 4, 6, 8, and 10. ***$p < 0.001$ for the comparison of the two groups on day 3. +++$p < 0.001$ for the comparison of day 3 to day 11 within the Cocaine Group and for the comparison of the two groups on day 11. †$p = 0.07$ for a statistical trend toward a difference between the two groups on day 12.

on day 12 of the Cocaine Group compared to the Vehicle Control (not quite significant in this case), suggesting some conditioned activation associated with repeated treatment with cocaine in the test environment.

For ethanol (*see* **Table 11.3**)

(a) On days 4, 5, 7, 8, 10, and 11, weigh mice, treat with saline or ethanol (in colony room or in test room), and return them to their home cages – holding cages are not used on these days.

(b) On days 6 and 9, weigh mice and place in holding cages, inject with saline or ethanol, and test as on day 3.

(c) On day 12, inject *all mice* with ethanol and test. Day 12 data will provide measures of between-group and within-group sensitization. A blood sample can be obtained on this day after testing to measure blood ethanol concentration in mice receiving ethanol for the first time versus those that have received ethanol repeatedly.

Table 11.3
Summary of test protocol for ethanol-induced locomotor sensitization

Day:	1	2	3	4	5	6	7	8	9	10	11	12	13
Vehicle Control	Veh	Veh	Veh	Veh	Veh	Veh	Veh	Veh	Veh	Veh	Veh	EtOH	Veh
Drug Group	Veh	Veh	EtOH	EtOH	EtOH	EtOH	EtOH	EtOH	EtOH	EtOH	EtOH	EtOH	Veh
		Test	Test	Test	–	–	Test	–	–	Test	–	–	Test	Test

Veh: vehicle treatment; EtOH: ethanol treatment; Test: locomotor test; –: no locomotor testing.

(d) On day 13, inject *all mice* with vehicle and test. Day 13 data will provide a measure of conditioned activation.

(e) If desired, wait for a period of 1–3 weeks and challenge the repeated Drug Group mice with drug to examine whether sensitization continues to be expressed.

Figure 11.2 gives an example of data from an ethanol sensitization study through step (d). Mice in this experiment were adult, male DBA/2 J strain mice and the dose of ethanol used on treatment and test days was 2 g/kg. Mice were tested for 20 min, with data collected in 5-min time intervals. Note (1) the similarity of the two groups in locomotor activity after saline on days 1 and 2, (2) the acute response to ethanol of the Ethanol Group on day 3, (3) the similarity of the acute ethanol response of the Vehicle Control on day 12 to that of the Ethanol Group on day 3, (4) the increase in ethanol stimulant response of the Ethanol Group from day 3 to day 12 (this is the measure of within-group sensitization), (5) the difference in response to ethanol between the two groups on day 12 (this is the measure of between-groups sensitization), and finally (6) the similarity of the two groups in locomotor behavior after saline on day 13, indicating no conditioned activation in the Ethanol Group in this study.

4. Notes

4.1. Treatment Issues

There is an array of factors to consider when designing an experiment to measure susceptibility to drug-induced sensitization or to identify underlying mechanisms. Because it may be desirable to examine both enhancement and attenuation of sensitization, it may be useful to have methods that produce submaximal as well as maximal levels.

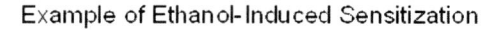

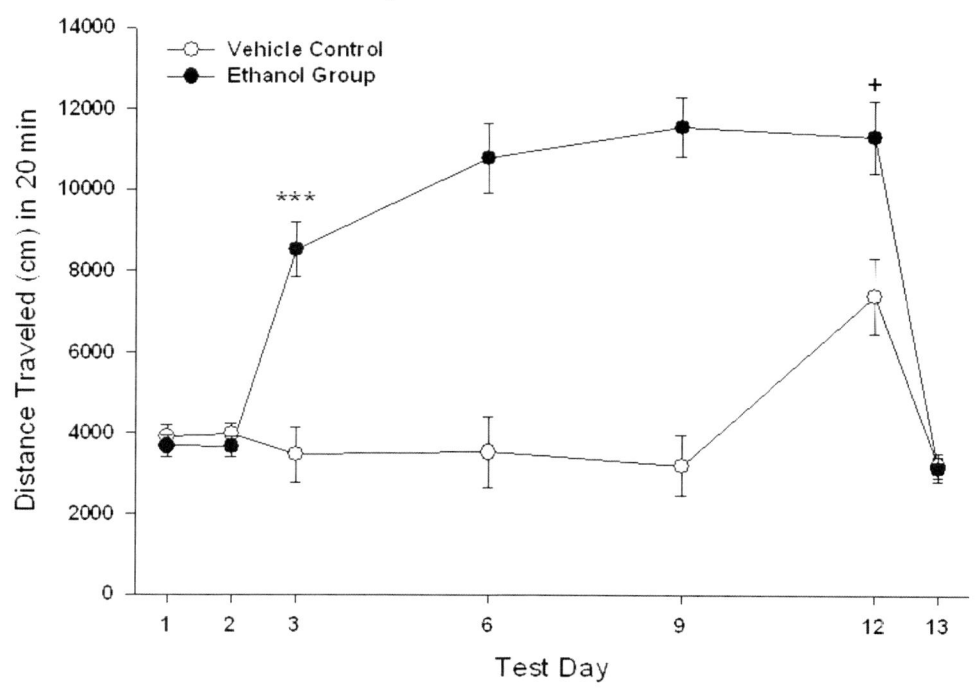

Fig. 11.2. Mean ± SEM for the distance traveled during 20-min tests following the protocol described in **Table 11.3**. Vehicle Control mice received 0.9% saline on all days shown except day 12, when they received an i.p. injection of 2 g/kg ethanol. They received saline injections in their colony room and were then returned to their home cages on intervening days 4, 5, 7, 8, 10, and 11. The Ethanol Group received 0.9% saline on days 1, 2, and 13, and 2 g/kg ethanol on all other days, including the intervening days between test days, when they were treated in the home cage. ***$p < 0.001$ for the comparison of the two groups on day 3. $^+p < 0.05$ for the comparison of day 3 to day 11 within the Ethanol Group and for the comparison of the two groups on day 11.

Factors to consider are dose, number of treatments, frequency or interval between treatments and challenge, and duration of testing. It was shown almost three decades ago that more intermittent treatment schedules produce a greater magnitude of locomotor sensitization to methamphetamine, but that the effect of interval between treatments was dependent upon dose (127); more robust sensitization was seen with longer intervals between treatments for higher methamphetamine doses. In a study that extended examination of treatment interval to other drugs of abuse, a treatment interval of 24 h or longer was needed to induce locomotor sensitization to methamphetamine, cocaine, and morphine (170). The same treatment schedule may produce different behavioral responses, even for drugs with similar mechanisms of action, like cocaine and methamphetamine (e.g., 175). These factors create some lack of clarity with regard to choosing a dose and treatment interval. However, for classical stimulant drugs like cocaine and amphetamines, it seems to be advisable for the induction of sensitization to utilize an intermittent treatment schedule in which drug is allowed to fully clear between treatments, rather than a more chronic treatment schedule (176).

For ethanol, we have found more reliable and robust sensitization with a 24-h treatment interval; however, both daily (38, 59, 74, 177) and less frequent (56, 178) treatments have been shown to induce sensitization to ethanol. A systematic study showed that in Swiss mice, treatment intervals of 24, 48, and 96 h induced similar degrees of ethanol-induced sensitization (179). However, it is currently unknown whether there are genotype-dependent effects of treatment interval, because this has not been systematically studied. On the other hand, dose of ethanol has been clearly shown to play an important role in both the development and expression of sensitization (179, 180). In our experience, sensitization to ethanol can be reliably induced using a 24-h treatment interval, with ethanol doses of 1.5 g/kg or greater, in strains that are susceptible to ethanol-induced sensitization. Higher doses for the induction of sensitization are advisable in less sensitive genotypes (e.g., 180). Sensitization can be induced by three or fewer exposures to ethanol in some mouse strains, but may require a larger number of exposures in others (181–183). Also, for ethanol, we have found that at least the sensitization-resistant C57BL/6 J strain is more likely to show sensitization using a repeated injection procedure that does not include test environment exposure during the treatment phase (182).

Pharmacokinetic/pharmacodynamic factors likely provide a partial explanation for the influence of treatment interval on magnitude of sensitization. Longer treatment intervals may be required for drugs like methamphetamine that have a longer half-life if a drug-free period between administrations is important in the sensitization process. In fact, at least for ethanol, sensitization has been described as a kindling-like process that could influence craving (184) and may require recurrent cycles of exposure, drug clearance, and withdrawal to fully develop (185). Repeated cycles of chronic ethanol exposure with intervening drug-free periods have been shown to result in increased ethanol intake compared to levels seen after a single episode of chronic exposure (186, 187). An escalating dose, binge-like model of cocaine administration has been advocated for studying sensitization, as an exposure model that might better reflect escalating drug use (188). This may be a good choice, depending upon the goal of the drug administration model. In a study of non-treatment-seeking cocaine users, the effects of escalating doses of smoked cocaine were studied. Sensitization of heart rate, blood pressure, positive drug effect ratings, and cocaine liking were found in the escalating dose group, but not in the fixed dose group (147).

4.2. Test Issues

Test frequency and duration should also be considered when designing a sensitization study. Tests after every drug administration or every few administrations allow the acquisition pattern to be tracked. However, acquisition pattern may vary from experiment to experiment even when identical procedures are used with the same type of mouse in the same laboratory. For example, results for two groups of CFW mice tested for the acquisition of ethanol-induced sensitization were presented in a single paper (189). In both groups, robust sensitization was demonstrated. However, in one group, maximum sensitization was present on the second test day, which was after the fourth ethanol treatment, whereas in the other group, a progressive increase in sensitization was evident across two additional tests, after the seventh and tenth ethanol treatments. One possible explanation for the differences in this case is individual differences in susceptibility to sensitization. CFW mice are a genetically heterogeneous stock and there could be genotype-dependent differences among individual animals that would be revealed in mean response differences. Another possible source of variation is environmental. Although the data for the two experiments were collected in a single laboratory, using common equipment and methods, it is not known if the same person collected both data sets or during what season of the year the data were collected and what impact such variables may have had. Others have shown that one cannot completely control environmental factors that may influence experimental results (190, 191), even when using genetically identical individuals (i.e., inbred strains).

Duration of the behavioral test should take into account the pharmacodynamics of the drug which affect the duration of the behavioral response. For example, for ethanol the behavioral stimulant effects are rapid and relatively short-lived (e.g., 87, 192, 193), so a shorter test duration (~15 min) may be appropriate. For methamphetamine, the stimulant effects occur rapidly after administration, but last longer (194, 195); thus, a somewhat longer test duration (~60 min) may be desirable if the goal is to measure behavior for the entire duration of the drug response; however, sensitization can clearly be detected during earlier time points for drugs with longer durations of action (e.g., 2, 196, 197), so shorter test periods may also suit the goals of the research. Clearly, time is an important factor to include in the analysis of sensitization data. If data are accumulated into a single, long test period, but effects occur and subside rapidly, these transient effects may be difficult to detect without consideration of time-dependent patterns of response.

For the measurement of expression of sensitization, dose, amount of time elapsed since the last drug treatment, and environmental factors must be considered. Most commonly, the dose used for pretreatment has also been used as the challenge dose when testing for the expression of sensitization. However, it may be desirable to treat with a higher dose and then test with a drug dose appropriate for measurement of psychomotor stimulation. Higher doses of some drugs in some genotypes may be more efficacious for inducing sensitization (although this has not been systematically studied), but would be behaviorally debilitating in locomotor challenge tests. Of course, if one wanted to examine the role of pre-treatment dose, the ideal design might be to treat different groups of animals with multiple doses and then test all with a common dose (e.g., 180). Some animal research has also indicated that an incubation period after the cessation of drug treatment may be important for the establishment of long-term central nervous system changes that accompany sensitization. However, an early examination of dopamine responses in comparison to behavioral sensitization to cocaine showed that there was no perfect correspondence between augmented behavior and augmented extracellular dopamine in response to cocaine challenge; effects were dependent upon dose and the time after cocaine treatment and withdrawal that the measurements were taken (198). This and other studies have shown that the augmented behavioral response can be seen both during the repeated treatment phase and upon challenge at various times after cessation of treatment. Therefore, the goals of the research may dictate the particular choice of whether an incubation period is included in the research design or not. For example, if one is interested in recording genotype-dependent differences in susceptibility to drug-induced sensitization, an incubation period may not be useful. However, if neurochemical or genetic changes associated with persistent sensitization-related changes are the goal of the investigation, an incubation period or even a time course after drug withdrawal may be an important facet of the design.

4.3. Environmental Factors

One environmental factor pertains specifically to the sensitization methods described in **Section 3**. As an alternative to using holding cages, mice can be weighed and returned to their home cages prior to treatment or weighed one at a time just prior to injection as they are placed immediately into an activity monitor. The use of holding cages allows for all injection syringes to be prepared for a set of mice, and mice to be placed into activity monitors within a short period of time so that testing is better synchronized. We have performed ethanol sensitization studies both with and without holding cages and have not found systematic effects of this short change in environment. However, we have not systematically studied the question of holding cage use.

Another environmental issue is the influence of the environmental cues that are associated with drug effects by virtue of being paired with drug treatment. Such cues have been shown to play an important role in the expression of sensitization (199–201). However, not all drug-induced sensitization is context-dependent; at least for ethanol, sensitization can be expressed in a novel environment after home cage treatment (38, 66, 176, 202). To demonstrate context-dependent sensitization, it has been shown that when animals consistently receive drug in association with a particular environment, the magnitude of sensitization expressed will be greater when testing occurs in the drug-associated environment, than when tested in another environment (203–205), although the affective nature of the environment can also play a role in the outcome (206). Likewise, neurochemical sensitization may not be fully expressed in an alternative environment (203). In addition, animals tested after vehicle treatment in the drug-paired environment will sometimes show "conditioned activation" – a response elevated above that of similarly treated animals that have not received the drug–environment pairing (e.g., 202). As they have in animal studies, in our opinion, associative cues need to be taken carefully into consideration in human studies of drug-induced sensitization. Described in **Section 1** was a study in which both behavioral and neurochemical sensitization to amphetamine were seen in men without a history of drug abuse (143). One potentially critical characteristic of that study was that each drug administration was paired with the same particular context and was consistent with the context used for the challenge tests. In other work by this research group, drug cues (in this case the drug context plus administration of a placebo pill) were also able to elicit a sensitized DA response in humans (144).

4.4. The Role of Stress in Drug-Induced Sensitization

Drugs that induce sensitization have been shown to mimic the effects of stressors by altering levels of HPA axis-associated peptides or receptor levels (4, 207–210). Activation of the HPA or stress axis may be critical to the development of sensitization (50, 211, 212). Therefore, when conducting studies of drug-induced sensitization, it is important to consider the potential stressors that may be present as part of the research design and could influence the results. For example, an effect of cage crowding on the magnitude of sensitization to ethanol may have been due to effects of stress (213). Also, even a seemingly simple change in design, such as a single exposure to the test environment prior to being tested for sensitization with a drug challenge, can have a profound effect on magnitude of sensitization (214); in one case it reduced magnitude of sensitization, suggesting that novelty of the environment is an important component in some sensitization-induction or expression procedures. In addition to possible effects of stress axis activation on drug response, the persistent neuroadaptations underlying drug-induced sensitization have been proposed to increase sensitivity to stressors, more readily activating mechanisms that lead to relapse (215, 216). Consistency of the technician handling the mice throughout the experiment (189) is likely an important consideration with regard to the role of stress in the development of drug-induced sensitization.

5. Summary

Behavioral and neurochemical sensitization are relevant to substance use disorders and play a particularly important oppositional role in the maintenance of sobriety. Methods for inducing sensitization and measuring susceptibility have been developed, and different methods are differentially influenced by stress and environmental factors. Consideration should be given not only to drug dose, number of treatments, treatment interval, and interval between sensitization induction and measurement of expression, but also to specifics of handling and characteristics of the environment in which animals are treated and tested. Translation of results from animal studies to human subjects is complicated by ethical issues associated with administering drugs to relatively drug-naïve individuals, which is necessary to obtain an accurate initial response measure, and also because the measurement of sensitization requires that drug be administered more than one time. However, there is evidence from human studies that drug-induced sensitization can be measured using behavioral rating scales and imaging tools and that the neural processes that support sensitization may play important roles in addiction-related processes, particularly relapse.

Acknowledgments

Supported by the Department of Veterans Affairs, NIDA grant P50 DA018165, and NIAAA grants P60 AA010760, U01 AA016655, and F31 AA018043.

References

1. Valjent E, Bertran-Gonzalez J, Aubier B, Greengard P, Hervé D, Girault JA (2010) Mechanisms of locomotor sensitization to drugs of abuse in a two-injection protocol. Neuropsychopharmacology 35:401–415

2. Phillips TJ, Dickinson S, Burkhart-Kasch S (1994) Behavioral sensitization to drug stimulant effects in C57BL/6 J and DBA/2 J inbred mice. Behav Neurosci 108: 789–803

3. Phillips TJ, Huson M, Gwiazdon C, Burkhart-Kasch S, Shen EH (1995) Effects of acute and repeated ethanol exposures on the locomotor activity of BXD recombinant inbred mice. Alcohol Clin Exp Res 19:269–278
4. Vanderschuren LJ, Schmidt ED, De Vries TJ, Van Moorsel CA, Tilders FJ, Schoffelmeer AN (1999) A single exposure to amphetamine is sufficient to induce long-term behavioral, neuroendocrine, and neurochemical sensitization in rats. J Neurosci 19:9579–9586
5. Paulson PE, Camp DM, Robinson TE (1991) Time course of transient behavioral depression and persistent behavioral sensitization in relation to regional brain monoamine concentrations during amphetamine withdrawal in rats. Psychopharmacology 103:480–492
6. Blum K, Chen TJ, Downs BW, Bowirrat A, Waite RL, Braverman ER, Madigan M, Oscar-Berman M, DiNubile N, Stice E, Giordano J, Morse S, Gold M (2009) Neurogenetics of dopaminergic receptor supersensitivity in activation of brain reward circuitry and relapse: proposing "deprivation-amplification relapse therapy" (DART). Postgrad Med 121:176–196
7. Chen JC, Chen PC, Chiang YC (2009) Molecular mechanisms of psychostimulant addiction. Chang Gung Med J 32:148–154
8. De Vries TJ, Schoffelmeer AN, Binnekade R, Mulder AH, Vanderschuren LJ (1998) Drug-induced reinstatement of heroin- and cocaine-seeking behaviour following long-term extinction is associated with expression of behavioural sensitization. Eur J Neurosci 10:3565–3571
9. De Vries TJ, Schoffelmeer AN, Binnekade R, Raasø H, Vanderschuren LJ (2002) Relapse to cocaine- and heroin-seeking behavior mediated by dopamine D2 receptors is time-dependent and associated with behavioral sensitization. Neuropsychopharmacology 26:18–26
10. Kalivas PW, Pierce RC, Cornish J, Sorg BA (1998) A role for sensitization in craving and relapse in cocaine addiction. J Psychopharmacol 12:49–53
11. Niwa M, Yan Y, Nabeshima T (2008) Genes and molecules that can potentiate or attenuate psychostimulant dependence: relevance of data from animal models to human addiction. Ann NY Acad Sci 1141:76–95
12. Thomas MJ, Kalivas PW, Shaham Y (2008) Neuroplasticity in the mesolimbic dopamine system and cocaine addiction. Br J Pharmacol 154:327–342
13. Kandel ER (2000) Cellular mechanisms of learning and the biological basis of individuality. In: Kandel ER, Schwartz JH, Jessell TM (eds) Principles of neural science, 4th edn. McGraw-Hill, San Francisco, CA, pp 1247–1279
14. Kandel ER, Schwartz JH (1982) Molecular biology of learning: modulation of transmitter release. Science 218:433–443
15. Pinsker HM, Hening WA, Carew TJ, Kandel ER (1973) Long-term sensitization of a defensive withdrawal reflex in Aplysia. Science 182:1039–1042
16. Bailey CH, Chen M (1983) Morphological basis of long-term habituation and sensitization in Aplysia. Science 220:91–93
17. Bliss TV, Lomo T (1973) Long-lasting potentiation of synaptic transmission in the dentate area of the anaesthetized rabbit following stimulation of the perforant path. J Physiol 232:331–356
18. Kandel ER, Spencer WA (1968) Cellular neurophysiological approaches in the study of learning. Physiol Rev 48:65–134
19. Feltenstein MW, See RE (2007) NMDA receptor blockade in the basolateral amygdala disrupts consolidation of stimulus-reward memory and extinction learning during reinstatement of cocaine-seeking in an animal model of relapse. Neurobiol Learn Mem 88:435–444
20. Hyman SE, Malenka RC, Nestler EJ (2006) Neural mechanisms of addiction: the role of reward-related learning and memory. Annu Rev Neurosci 29:565–598
21. Lee JL, Di Ciano P, Thomas KL, Everitt BJ (2005) Disrupting reconsolidation of drug memories reduces cocaine-seeking behavior. Neuron 47:795–801
22. Robbins TW, Ersche KD, Everitt BJ (2008) Drug addiction and the memory systems of the brain. Ann NY Acad Sci 1141:1–21
23. Russo SJ, Mazei-Robison MS, Ables JL, Nestler EJ (2009) Neurotrophic factors and structural plasticity in addiction. Neuropharmacology 56:73–82
24. von der Goltz C, Vengeliene V, Bilbao A, Perreau-Lenz S, Pawlak CR, Kiefer F, Spanagel R (2009) Cue-induced alcohol-seeking behaviour is reduced by disrupting the reconsolidation of alcohol-related memories. Psychopharmacology (Berl) 205:389–397
25. Karler R, Calder LD, Chaudhry IA, Turkanis SA (1989) Blockade of "reverse tolerance" to cocaine and amphetamine by MK-801. Life Sci 45:599–606
26. Davis S, Butcher SP, Morris RG (1992) The NMDA receptor antagonist D-2-amino-5-

phosphonopentanoate (D-AP5) impairs spatial learning and LTP in vivo at intracerebral concentrations comparable to those that block LTP in vitro. J Neurosci 12:21–34
27. Vanderschuren LJ, Kalivas PW (2000) Alterations in dopaminergic and glutamatergic transmission in the induction and expression of behavioral sensitization: a critical review of preclinical studies. Psychopharmacology 151:99–120
28. Wolf ME (1998) The role of excitatory amino acids in behavioral sensitization to psychomotor stimulants. Prog Neurobiol 54:679–720
29. Wise RA, Bozarth MA (1987) A psychomotor stimulant theory of addiction. Psych Rev 94:469–492
30. Post RM, Rose H (1976) Increasing effects of repetitive cocaine administration in the rat. Nature 260:731–732
31. Segal DS, Mandell AJ (1974) Long-term administration of *d*-amphetamine: progressive augmentation of motor activity and stereotypy. Pharmacol Biochem Behav 2:249–255
32. Boutrel B, de Lecea L (2008) Addiction and arousal: the hypocretin connection. Physiol Behav 93:947–951
33. Haile CN, Hiroi N, Nestler EJ, Kosten TA (2001) Differential behavioral responses to cocaine are associated with dynamics of mesolimbic dopamine proteins in Lewis and Fischer 344 rats. Synapse 41: 179–190
34. Phillips TJ (1997) Behavior genetics of drug sensitization. Crit Rev Neurobiol 11:21–33
35. Segal DS, Weinberger SB, Cahill J, McCunney SJ (1980) Multiple daily amphetamine administration: behavioral and neurochemical alterations. Science 207:904–907
36. Yang PB, Swann AC, Dafny N (2006) Chronic methylphenidate modulates locomotor activity and sensory evoked responses in the VTA and NAc of freely behaving rats. Neuropharmacology 51:546–556
37. Wills TA, Knapp DJ, Overstreet DH, Breese GR (2009) Sensitization, duration, and pharmacological blockade of anxiety-like behavior following repeated ethanol withdrawal in adolescent and adult rats. Alcohol Clin Exp Res 33:455–463
38. Pastor R, McKinnon CS, Scibelli AC, Burkhart-Kasch S, Reed C, Ryabinin AE, Coste SC, Stenzel-Poore MP, Phillips TJ (2008) Corticotropin-releasing factor 1 receptor involvement in behavioral neuroadaptation to ethanol: a urocortin1-independent mechanism. Proc Natl Acad Sci USA 105:9070–9075
39. Pierce RC, Kalivas PW (1997) A circuitry model of the expression of behavioral sensitization to amphetamine-like psychostimulants. Brain Res Rev 25:192–216
40. Robinson TE, Becker JB (1986) Enduring changes in brain and behavior produced by chronic amphetamine administration: a review and evaluation of animal models of amphetamine psychosis. Brain Res 396:157–198
41. Stewart J, Badiani A (1993) Tolerance and sensitization to the behavioral effects of drugs. Behav Pharmacol 4:289–312
42. Boehm SL II, Goldfarb KJ, Serio KM, Moore EM, Linsenbardt DN (2008) Does context influence the duration of locomotor sensitization to ethanol in female DBA/2 J mice? Psychopharmacology 197:191–201
43. Castner SA, Goldman-Rakic PS (1999) Long-lasting psychotomimetic consequences of repeated low-dose amphetamine exposure in rhesus monkeys. Neuropsychopharmacology 20:10–28
44. Lessov CN, Phillips TJ (1998) Duration of sensitization to the locomotor stimulant effects of ethanol in mice. Psychopharmacology (Berl) 135:374–382
45. Paulson PE, Robinson TE (1995) Amphetamine-induced time-dependent sensitization of dopamine neurotransmission in the dorsal and ventral striatum: a microdialysis study in behaving rats. Synapse 19:56–65
46. Paulson PE, Camp DM, Robinson TE (1991) Time course of transient behavioral depression and persistent behavioral sensitization in relation to regional brain monoamine concentrations during amphetamine withdrawal in rats. Psychopharmacology (Berl) 103:480–492
47. Brodkin ES, Kosten TA, Haile CN, Heninger GR, Carlezon WA Jr, Jatlow P, Remmers EF, Wilder RL, Nestler EJ (1999) Dark Agouti and Fischer 344 rats: differential behavioral responses to morphine and biochemical differences in the ventral tegmental area. Neuroscience 88:1307–1315
48. Correa M, Sanchis-Segura C, Pastor R, Aragon CM (2004) Ethanol intake and motor sensitization: the role of brain catalase activity in mice with different genotypes. Physiol Behav 82:231–240
49. Orsini C, Bonito-Oliva A, Conversi D, Cabib S (2005) Susceptibility to conditioned place preference induced by addictive drugs in mice of the C57BL/6 and DBA/2 inbred strains. Psychopharmacology (Berl) 181:327–336
50. Phillips TJ, Roberts AJ, Lessov CN (1997) Behavioral sensitization to ethanol: genetics

and the effects of stress. Pharmacol Biochem Behav 57:487–493
51. Phillips TJ, Huson MG, McKinnon CS (1998) Localization of genes mediating acute and sensitized locomotor responses to cocaine in BXD/Ty recombinant inbred mice. J Neurosci 18:3023–3034
52. Szumlinski KK, Lominac KD, Frys KA, Middaugh LD (2005) Genetic variation in heroin-induced changes in behaviour: effects of B6 strain dose on conditioned reward and locomotor sensitization in 129-B6 hybrid mice. Genes Brain Behav 4: 324–336
53. Tolliver BK, Belknap JK, Woods WE, Carney JM (1994) Genetic analysis of sensitization and tolerance to cocaine. J Pharmacol Exp Ther 270:1230–1238
54. Yang PB, Amini B, Swann AC, Dafny N (2003) Strain differences in the behavioral responses of male rats to chronically administered methylphenidate. Brain Res 971:139–152
55. Balda MA, Anderson KL, Itzhak Y (2009) Development and persistence of long-lasting behavioral sensitization to cocaine in female mice: role of the *nNOS* gene. Neuropharmacology 56:709–715
56. Morice E, Denis C, Giros B, Nosten-Bertrand M (2010) Evidence of long-term expression of behavioral sensitization to both cocaine and ethanol in dopamine transporter knockout mice. Psychopharmacology 208:57–66
57. Phillips TJ, Kamens HM, Wheeler JM (2008) Behavioral genetic contributions to the study of addiction-related amphetamine effects. Neurosci Biobehav Rev 32:707–759
58. Schmidt LS, Miller AD, Lester DB, Bay-Richter C, Schülein C, Frikke-Schmidt H, Wess J, Blaha CD, Woldbye DP, Fink-Jensen A, Wortwein G (2010) Increased amphetamine-induced locomotor activity, sensitization, and accumbal dopamine release in M(5) muscarinic receptor knockout mice. Psychopharmacology (Berl) 207: 547–558
59. Sharpe AL, Low MJ (2009) Proopiomelanocortin peptides are not essential for development of ethanol-induced behavioral sensitization. Alcohol Clin Exp Res 33:1202–1207
60. de Jong IE, Steenbergen PJ, de Kloet ER (2009) Behavioral sensitization to cocaine: cooperation between glucocorticoids and epinephrine. Psychopharmacology 204:693–703
61. Marinelli M, Piazza PV (2002) Interaction between glucocorticoid hormones, stress and psychostimulant drugs. Eur J Neurosci 16:387–394
62. Antelman SM, Eichler AJ, Black CA, Kocan D (1980) Interchangeability of stress and amphetamine in sensitization. Science 207:329–331
63. Diaz-Otanez CS, Capriles ND, Cancela LM (1997) D1 and D2 dopamine and opiate receptors are involved in the restraint-stress induced sensitization to the psychostimulant effects of amphetamine. Pharmacol Biochem Behav 58:9–14
64. Goeders NE (2002) Stress and cocaine addiction. J Pharmacol Exp Ther 301:785–789
65. Lepsch LB, Gonzalo LA, Magro FJ, Delucia R, Scavone C, Planeta CS (2005) Exposure to chronic stress increases the locomotor response to cocaine and the basal levels of corticosterone in adolescent rats. Addict Biol 10:251–256
66. Roberts AJ, Lessov CN, Phillips TJ (1995) Critical role for glucocorticoid receptors in stress- and ethanol-induced locomotor sensitization. J Pharmacol Exp Ther 275:790–797
67. Davidson C, Lee TH, Ellinwood EH (2005) Acute and chronic continuous methamphetamine have different long-term behavioral and neurochemical consequences. Neurochem Int 46:189–203
68. Chefer VI, Shippenberg TS (2002) Changes in basal and cocaine-evoked extracellular dopamine uptake and release in the rat nucleus accumbens during early abstinence from cocaine: quantitative determination under transient conditions. Neuroscience 112:907–919
69. Ding ZM, Rodd ZA, Engleman EA, McBride WJ (2009) Sensitization of ventral tegmental area dopamine neurons to the stimulating effects of ethanol. Alcohol Clin Exp Res 33:1571–1581
70. Domino EF, Tsukada H (2009) Nicotine sensitization of monkey striatal dopamine release. Eur J Pharmacol 607:91–95
71. Janowsky AJ, Mah C, Johnson RA, Cunningham CL, Phillips TJ, Crabbe JC, Eshleman AJ, Belknap JK (2001) Mapping genes that regulate density of dopamine transporters and correlated behaviors in recombinant inbred mice. J Pharma Exp Ther 298:634–643
72. Borgkvist A, Valjent E, Santini E, Herve D, Girault JA, Fisone G (2008) Delayed, context- and dopamine D1 receptor-dependent activation of ERK in morphine-sensitized mice. Neuropharmacol 55: 230–237
73. Broadbent J, Kampmueller KM, Koonse SA (2003) Expression of behavioral sensitization

to ethanol by DBA/2 J mice: the role of NMDA and non-NMDA glutamate receptors. Psychopharmacology (Berl) 167:225–234
74. Meyer PJ, Phillips TJ (2003) Bivalent effects of MK-801 on ethanol-induced sensitization do not parallel its effects on ethanol-induced tolerance. Behav Neurosci 117:641–649
75. Sofuoglu M, Sewell RA (2009) Norepinephrine and stimulant addiction. Addict Biol 14:119–129
76. Kotlinska J, Bochenski M (2009) Pretreatment with group I metabotropic glutamate receptors antagonists attenuates lethality induced by acute cocaine overdose and expression of sensitization to hyperlocomotor effect of cocaine in mice. Neurotox Res (Published Online Nov 21, 2009). doi:10.1007/s12640-009-9136-8
77. Szumlinski KK, Abernathy KE, Oleson EB, Klugmann M, Lominac KD, He DY, Ron D, During M, Kalivas PW (2006) Homer isoforms differentially regulate cocaine-induced neuroplasticity. Neuropsychopharmacology 31:768–777
78. Zweifel LS, Argilli E, Bonci A, Palmiter RD (2008) Role of NMDA receptors in dopamine neurons for plasticity and addictive behaviors. Neuron 59:486–496
79. Bahi A, Boyer F, Chandrasekar V, Dreyer JL (2008) Role of accumbens BDNF and TrkB in cocaine-induced psychomotor sensitization, conditioned-place preference, and reinstatement in rats. Psychopharmacology 199:169–182
80. Narendran R, Martinez D (2008) Cocaine abuse and sensitization of striatal dopamine transmission: a critical review of the preclinical and clinical imaging literature. Synapse 62:851–869
81. Holstein SE, Dobbs L, Phillips TJ (2009) Attenuation of the stimulant response to ethanol is associated with enhanced ataxia for a GABA, but not a GABA, receptor agonist. Alcohol Clin Exp Res 33:108–120
82. Meyer PJ, Phillips TJ (2003) Sensitivity to ketamine, alone or in combination with ethanol, is altered in mice selectively bred for sensitivity to ethanol's locomotor effects. Alcohol Clin Exp Res 27:1701–1709
83. Davidson C, Lazarus C, Xiong X, Lee TH, Ellinwood EH (2002) 5-HT2 receptor antagonists given in the acute withdrawal from daily cocaine injections can reverse established sensitization. Eur J Pharmacol 453:255–263
84. Doetschman T (2009) Influence of genetic background on genetically engineered mouse phenotypes. Methods Mol Biol 530:423–433
85. Eisener-Dorman AF, Lawrence DA, Bolivar VJ (2009) Cautionary insights on knockout mouse studies: the gene or not the gene? Brain Behav Immun 23:318–324
86. Gerlai R (1996) Gene targeting in neuroscience: the systemic approach. Trends Neurosci 19:188–189
87. Harrison SJ, Nobrega JN (2009) A functional role for the dopamine D3 receptor in the induction and expression of behavioural sensitization to ethanol in mice. Psychopharmacology 207:47–56
88. Robinson TE, Berridge KC (1993) The neural basis of drug craving: an incentive-sensitization theory of addiction. Brain Res Rev 18:247–291
89. Robinson TE, Berridge KC (2008) The incentive sensitization theory of addiction: some current issues. Philos Trans R Soc Lond B Biol Sci 363:3137–3146
90. Carelli RM (2004) Nucleus accumbens cell firing and rapid dopamine signaling during goal-directed behaviors in rats. Neuropharmacology 47:180–189
91. Robinson TE, Berridge KC (2003) Addiction. Annu Rev Psychol 54:25–53
92. Salamone JD, Correa M, Mingote SM, Weber SM (2005) Beyond the reward hypothesis: alternative functions of nucleus accumbens dopamine. Curr Opin Pharmacol 5:34–41
93. Schultz W (2007) Multiple dopamine functions at different time courses. Annu Rev Neurosci 30:259–288
94. Wise RA (2008) Dopamine and reward: the anhedonia hypothesis 30 years on. Neurotox Res 14:169–183
95. Berridge KC (2007) The debate over dopamine's role in reward: the case for incentive salience. Psychopharmacology 191:391–431
96. Koob GF (2009) Dynamics of neuronal circuits in addiction: reward, antireward, and emotional memory. Pharmacopsychiatry 42:S32–S41
97. Blaszczynski A, Nower LA (2002) Pathways model of problem and pathological gambling. Addiction 97:487–499
98. Carlezon WA Jr, Thomas MJ (2009) Biological substrates of reward and aversion: a nucleus accumbens activity hypothesis. Neuropharmacology 56:122–132
99. Davis C, Carter JC (2009) Compulsive overeating as an addiction disorder. A review of theory and evidence. Appetite 53:1–8

100. Fenu S, Wardas J, Morelli M (2009) Impulse control disorders and dopamine dysregulation syndrome associated with dopamine agonist therapy in Parkinson's disease. Behav Pharmacol 20:363–379
101. Fiorino DF, Phillips AG (1999) Facilitation of sexual behavior and enhanced dopamine efflux in the nucleus accumbens of male rats after d-amphetamine-induced behavioral sensitization. J Neurosci 19:456–463
102. Mathes WF, Brownley KA, Mo X, Bulik CM (2009) The biology of binge eating. Appetite 52:545–553
103. Roitman MF, Na E, Anderson G, Jones TA, Bernstein IL (2002) Induction of a salt appetite alters dendritic morphology in nucleus accumbens and sensitizes rats to amphetamine. J Neurosci 22:RC225
104. Zack M, Poulos CX (2004) Amphetamine primes motivation to gamble and gambling-related semantic networks in problem gamblers. Neuropsychopharmacology 29:195–207
105. Zack M, Poulos CX (2009) Parallel roles for dopamine in pathological gambling and psychostimulant addiction. Curr Drug Abuse Rev 2:11–25
106. Abercrombie ED, Keefe KA, Di Frischia DS, Zigmend MJ (1989) Differential effect of stress on in vivo dopamine release in striatum, nucleus accumbens and medial frontal cortex. J Neurochem 52:1655–1658
107. Kalivas PW, Stewart J (1991) Dopamine transmission in the initiation and expression of drug- and stress-induced sensitization of motor activity. Brain Res Rev 16:223–244
108. Thierry AM, Tassin JP, Blanc G, Glowinski J (1976) Selective activation of the mesocortical DA system by stress. Nature 263:242–243
109. Everitt BJ, Dickinson A, Robbins TW (2001) The neuropsychological basis of addictive behaviour. Brain Res Brain Res Rev 36:129–138
110. Hyman SE (2005) Addiction: a disease of learning and memory. Am J Psychiatry 162:1414–1422
111. Bechara A, Dolan S, Hindes A (2002) Decision-making and addiction (part II): myopia for the future or hypersensitivity to reward? Neuropsychologia 40:1690–1705
112. Jentsch JD, Taylor JR (1999) Impulsivity resulting from frontostriatal dysfunction in drug abuse: implications for the control of behavior by reward-related stimuli. Psychopharmacology (Berl) 146:373–390
113. Rogers RD, Robbins TW (2001) Investigating the neurocognitive deficits associated with chronic drug misuse. Curr Opin Neurobiol 11:250–257
114. Schoenbaum G, Shaham Y (2008) The role of orbitofrontal cortex in drug addiction: a review of preclinical studies. Biol Psychiatry 63:256–262
115. Lett BT (1989) Repeated exposures intensify rather than diminish the rewarding effects of amphetamine, morphine, and cocaine. Psychopharmacology (Berl) 98:357–362
116. Lorrain DS, Arnold GM, Vezina P (2000) Previous exposure to amphetamine increases incentive to obtain the drug: long-lasting effects revealed by the progressive ratio schedule. Behav Brain Res 107:9–19
117. McSweeney FK, Murphy ES, Kowal BP (2005) Regulation of drug taking by sensitization and habituation. Exp Clin Psychopharmacol 13:163–184
118. Piazza PV, Deminière JM, Le Moal M, Simon H (1989) Factors that predict individual vulnerability to amphetamine self-administration. Science 245:1511–1513
119. Vezina P (2004) Sensitization of midbrain dopamine neuron reactivity and the self-administration of psychomotor stimulant drugs. Neurosci Biobehav Rev 27:827–839
120. Cardinal RN, Parkinson JA, Hall J, Everitt BJ (2002) Emotion and motivation: the role of the amygdala, ventral striatum, and prefrontal cortex. Neurosci Biobehav Rev 26:321–352
121. Di Ciano P, Benham-Hermetz J, Fogg AP, Osborne GE (2007) Role of the prelimbic cortex in the acquisition, re-acquisition or persistence of responding for a drug-paired conditioned reinforcer. Neuroscience 150:291–298
122. Taylor JR, Horger BA (1999) Enhanced responding for conditioned reward produced by intra-accumbens amphetamine is potentiated after cocaine sensitization. Psychopharmacology (Berl) 142:31–40
123. Uslaner JM, Acerbo MJ, Jones SA, Robinson TE (2006) The attribution of incentive salience to a stimulus that signals an intravenous injection of cocaine. Behav Brain Res 169:320–324
124. Wyvell CL, Berridge KC (2001) Incentive sensitization by previous amphetamine exposure: increased cue-triggered "wanting" for sucrose reward. J Neurosci 21:7831–7840
125. Tindell AJ, Berridge KC, Zhang J, Pecina S, Aldridge JW (2005) Ventral pallidal neurons code incentive motivation: amplification by mesolimbic sensitization and amphetamine. Eur J Neurosci 22:2617–2634
126. Bradberry CW (2008) Comparison of acute and chronic neurochemical effects of

cocaine and cocaine cues in rhesus monkeys and rodents: focus on striatal and cortical dopamine systems. Rev Neurosci 19:113–128
127. Hirabayashi M, Alam MR (1981) Enhancing effect of methamphetamine on ambulatory activity produced by repeated administration in mice. Pharmacol Biochem Behav 15:925–932
128. Kalivas PW, Duffy P, Dilts R, Abhold R (1988) Enkephalin modulation of A10 dopamine neurons: a role in dopamine sensitization. Ann NY Acad Sci 537:405–414
129. Nakamura H, Hishinuma T, Tomioka Y, Ido T, Iwata R, Funaki Y, Itoh M, Fujiwara T, Yanai K, Sato M, Numachi Y, Yoshida S, Mizugaki M (1996) Positron emission tomography study of the alterations in brain distribution of [11C]methamphetamine in methamphetamine-sensitized dog. Ann NY Acad Sci 801:401–408
130. Schwandt ML, Higley JD, Suomi SJ, Heilig M, Barr CS (2008) Rapid tolerance and locomotor sensitization in ethanol-naïve adolescent rhesus macaques. Alcohol Clin Exp Res 32:1217–1228
131. Wallach MB, Gershon S (1971) Sensitization to amphetamines. Psychopharmacol Bull 7:30–31
132. Weiner WJ, Goetz CG, Nausieda PA, Klawans HL (1979) Amphetamine-induced hypersensitivity in guinea pigs. Neurology 29:1054–1057
133. Heberlein U, Tsai LT, Kapfhamer D, Lasek AW (2009) Drosophila, a genetic model system to study cocaine-related behaviors: a review with focus on LIM-only proteins. Neuropharmacology 56(Suppl 1):97–106
134. McClung C, Hirsh J (1998) Stereotypic behavioral responses to free-base cocaine and the development of behavioral sensitization in *Drosophila*. Curr Biol 8:109–112
135. Benwell MEM, Balfour DJK (1992) The effects of acute and repeated nicotine treatment on nucleus accumbens dopamine and locomotor activity. Br J Pharmacol 105:849–856
136. Joyce EM, Iversen SD (1979) The effect of morphine applied locally to mesencephalic dopamine cell bodies on spontaneous motor activity in the rat. Neurosci Lett 14:207–212
137. Kalivas PW, Duffy P (1993) Time course of extracellular dopamine and behavioral sensitization to cocaine. I. Dopamine axon terminals. J Neurosci 13:266–275
138. Kapasova Z, Szumlinski KK (2008) Strain differences in alcohol-induced neurochemical plasticity: a role for accumbens glutamate in alcohol intake. Alcohol Clin Exp Res 32:617–631
139. Kazahaya Y, Akimoto K, Otsuki S (1989) Subchronic methamphetamine treatment enhances methamphetamine- or cocaine-induced dopamine efflux in vivo. Biol Psychiatry 25:903–912
140. Zapata A, Chefer VI, Ator R, Shippenberg TS, Rocha BA (2003) Behavioural sensitization and enhanced dopamine response in the nucleus accumbens after intravenous cocaine self-administration in mice. Eur J Neurosci 17:590–596
141. Zapata A, Gonzales RA, Shippenberg TS (2006) Repeated ethanol intoxication induces behavioral sensitization in the absence of a sensitized accumbens dopamine response in C57BL/6 J and DBA/2 J mice. Neuropsychopharmacology 31:396–405
142. Berger SP, Hall S, Mickalian JD, Reid MS, Crawford CA, Delucchi K, Carr K, Hall S (1996) Haloperidol antagonism of cue-elicited cocaine craving. Lancet 347:504–508
143. Boileau I, Dagher A, Leyton M, Gunn RN, Baker GB, Diksic M, Benkelfat C (2006) Modeling sensitization to stimulants in humans: an [11C]raclopride/positron emission tomography study in healthy men. Arch Gen Psychiatry 63:1386–1395
144. Boileau I, Dagher A, Leyton M, Welfeld K, Booij L, Diksic M, Benkelfat C (2007) Conditioned dopamine release in humans: a positron emission tomography [11C]raclopride study with amphetamine. J Neurosci 27:3998–4003
145. Breier A, Su T-P, Saunders R, Carson RE, Kolachana BS, de Bartolomeis A, Weinberger DR, Weisenfeld N, Malhotra AK, Eckelman WC, Pickar D (1997) Schizophrenia is associated with elevated amphetamine-induced synaptic dopamine concentrations: evidence from a novel positron emission tomography method. Proc Natl Acad Sci USA 94:2569–2574
146. Cox SM, Benkelfat C, Dagher A, Delaney JS, Durand F, McKenzie SA, Kolivakis T, Casey KF, Leyton M (2006) Striatal dopamine responses to intranasal cocaine self-administration in humans. Biol Psychiatry 65:846–850
147. Foltin RW, Ward AS, Haney M, Hart CL, Collins ED (2003) The effects of escalating doses of smoked cocaine in humans. Drug Alcohol Depend 70:149–157
148. Johanson CE, Uhlenhuth EH (1981) Drug preference and mood in humans: repeated assessment of *d*-amphetamine. Pharmacol Biochem Behav 14:159–163

149. Kegeles LS, Zea-Ponce Y, Abi-Dargham A, Rodenhiser J, Wang T, Weiss R, Van Heertum RL, Mann JJ, Laruelle M (1999) Stability of [123I]IBZM SPECT measurement of amphetamine-induced striatal dopamine release in humans. Synapse 31:302–308
150. Kelly TH, Foltin RW, Fischman MW (1991) The effects of repeated amphetamine exposure on multiple measures of human behavior. Pharmacol Biochem Behav 38:417–426
151. Leyton M, Casey KF, Delaney JS, Kolivakis T, Benkelfat C (2005) Cocaine craving, euphoria, and self-administration: a preliminary study of the effect of catecholamine precursor depletion. Behav Neurosci 119:1619–1627
152. Martinez D, Narendran R, Foltin RW, Slifstein M, Hwang DR, Broft A, Huang Y, Cooper TB, Fischman MW, Kleber HD, Laruelle M (2007) Amphetamine-induced dopamine release: markedly blunted in cocaine dependence and predictive of the choice to self-administer cocaine. Am J Psychiatry 164:622–629
153. Nagoshi C, Kumor KM, Muntaner C (1992) Test–retest stability of cardiovascular and subjective responses to intravenous cocaine in humans. Br J Addict 87:591–599
154. Newlin DB, Thomson JB (1991) Chronic tolerance and sensitization to alcohol in sons of alcoholics. Alcohol Clin Exp Res 15:399–405
155. Newlin DB, Thomson JB (1999) Chronic tolerance and sensitization to alcohol in sons of alcoholics: II. Replication and reanalysis. Exp Clin Psychopharmacol 7:234–243
156. Rothman RB, Gorelick DA, Baumann MH, Guo XY, Herning RI, Pickworth WB, Gendron TM, Koeppl B, Thomson LE III, Henningfield JE (1994) Lack of evidence for context-dependent cocaine-induced sensitization in humans: preliminary studies. Pharmacol Biochem Behav 49:583–588
157. Sax KW, Strakowski SM (1998) Enhanced behavioral response to repeated d-amphetamine and personality traits in humans. Biol Psychiatry 44:1192–1195
158. Strakowski SM, Sax KW (1998) Progressive behavioral response to repeated d-amphetamine challenge: further evidence for sensitization in humans. Biol Psychiatry 44:1171–1177
159. Strakowski SM, Sax KW, Setters MJ, Keck PE Jr (1996) Enhanced response to repeated d-amphetamine challenge: evidence for behavioral sensitization in humans. Biol Psychiatry 40:872–880
160. Strakowski SM, Sax KW, Rosenberg HL, DelBello MP, Adler CM (2001) Human response to repeated low dose d-amphetamine: evidence for behavioral enhancement and tolerance. Neuropsychopharmacol 25:548–554
161. Szechtman H, Cleghorn JM, Brown GM, Kaplan RD, Franco S, Rosenthal K (1998) Sensitization and tolerance to apomorphine in men: yawning, growth hormone, nausea, and hyperthermia. Psychiatry Res 23:245–255
162. Volkow ND, Wang G-J, Telang F, Fowler JS, Logan J, Childress A-R, Jayne M, Ma Y, Wong C (2006) Cocaine cues and dopamine in dorsal striatum: mechanism of craving in cocaine addiction. J Neurosci 26:6583–6588
163. Wong DF, Kuwabara H, Schretlen DJ, Bonson KR, Zhou Y, Nandi A, Brasić JR, Kimes AS, Maris MA, Kumar A, Contoreggi C, Links J, Ernst M, Rousset O, Zukin S, Grace AA, Lee JS, Rohde C, Jasinski DR, Gjedde A, London ED (2006) Increased occupancy of dopamine receptors in human striatum during cue-elicited cocaine craving. Neuropsychopharmacol 31:2716–2727
164. Leyton M (2007) Conditioned and sensitized responses to stimulant drugs in humans. Prog Neuropsychopharmacol. Biol Psychiatry 31:1601–1613
165. Ellinwood EH Jr (1968) Amphetamine psychosis. II. Theoretical implications. Int J Neuropsychiatry 4:45–54
166. Ellinwood EH Jr, Balster RL (1974) Rating the behavioral effects of amphetamine. Eur J Pharmacol 28:35–41
167. Berridge KC, Robinson TE (1998) What is the role of dopamine in reward: hedonic impact, reward learning, or incentive salience? Brain Res Brain Res Rev 28:309–369
168. Blackburn JR, Pfaus JG, Phillips AG (1992) Dopamine functions in appetitive and defensive behaviours. Prog Neurobiol 39:247–279
169. Ikemoto S, Panksepp J (1999) The role of nucleus accumbens dopamine in motivated behavior: a unifying interpretation with special reference to reward-seeking. Brain Res Brain Res Rev 31:6–41
170. Kuribara H (1996) Effects of interdose interval on ambulatory sensitization to methamphetamine, cocaine and morphine in mice. Eur J Pharmacol 316:1–5
171. Grimm JW, Hope BT, Wise RA, Shaham Y (2001) Neuroadaptation. Incubation of cocaine craving after withdrawal. Nature 412:141–142
172. Lu L, Hope BT, Dempsey J, Liu SY, Bossert JM, Shaham Y (2005) Central amygdala ERK signaling pathway is critical to incubation of cocaine craving. Nat Neurosci 8:212–219

173. Biagioni F, Pellegrini A, Ruggieri S, Murri L, Paparelli A, Fornai F (2009) Behavioural sensitisation during dopamine replacement therapy in Parkinson's disease is reminiscent of the addicted brain. Curr Top Med Chem 9:894–902
174. Fornai F, Biagioni F, Fulceri F, Murri L, Ruggieri S, Paparelli A (2009) Intermittent dopaminergic stimulation causes behavioral sensitization in the addicted brain and parkinsonism. Int Rev Neurobiol 88:371–398
175. Hirabayashi M, Okada S, Tadokoro S (1991) Comparison of sensitization to ambulation-increasing effects of cocaine and methamphetamine after repeated administration in mice. J Pharm Pharmacol 43:827–830
176. King GR, Xiong Z, Ellinwood EH Jr (1998) Blockade of the expression of sensitization and tolerance by ondansetron, a 5-HT3 receptor antagonist, administered during withdrawal from intermittent and continuous cocaine. Psychopharmacology 135:263–269
177. Meyer PJ, Phillips TJ (2007) Behavioral sensitization to ethanol does not result in cross-sensitization to NMDA receptor antagonists. Psychopharmacology (Berl) 195:103–115
178. Broadbent J, Kampmueller KM, Koonse SA (2005) Role of dopamine in behavioral sensitization to ethanol in DBA/2 J mice. Alcohol 35:137–148
179. Didone V, Quoilin C, Tirelli E, Quertemont E (2008) Parametric analysis of the development and expression of ethanol-induced behavioral sensitization in female Swiss mice: effects of dose, injection schedule, and test context. Psychopharmacology (Berl) 201:249–260
180. Grahame NJ, Rodd-Henricks K, Li T-K, Lumeng L (2000) Ethanol locomotor sensitization, but not tolerance correlates with selection for alcohol preference in high- and low-alcohol preferring mice. Psychopharmacology (Berl) 151:252–260
181. Kayir H, Uzbay IT (2002) Investigation of a possible sensitization development to a challenge dose of ethanol after 2 weeks following the single injection in mice. Pharmacol Biochem Behav 73:551–556
182. Lessov CN, Palmer AA, Quick EA, Phillips TJ (2001) Voluntary ethanol drinking in C57BL/6 J and DBA/2 J mice before and after sensitization to the locomotor stimulant effects of ethanol. Psychopharmacology (Berl) 155:91–99
183. Quadros IM, Hipólide DC, Frussa-Filho R, De Lucca EM, Nobrega JN, Souza-Formigoni ML (2002) Resistance to ethanol sensitization is associated with increased NMDA receptor binding in specific brain areas. Eur J Pharmacol 442:55–61
184. Jasova D, Bob P, Fedor-Freybergh P (2007) Alcohol craving, limbic irritability, and stress. Med Sci Monit 13:CR543–CR547
185. Becker HC, Diaz-Granados JL, Weathersby RT (1997) Repeated ethanol withdrawal experience increases the severity and duration of subsequent withdrawal seizures in mice. Alcohol 14:319–326
186. Becker HC, Lopez MF (2004) Increased ethanol drinking after repeated chronic ethanol exposure and withdrawal experience in C57BL/6 mice. Alcohol Clin Exp Res 28:1829–1838
187. O'Dell LE, Roberts AJ, Smith RT, Koob GF (2004) Enhanced alcohol self-administration after intermittent versus continuous alcohol vapor exposure. Alcohol Clin Exp Res 28:1676–1682
188. Bailey A, Metaxas A, Yoo JH, McGee T, Kitchen I (2008) Decrease of D2 receptor binding but increase in D2-stimulated G-protein activation, dopamine transporter binding and behavioural sensitization in brains of mice treated with a chronic escalating dose 'binge' cocaine administration paradigm. Eur J Neurosci 28:759–770
189. Fish EW, DeBold JF, Miczek KA (2002) Repeated alcohol: behavioral sensitization and alcohol-heightened aggression in mice. Psychopharmacology (Berl) 160:39–48
190. Crabbe JC, Wahlsten D, Dudek BC (1999) Genetics of mouse behavior: interactions with laboratory environment. Science 284:1670–1672
191. Wahlsten D, Bachmanov A, Finn DA, Crabbe JC (2006) Stability of inbred mouse strain differences in behavior and brain size between laboratories and across decades. Proc Natl Acad Sci USA 103:16364–16369
192. Phillips TJ, Burkhart-Kasch S, Terdal ES, Crabbe JC (1991) Response to selection for ethanol-induced locomotor activation: genetic analyses and selection response characterization. Psychopharmacology 103:557–566
193. Nuutinen S, Karlstedt K, Aitta-Aho T, Korpi ER, Panula P (2010) Histamine and H3 receptor dependent mechanisms regulate ethanol stimulation and conditioned place preference in mice. Psychopharmacology 208:75–86
194. Fujii H, Ishihama T, Ago Y, Shintani N, Kakuda M, Hashimoto H, Baba A, Matsuda T (2007) Methamphetamine-induced hyperactivity and behavioral sensitization in PACAP deficient mice. Peptides 28:1674–1679

195. Kelly MA, Low MJ, Rubinstein M, Phillips TJ (2008) Role of dopamine D1-like receptors in methamphetamine locomotor responses of D2 receptor knockout mice. Genes Brain Behav 7:568–577
196. Ago Y, Nakamura S, Kajita N, Uda M, Hashimoto H, Baba A, Matsuda T (2007) Ritanserin reverses repeated methamphetamine-induced behavioral and neurochemical sensitization in mice. Synapse 61:757–763
197. Kamens HM, Burkhart-Kasch S, McKinnon CS, Li N, Reed C, Phillips TJ (2005) Sensitivity to psychostimulants in mice bred for high and low stimulation to methamphetamine. Genes Brain Behav 4:110–125
198. Kalivas PW, Duffy P (1993) Time course of extracellular dopamine and behavioral sensitization to cocaine. II. Dopamine perikarya. J Neurosci 13:276–284
199. Faria RR, Lima Rueda AV, Sayuri C, Soares SL, Malta MB, Carrara-Nascimento PF, da Silva Alves A, Marcourakis T, Yonamine M, Scavone C, Giorgetti Britto LR, Camarini R (2008) Environmental modulation of ethanol-induced locomotor activity: correlation with neuronal activity in distinct brain regions of adolescent and adult Swiss mice. Brain Res 1239:127–140
200. Vezina P, Leyton M (2009) Conditioned cues and the expression of stimulant sensitization in animals and humans. Neuropharmacology 56(Suppl 1):160–168
201. Quadros IM, Souza-Formigoni ML, Fornari RV, Nobrega JN, Oliveira MG (2003) Is behavioral sensitization to ethanol associated with contextual conditioning in mice? Behav Pharmacol 14:129–136
202. Singer BF, Tanabe LM, Gorny G, Jake-Matthews C, Li Y, Kolb B, Vezina P (2009) Amphetamine-induced changes in dendritic morphology in rat forebrain correspond to associative drug conditioning rather than nonassociative drug sensitization. Biol Psychiatry 65:835–840
203. Marin MT, Berkow A, Golden SA, Koya E, Planeta CS, Hope BT (2009) Context-specific modulation of cocaine-induced locomotor sensitization and ERK and CREB phosphorylation in the rat nucleus accumbens. Eur J Neurosci 30:1931–1940
204. Robinson TE, Browman KE, Crombag HS, Badiani A (1998) Modulation of the induction or expression of psychostimulant sensitization by the circumstances surrounding drug administration. Neurosci Biobehav Rev 22:347–354
205. Vezina P, Giovino AA, Wise RA, Stewart J (1989) Environment-specific cross-sensitization between the locomotor activating effects of morphine and amphetamine. Pharmacol Biochem Behav 32:581–584
206. Yetnikoff L, Arvanitogiannis A (2005) A role for affect in context-dependent sensitization to amphetamine. Behav Neurosci 119:1678–1681
207. Knych ET, Eisenberg RM (1979) Effect of amphetamine on plasma corticosterone in the conscious rat. Neuroendocrinology 29:110–118
208. Lowy MT, Novotney S (1994) Methamphetamine-induced decrease in neural glucocorticoid receptors: relationship to monoamine levels. Brain Res 638:175–181
209. Rivier C (1996) Alcohol stimulates ACTH secretion in the rat: mechanisms of action and interactions with other stimuli. Alcohol Clin Exp Res 20:240–254
210. Zhou Y, Spangler R, Schlussman SD, Yuferov VP, Sora I, Ho A, Uhl GR, Kreek MJ (2002) Effects of acute "binge" cocaine on preprodynorphin, preproenkephalin, proopiomelanocortin, and corticotropin-releasing hormone receptor mRNA levels in the striatum and hypothalamic–pituitary–adrenal axis of mu-opioid receptor knockout mice. Synapse 45:220–229
211. de Jong IE, Oitzl MS, de Kloet ER (2007) Adrenalectomy prevents behavioural sensitisation of mice to cocaine in a genotype-dependent manner. Behav Brain Res 177:329–339
212. Deroche V, Marinelli M, Maccari S, Le Moal M, Simon H, Piazza PV (1995) Stress-induced sensitization and glucocorticoids. I. Sensitization of dopamine-dependent locomotor effects of amphetamine and morphine depends on stress-induced corticosterone secretion. J Neurosci 15:7181–7188
213. Araujo NP, Camarini R, Souza-Formigoni ML, Carvalho RC, Abílio VC, Silva RH, Ricardo VP, Ribeiro Rde A, Frussa-Filho R (2005) The importance of housing conditions on behavioral sensitization and tolerance to ethanol. Pharmacol Biochem Behav 82:40–45
214. Meyer PJ, Palmer AA, McKinnon CS, Phillips TJ (2005) Behavioral sensitization to ethanol is modulated by environmental conditions, but is not associated with cross-sensitization to allopregnanolone or pentobarbital in DBA/2J mice. Neuroscience 131:263–273
215. Chauvet C, Lardeux V, Goldberg SR, Jaber M, Solinas M (2009) Environmental enrichment reduces cocaine seeking and rein-

statement induced by cues and stress but not by cocaine. Neuropsychopharmacology 34:2767–2778
216. Yui K, Goto K, Ikemoto S, Ishiguro T, Angrist B, Duncan GE, Sheitman BB, Lieberman JA, Bracha SH, Ali SF (1999) Neurobiological basis of relapse prediction in stimulant-induced psychosis and schizophrenia: the role of sensitization. Mol Psychiatry 4:512–523

Chapter 12

Evaluating Behavioral Outcomes from Ischemic Brain Injury

Paco S. Herson, Julie Palmateer, Patricia D. Hurn, and A. Courtney DeVries

Abstract

Brain injury resulting from cerebral ischemia is a significant clinical problem. Stroke (focal cerebral ischemia) is the third leading cause of death in the United States. In addition, approximately 500,000 people annually in the United States suffer brain injury after global brain ischemia consequent to cardiac arrest and cardiopulmonary resuscitation (CPR). Despite intensive research over the past few decades, survival and neurological outcome for both types of ischemic injury remain poor. Therefore, the need for therapies to protect the brain during ischemic episodes and to enhance its potential for plasticity and repair after ischemia remains paramount. This chapter discusses behavioral techniques and considerations that are vital to brain injury studies of experimental stroke or cardiac arrest. Testing protocols are focused on the mouse, as this species is readily amenable to genetic alteration and so has rapidly become the dominant species employed in most laboratories.

1. Introduction

1.1. Overview

Brain injury resulting from cerebral ischemia is a significant clinical problem. Indeed, stroke remains the third leading cause of death in the United States, and a similar epidemiology has been reported in most countries around the world. Despite considerable efforts in clinical and basic science research, thrombolytic tissue plasminogen activator (tPA) is currently the only agent approved for acute stroke treatment, although its use is limited by a narrow treatment window. Therefore, the need for therapies to protect the brain during ischemic episodes and to enhance its potential for plasticity and repair after ischemia remains paramount. From 1999 to 2009, a series of expert panels were convened to recommend best practices for pre-clinical animal studies (for most recent study *see* 1). The resulting guidelines, known as the Stroke Therapy Academic Industry Roundtable (STAIR) recommendations, specifically include short-term and long-term behavioral variables. The STAIR recommendations, and other expert sources, emphasize that behavioral outcomes, rather than histologic or molecular endpoints alone, are essential for moving experimental therapies forward to human trials, where functional outcomes remain the gold standard for measuring treatment efficacy.

Accordingly, this chapter discusses techniques for evaluating behavioral outcomes and other considerations that are vital to brain injury studies of experimental stroke (focal cerebral ischemia) or cardiac arrest (global cerebral ischemia). A variety of animal models have been employed for pre-clinical drug and therapy studies, including higher order, gyrencephalic species such as dog, cat, and non-human primates. These models, however, pose significant challenges including difficulty in assessing behavior over time and coherently among laboratories, high cost, and lack of ready animal availability. In contrast, behavioral outcomes in rodent models such as gerbil, rat, and mouse have been well studied. This chapter will focus on behavioral testing in mouse since this species is readily amenable to genetic alteration and has therefore rapidly become the dominant species used in most laboratories.

1.2. Natural Mouse Characteristics to Be Considered in Behavioral Assessment

Over the past two decades, the popularity of using mice in behavioral studies has grown dramatically with the mainstream availability of genetically altered mice. Most of the rodent behavioral tests that are currently used were originally developed for rats, and then modified for testing mice. Although there remains a tendency to think of mice as miniature rats, basic behavioral differences between the most popular laboratory strains of rats and mice include baseline locomotor behavior, rearing, exploratory behavior, aggressive behavior, stress responsivity, learning of cognitive tasks, and social behavior. These differences can affect the mouse's performance in the behavioral task and obscure data interpretation if not taken into consideration when adopting a protocol. Thus, it is crucial to choose behavioral protocols that have been well characterized and validated for use with mice. The large size differential between mice and rats also means that few testing apparati are the optimal size for both species. Finally, it is preferable to maintain separate testing facilities for mice and rats since rats are mouse predators, and lingering rat odors may affect mouse behavior.

1.3. Importance of Biological Sex and Genetic Strain

Knowledge of molecular and cellular mechanisms of neuroprotection or plasticity is important in both sexes. Accordingly, optimal injury-related behavioral studies consider potential gender differences in response to ischemia and repair. In humans, clear sex-linked patterns of cerebrovascular disease and stroke have been observed. For example, overall incidence of stroke is higher in men compared to women in all countries and across ethnic backgrounds (2). However, the incidence of stroke increases in both sexes with age and evidence now suggests that *outcome* from stroke is worse in aged women than in men. Not surprisingly, similar sex-linked sensitivity is readily apparent in rodent injury models. The animal's sex impacts injury assessment in two ways. First, the amount of tissue damage resulting from an ischemic event is typically greater in the adult male versus female until the middle years of life. These differences are due in part to cellular actions of sex steroids: 17β-estradiol, testosterone, and the progesterone–neurosteroid family. Although reported behavioral studies in both sexes are rare, it logically follows that differences in the extent of tissue damage will influence, and may correlate with, functional measures of sensory–motor performance and cognition. Second, emerging evidence suggests that the molecular mechanisms of cell death or repair are not necessarily identical in both sexes (for recent review, *see* 3). Therefore, studies of genetically engineered animals are best constructed to include both sexes despite the additional cost and increased animal use.

Genetic strain also strongly influences outcomes from ischemic injury. Common mouse strains such as C57BL/6, SV129, CD1, and BALBc display large inter-strain differences in cerebrovascular anatomy, hemodynamic variability, and susceptibility to tissue damage from a controlled ischemic insult (4–6). Accordingly, outcomes from experimental focal and global ischemia may be largely determined by background strain or confluence of genetic strain. Recovery factors such as susceptibility to post-insult inflammation and systemic infection are also known to be strain dependent (7), which may greatly impact plasticity, neurogenesis, and consequent functional recovery.

2. General Primer for Establishing Mouse Behavioral Testing

2.1. Housing and Animal Issues

The housing environment during both development and adulthood can influence physiological and behavioral responses in mice. For any given study, all mice that are purchased as adults and shipped to the testing facility should be procured from a single vendor to minimize variability due to different developmental exposures. Subtle differences in breeding facilities can have long-term effects on stress responsivity, which, in turn, can add variability to behavioral data sets. Furthermore, significant vendor effects on cerebral arteries and infarct volumes following cerebral ischemia have been reported (8).

The mice within a study should be the same sex (unless sex differences are being examined, see above) and the same approximate size at time of ischemia. They should also be allowed to acclimate to the vivarium for at least 1 week prior to being enrolled in the study. Transporting animals is stressful to them, and stress can alter both behavioral responses and ischemic outcome (9, 10). The mice should be housed on a set light cycle in cages that are cleaned on a schedule. However, to minimize differences in stress during testing, it is ideal to avoid cage cleaning within the 24 h that precede behavioral testing.

Good communication with the animal care staff should be established to ensure the most consistent possible environment. For example, the staff must notify investigators about facility-wide changes that could affect ischemic outcomes, such as a change in the brand of rodent chow. Indeed, rodent chow has high and often variable concentrations of phytoestrogens, which can alter physiological responses to estrogens and, in turn, can affect behavior and infarct size; the type and amount of food should remain consistent for the duration of the study.

The number of mice housed together in cages also should remain consistent across groups. Efforts to increase efficiency within animal colonies have led many research institutions to shift from charging per diem based on the number of mice to charging per diem based on the number of cages (allowing up to five mice per cage). Thus, the more mice housed per cage, the lower the overall per diem charge. Although there may be clear fiscal benefits to group housing, male mice tend to be territorial and will fight to establish dominance hierarchies. The result is a stratification of cutaneous wounding, testosterone concentrations, and corticosterone concentrations across animals within the cage (11). Each of these factors can influence behavior and ischemic outcome.

2.2. Testing Environment

The testing environment can influence the magnitude of behavioral effects (12); so, it is imperative to maintain a consistent testing environment in order to reduce variability and increase reproducibility of the data. An ideal testing environment is located close to the vivarium to reduce transport stress and has steady temperature and lighting, as well as low noise and vibration. The mice awaiting testing should be maintained in an adjacent room so that they are not influenced by the vocalizations or olfactory alarm cues emitted by other mice during routine handling or behavioral testing. Olfaction is the primary modality by which rodents communicate with other rodents and detect predators. Indeed, clothing can retain odors that can influence a mouse's behavior, such as odors from pet cats and dogs, i.e., predators. Wearing hospital scrubs that are laundered daily and avoiding scented toiletries will reduce this concern. Cleaning the behavioral apparati between each animal with 70% ethanol is also important for reducing the potential effects of odor cues from other mice.

2.3. Steps to Reduce Variability and Maximize Effect Size

Maintaining consistent vivarium and behavioral testing environments and strictly adhering to the study protocol are important steps toward decreasing variability and achieving sufficient statistical power with the fewest number of animals. All essential procedures should be performed by a single, well-trained individual. Although it is acceptable for several people to work on a single study, all surgeries should be performed by the same individual; additionally all data for a particular behavioral task should be collected by one individual who is not aware of the experimental assignments of the animals they are testing. This is particularly true not only for outcome measures that are somewhat subjective, such as neuroscore, but it is also important for automated tasks, such as measuring locomotor activity, because the style of handling and other personal techniques that influence placement of the mouse in the apparatus can also impact behavior.

The proper timing of ischemia and behavioral testing is also vital because rodents exhibit strong circadian patterns in physiology and behavior. Indeed, a shift of as little as 4 h in the timing of cerebral ischemia relative to the onset of the light phase can have a significant effect on the extent of neuronal damage and subsequent behavioral alterations (13). Likewise, each behavioral task should be conducted during the same time of day, whether during the dark cycle, which is the active phase for mice, or during the light cycle, which is their inactive phase. Testing during the active phase is preferable if the testing can be conducted in darkness or under red light, such as measuring general locomotor activity; however, tasks that require bright lights, such as the elevated plus maze, water maze tracking system, or passive avoidance (*see* below), should be conducted during the light cycle to avoid disrupting the animal's circadian rhythm.

Other crucial factors to consider are the number of mice to house per cage and whether to include objects for environmental enrichment. Although social and environmental enrichment have been reported to improve animal welfare, they have also been shown to reduce infarct size and/or behavioral deficits in several

2.4. Choosing Ischemic Duration and a Time Course for Behavioral Testing

models of cerebral ischemia (14–16) and may therefore ultimately necessitate longer durations of ischemia and a greater number of animals to achieve appropriate statistical power.

Focal cerebral ischemia models are appropriate for studying sensory-motor, cognitive, and affective behaviors. In general, longer durations of focal ischemia and earlier testing time points should be used for assessing gross motor deficits because rodents are adept at recovering basic skills within days or weeks of stroke. In contrast, persistent effects on fine motor control, such as skilled reaching, can be observed for several months following stroke (17). Cognitive and affective deficits are also persistent for weeks to months after stroke. If the goal is to study cognitive function or another complex behavior, the investigator should choose (1) a short duration (30–60 min) of transient middle cerebral artery occlusion (MCAO) and (2) a post-surgical recovery period that is sufficiently long to allow full sensory-motor recovery prior to initiating additional testing. Delaying complex testing until ischemia-induced sensory-motor deficits are no longer detected will minimize the possibility that sensory-motor deficits will become a confounding factor in the interpretation of data from cognitive and affective tasks. In contrast to stroke, the duration of CA/CPR is not typically altered from study to study because the only major sensory-motor change is hyperactivity, and this typically resolves within 4–5 days after CA/CPR.

Choosing when and how often to conduct behavioral testing will depend on the goals and expected length of the experiment. Baseline testing is beneficial because it allows statistical analyses within animal as well as across groups, and it is a particularly important time point when the mice are pre-treated with a drug or manipulation that could have effects on behavior that are independent of its effects on ischemia. The frequency of testing mice after ischemia should be determined based on the duration of the study and the likelihood that repeated testing will obscure the behavioral results. For example, in order for mice to perform well consistently, water maze training may require as many as three trials per day for approximately 10 consecutive days. In contrast, certain other tasks, such as the open field and elevated plus maze, require environmental novelty, and the data can be obscured by repeated testing because the apparatus becomes familiar to the mouse and no longer evokes anxiety. If a cohort of mice is to be tested using multiple tasks that are likely to evoke the same low level of stress or anxiety, the presentation of tasks should be counterbalanced. In contrast, if the tasks are invasive or potentially stress-provoking, such as passive avoidance, the tasks should be presented in a consistent sequence from least to most stress-provoking or should be conducted in separate cohorts of mice.

3. Focal Cerebral Ischemia and Stroke

3.1. Clinical Connections

Stroke is the leading cause of serious long-term disability in the United States. Estimates suggest that approximately 15–30% of stroke survivors are permanently disabled. Although advancing age is clearly related to stroke incidence, it must be emphasized that stroke also occurs in the newborn and pediatric populations. Because approximately 25% of stroke victims are under the age of 65 years, large economic costs are associated with the disease and its consequent disability. Stroke survivors are routinely assessed for acute severity of stroke through neurological deficit as well as imaging modalities. During recovery, the level of disability and potential for recovery are frequently assessed as means of assessing long-term prognosis and effects of therapy. In general, functional recovery reaches a maximum level by approximately 3–6 months post-stroke onset, and a number of studies suggest that there is little further recovery beyond this point. Many patients reach their maximum functional return to activities of daily living within 6 weeks of onset. Accordingly, steady functional re-assessment is an active part of the survivor's treatment plan during this period.

Notably, phase III clinical trials also use disability and neurological deficit scales as primary outcomes. The most commonly used outcomes scales are the NIH Stroke Scale, modified Rankin Scale, and the Barthel Index. These scales cover a wide range of disabilities: motor control difficulties ranging from paralysis to varying types of incoordination; sensory loss or pain; language disturbances involving comprehension or speech execution; emotional distress such as depression; and memory and learning disorders (for recent review, *see* 18). Clearly, some of these variables can be recapitulated in experimental animal studies, and this is one basis for conducting functional assessment over the recovery period of animal survivors.

3.2. Methods and Models

The most frequently used animal model to study focal cerebral ischemia is the middle cerebral artery occlusion (MCAO) model. Several different MCAO methods/models have been developed, resulting in either temporary or permanent loss of perfusion to the middle cerebral artery (MCA). These MCAO models have been extensively used because all result in significant reduction of cerebral blood flow (CBF) to both the striatum and cortex, making them a reasonable model of human thrombo-embolic stroke. The most common method for inducing focal cerebral ischemia uses an intraluminal filament to occlude the MCA (19, 20). A nylon suture is inserted into the internal carotid artery of mice or rats and advanced until the MCA is blocked, as determined by laser Doppler flowmetry. Reperfusion occurs when the suture is removed, making this a very convenient method for inducing temporary focal ischemia. This method is well suited for neurobehavioral assessment studies since the filament can be removed after varying times of occlusion to alter severity and long-term outcome.

Details of the intraluminal filament method are described as follows: Anesthesia is induced with 3% isoflurane and maintained at 1–1.5%. A temperature probe is inserted into the left temporalis muscle, and head temperature is maintained at 35.5–37.5°C throughout surgery with water pads. A small laser Doppler probe is affixed to the skull to monitor cortical perfusion and to verify vascular occlusion and reperfusion. An incision is made to expose the external carotid artery, and a nylon monofilament with a silicone-coated tip is inserted into the right internal carotid artery via the external carotid artery until the laser Doppler flowmetry value drops to <20% of baseline. After securing the filament in place, the surgical site is closed with Vicryl sutures. Each mouse is then placed in a separate cage with a warm water pad under the cage for 60–120 min. Mice are re-anesthetized, and the laser Doppler probe is re-positioned over the same site on the skull. Then, the occluding filament is withdrawn for reperfusion. In the sham MCAO animal group, external and internal carotid arteries are located but not disturbed. Sham-operated animals are important surgical controls in behavioral studies because anesthetic exposure and the potential stress or pain associated with all surgeries can affect behavior.

3.3. Behavioral Outcome Assessment – Methods

Using mice in neurobehavioral experiments after exposure to cerebral ischemia creates some unique complications that must be considered in order to generate high quality, interpretable data. Cerebral ischemia not only causes a core of neuronal death in the striatum and cortex, but it also results in widespread neuroinflammation and peripheral immune responses that can substantially alter behavior and compromise the health of the animals; the survival rate is generally 80–90%. Therefore, a strategy for replacing animals that die prematurely is critical. The simplest approach to this problem is a double-blind randomized study design with targeted replacement whereby a separate researcher who is not involved in the surgery or behavioral assessment randomly assigns mice to treatment groups and randomly inserts replacement animals as mice drop out due to mortality, maintaining approximately uniform group sizes of 10–15 animals. In addition to standard post-operative care, such as maintaining hydration and body temperature, the general health of the animals must be monitored regularly using a neuroscore (*see* below). Sufficient recovery is critical in order to engage the animals in cognitive tasks that measure more subtle behavioral deficits such as memory, anxiety, and affective function. To maximize the use of experimental animals, a small battery of neurobehavioral tasks is recommended for each mouse. Described below are selected neurobehavioral tests that have proved to be consistent across experimental rodent focal cerebral ischemia studies in our laboratories (17, 21, 22). In addition, we will describe the Morris water maze task because it is a subtle learning and memory task commonly used in ischemia studies.

Timeline for behavioral testing in MCAO mice: Behavioral and functional tests are performed in the same order on all mice undergoing MCAO or sham MCAO. Tests are administered in the order of least to most stress-provoking so that the outcome of later tests will not be skewed. Each mouse is handled and weighed daily in the lab throughout the testing period. The testing schedule before and after MCAO is shown in **Fig. 12.1**. Baseline performance is assessed using the open field (OF) and paw preference (cylinder) tests (PPT) on Day 1, immediately preceding experimental brain injury (MCAO). A combination of behavioral tasks is chosen to assess overall general health (neuroscore), sensory-motor impairment (open field), forelimb asymmetry (paw preference or cylinder test), and cognitive function (novel object recognition, passive/active avoidance) during the 9-day reperfusion period. Specifically, a neuroscore (NS) is performed on Days 1 and 3 after MCAO. PPT is repeated on Days 3 and 7 following MCAO, and OF is performed again on Day 5. The novel object recognition test (ORT) is performed on Days 6 and 7; and finally, passive avoidance (PA) is performed on Days 8 and 9 of the reperfusion period.

Experimental Timeline For Focal Cerebral Ischemia

MCAO - 10 day experiment

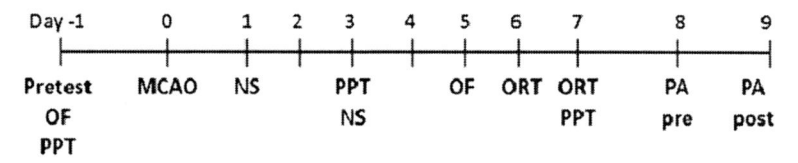

Day -1	0	1	2	3	4	5	6	7	8	9
Pretest OF PPT	MCAO	NS		PPT NS		OF	ORT	ORT PPT	PA pre	PA post

Fig. 12.1. Representative behavioral testing schedule for MCAO experiment. NS, neuroscore; OF, open field; PPT, paw preference; ORT, object recognition; PA, passive avoidance.

3.3.1. Sensorimotor

Neuroscore: The neuroscore described here for behavioral experiments is somewhat more thorough than other neuroscore methods described in the literature and used in combination with standard histology outcomes (23). The neuroscore covers nine measures of recovery: consciousness, interaction, grooming (eye appearance), respiratory rate, food intake, forelimb strength (ability to grab and hold onto wire cage top), nest building, motor function (standing and walking without leaning or circling), and general activity level. Each of these factors is assigned a different scale ranging from 0 to 3–5, depending on the factor; a maximum score is 23.

The mice are tested for latency to move (*see* below), weighed, and given subcutaneous fluids as part of the neuroscore testing. Neuroscore is performed on Days 1 and 3 post-operatively, but mice are weighed and given fresh food and fluids daily. Food intake is measured using a dish with a food pellet mixed into mush with water daily.

- *Outcomes*: A control healthy mouse will score 0, while a more impaired animal will score higher. Mice recover well from MCAO and generally by Day 4 or 5 have returned to a normal neuroscore value (<4). Since the animals are being compared to normal via first-hand observation, it is important that the tester has extensive experience including proficient handling skills with both healthy and impaired animals. Our experience is that statistically significant differences in neuroscores across treatment groups are not commonly observed. However, differences are often revealed with more subtle behavioral tasks described here. Nonetheless, performing the neuroscore is important in order to observe gross general health changes and to help habituate the mice to handling by the tester.

Latency to move: The latency to move test measures how quickly an animal moves one body length when placed in an open area. Exposure to the open is stressful to a mouse, and the normal response is to freeze for a few seconds to assess immediate danger and then move to a more protected area (near a wall or crevice). The latency to move test is performed on an open countertop approximately 2 feet square that is designated for this test only. A circle with a radius of 12 cm is drawn in the center of the countertop. The animal is placed by gently sliding it out of the container used to weigh the animals into the center of the circle. Every animal is placed facing the same direction. The animal is timed manually with a stopwatch to measure how long it takes for it to move out of the circle; timing is stopped when all four feet have left the circle. The maximum latency is 60 s. The area is fully cleaned with 70% ethanol after every animal to remove scent trails, which can affect this behavior.

- *Outcomes*: Latency to move in a healthy mouse is roughly 5–10 s. MCAO-injured animals have higher latencies due to apathy or physical inability. Significant improvement (decrease) in latency to move is commonly observed between Days 1 and 3 after MCAO.

Paw preference: Forelimb bias is tested using the paw preference or cylinder test. Focal ischemia generates unilateral damage, resulting in decreased use of the forelimb on the contralateral side to the injury. To track recovery throughout a longer period, this test is performed at Days 3 and 7. Animals are pre-tested on Day 1 to identify any naturally occurring "handedness." An animal is placed in an upright, clear, plastic cylinder with a diameter approximately equal to the length of the animal's body. (The restricted space encourages the animal to rear up to explore the cylinder but still gives the animal enough room to turn around freely.) The cylinder sits inside a white plastic box that is similar to the open field test. This blocks out the tester's movement and provides a familiar environment. Four video cameras are positioned within the box so that all four directions are visualized simultaneously (**Fig. 12.2**). The video is analyzed in slow motion to ensure accuracy. Only the initial touch is counted each time the mouse rears. A new touch will not be counted until the animal "resets" by returning all four limbs to the floor. If the animal touches the cylinder simultaneously with both forepaws, a count of "both" is entered. The first 20 rears are recorded for each trial.

Paw Preference

Fig. 12.2. Paw preference apparatus.

Outcomes: The final score is generally calculated as percent usage of the non-impaired limb. The advantage of this task is that it is sensitive enough to measure impairment at multiple time points for up to 2–3 weeks post-MCAO. However, caution should be taken because over time the mouse may become habituated to the cylinder and stop exploring and rearing. Testing during the dark cycle, when mice are naturally more active, may encourage more rearing behavior. The floor and cylinder are cleaned with 70% ethanol after each trial.

Open field: The open field test is a robust measure of recovery, assessing both anxiety/exploration behavior and recovery of spontaneous locomotor activity. Each animal is tested 1 day before injury and on Day 5 after injury in order to make a direct paired comparison of how the animal has recovered with respect to its prior mobility and willingness to explore. The mouse is placed in a white plastic box measuring 20.5 cm × 20.5 cm × 20.5 cm. An overhead camera records the animal's movement in four open fields simultaneously. The video is analyzed for average velocity and total distance moved and can also be analyzed for percent time spent near outer walls, as a measure of anxiety. The duration of each trial is 30 min. For all open field trials, each animal is placed in the same box that was used for its original trial to negate any slight differences in lighting or orientation. Each field is cleaned with 70% ethanol after each trial.

- *Outcomes*: All animals should have recovered sufficiently by Day 5 to have equivalent exploration and velocity compared to their baseline scores. If an animal exhibits significantly impaired activity/movement, it will be excluded from all future behavioral testing. The open field task has the added benefit of habituating mice to the field that is also used for novel object recognition.

3.3.2. Cognitive

Novel object recognition: The novel object recognition test is designed to assess hippocampal and cortical damage affecting memory. The test field is a box identical to the one used in the open field test except for two identical objects (e.g., padlock) placed in opposite corners, 1″ away from the walls so that the animals can move fully around them. Objects are chosen in advance to have irregular shapes or defects to appear more interesting and to be approximately the same size and of equal interest to the animals. Objects used frequently are a Masterlock padlock and a plastic clamp, each approximately 1.5″ long. On Day 6 (acquisition/training), the mouse is allowed to explore two identical objects. The trial is stopped when the total exploration time is 38 s, with 10 min maximum duration. Exploration time is measured with two stopwatches (one dedicated to each object), and video is recorded for confirmatory analysis off-line. On Day 7 (test), a novel object (e.g., 1 padlock replaced with plastic clamp) replaces the object the animal spent the least time exploring during acquisition. Time spent exploring the familiar versus novel object is recorded during a 5-min test.

- *Outcomes*: The percent of time spent exploring the new object during the 5-min test trial is used as the final outcome. Healthy animals with intact memory spend approximately 70% of the time exploring the novel object during the test trial. In contrast, injured animals spend significantly less time exploring the novel object, indicative of impaired memory. If an animal does not explore for the required 38 s during the training trial or 16 s during the test trial, then that animal is excluded from the test data.

Passive avoidance: The passive avoidance test is performed on Days 8 and 9 post-injury and is used to assess cognitive function by linking a punishment (short 2-mA shock to the paws) with moving to a preferred environment (dark chamber) during an acquisition/training period. The passive avoidance apparatus (Gemini Avoidance System, SD Instruments) consists of two chambers separated by an automated door. At the start of the acquisition period, both chambers are dark; the animal is placed in the right-hand chamber, and the door is closed. After a 10-s habituation, an overhead light turns on, and the door between the chambers opens. When the animal moves into the dark left-hand chamber, the door closes, and a 2-mA shock is delivered. Ten seconds after the shock is delivered, the animal is removed from the apparatus. Twenty-four hours later, the same protocol is repeated to measure retention of the negative association.

- *Outcomes*: After experiencing a single shock exposure in the dark chamber during the acquisition period, a fully healthy mouse will stay in the non-preferred environment (light chamber) until the maximum latency of 5 min is reached during the test trial 24 h later. Injured animals will enter the dark chamber again on the Day 2, ending the test with their latencies recorded. Animals that do not enter the dark chamber within 40 s on the acquisition day are removed from the test.

3.3.3. Affective

Post-stroke anxiety and depression are common in humans and can be demonstrated in rodents (24, 25). The two most commonly used tasks for assessing anxiety-like behavior after focal cerebral ischemia are the elevated plus maze (EPM) and the open field (OF) test. Both tasks require that the mice are fully mobile and have recovered any gross motor deficits that emerged following stroke.

Elevated plus maze: The EPM apparatus is raised approximately a meter above the floor and consists of two open arms (without walls) and two closed arms arranged in a "+" orientation. The maze is brightly lit to increase contrast between the open and closed arms. All four arms of the maze should be approximately 65 cm long and 5 cm wide. The walls around the closed arms should be approximately 15 cm high. The mouse is placed in the center of the apparatus facing an open arm, and the following measures are recorded: latency to enter arms, duration of time spent in closed and open arms, and frequency of arm entries. The total number of fecal boli excreted during the 5-min test is also recorded.

- *Outcomes*: A significant decrease in the percentage of open arm entries (open arm entries/total arm entries) is indicative of increased anxiety. The rationale is that anxious mice will prefer the relative safety of the dark enclosed arms to the bright open arms. A minimum of six arm entries are necessary for the task to be valid, and significant group differences in the total number of arm entries raise the concern that group differences in general locomotor activity could be a confounding factor in interpreting the EPM data as an index of anxiety-like behavior. An increase in the latency to enter an open arm and an increase in the number of fecal boli expressed can also be indicative of anxiety but tend to be less robust measures than percent open arm entries. Also, increased rearing in the enclosed arms is indicative of increased exploratory behavior.

Open field: In the open field test, anxiety-like behavior is assessed by comparing the amount of activity that occurs at the periphery versus the 100-sq cm zone in the middle of the apparatus. The open field apparatus is 40 cm × 40 cm × 38 cm (l × w × h) and consists of a sound-attenuating chamber with a light and a fan; movement of the mouse within the apparatus is recorded via a digital camera and computer. The bottom of the chamber is covered with clean bedding that is different in texture from the bedding used in the mouse's home cage. The mouse is placed in the center of the open field chamber at the beginning of the 5-min test. When the test is complete, the bedding is discarded, and all surfaces of the apparatus are cleaned with 70% ethanol.

- *Outcomes*: Decreased percentage of activity in the center of the apparatus (center activity/total activity) indicates increased anxiety-like behavior; the rationale is that anxious rodents avoid such open spaces.

The three most common tests used to measure depressive-like behavior after focal cerebral ischemia are the sucrose consumption test, the Porsolt swim test, and the tail suspension test.

Sucrose consumption test: The sucrose consumption test is based on the observation that rodents find sweet liquids, such as sucrose solutions, rewarding and will drink them preferentially when offered a choice between water and a sucrose solution. A decrease in preference for sucrose is interpreted as anhedonia, a decreased sensitivity to reward. Anhedonia is a core symptom of depression and can be effectively evaluated in rodents.

For this test, the mice must be housed individually. The first step is to familiarize the mice with the water bottles and sucrose solution. The mice should receive all of their water via water bottles beginning 2 weeks prior to stroke. (Be careful to monitor the mice for dehydration if they have been raised on an automatic watering system). One week prior to stroke, remove all water from the cage for the last 6 h of the light cycle. Then, at the onset of the dark cycle, place a bottle containing a pre-measured amount of water and a bottle containing a pre-measured amount of 3% sucrose in the cage. At the end of the first 6 h of the dark cycle, remove the bottles and record the amounts of water and sucrose that were consumed. Repeat this protocol the following night, and calculate the mean water and sucrose consumption for these two nights to determine baseline consumption volumes for each liquid. Following stroke, continue to monitor water consumption. When water consumption has returned to baseline levels, the post-ischemic component of the sucrose consumption test should be initiated using the same protocol as at baseline.

- *Outcomes*: In our experience, mice have a 2:1 preference for sucrose solution over water prior to stroke and no preference for sucrose after stroke. One precaution is to rule out drug effects on taste perception in mice treated with a drug during the sucrose consumption test.

The Porsolt forced swim test and the tail suspension test have similar rationales for measuring depressive-like behavior. The swim test assesses persistence of swimming, and the tail suspension test assesses persistence of struggling when the potential for escape is absent. Mice exhibiting depressive-like behavior are significantly more passive, i.e., floating rather than swimming or hanging rather than attempting to twist loose.

Porsolt swim test: This test requires pre-exposure to the task 1 week prior to stroke. The mouse is placed into an opaque cylinder tank measuring 24 cm in diameter and 53 cm high and filled to a depth of 30 cm with water maintained at $29 \pm 1°C$. The pre-stroke swim lasts 5 min, and the post-stroke swim lasts 3 min. The mice are scored for time spent actively swimming versus floating, i.e., no leg or tail movement that contributes to forward motion. The post-stroke swim test should not be conducted until the mice are fully alert and mobile. (Testing on post-surgical Day 7 has worked well for us.) Lethargy and motor deficits that hamper swimming are potential confounding factors and should be monitored closely during the recovery period prior to swim testing. Mice are excellent swimmers, even after stroke, if allowed sufficient recovery time. Unlike some strains of rats, mice do not typically dive under water, therefore, if they appear to be struggling or dipping below the surface of the water, remove them immediately to prevent possible drowning and re-assess the length of ischemia and the timing of the task post-surgically. After testing each animal, the water must be changed and the tank thoroughly cleaned with 70% ethanol prior to refilling.

- *Outcomes*: An increase in time spent floating compared to swimming is suggestive of increased depressive-like behavior.

Tail suspension test: Although it is not as commonly used as the Porsolt swim test for assessing depressive-like behavior, the tail suspension test has several advantages: (1) it does not require special equipment, (2) it is not confounded by mild motor deficits, and (3) it can be performed soon after ischemia. To perform this test, hang a mouse by its tail for 5 min and record the amount of time it spends immobile versus struggling (includes passive swinging).

- *Outcomes*: An increase in immobility is suggestive of increased depressive-like behavior.

4. Global Cerebral Ischemia Due to Cardiac Arrest/CPR

4.1. Clinical Connection

Each year in the United States, approximately 500,000 people suffer from cardiac arrest (CA) with many receiving cardiopulmonary resuscitation (CPR). Despite intense research over the past few decades, survival and neurological outcome remain poor. Only a small percentage of all patients who experience CA are successfully resuscitated and return to productive lives (<5%). Cognitive impairment is very common after CA, and it is estimated that 40–75% of all survivors suffer from some form of cognitive deficit (for reviews *see* 26, 27). The most common deficit is impaired short-term and long-term memory, followed by attention and executive functioning deficits. Recent advances in resuscitation techniques and response times have dramatically improved the success rate of resuscitation but have simultaneously increased the number of patients with debilitating long-term cognitive deficits. During CA, the brain temporarily loses its blood supply, leading to hypoxic/ischemic brain injury. The high prevalence of cognitive impairment is consistent with ischemic brain injury. Interestingly, the rate of return to spontaneous circulation is dramatically greater (~30%) than the rate of survival and hospital discharge (5–10%). This post-resuscitation loss of life is attributed to failure of the central nervous system to recover. If ischemic brain damage during CA could be prevented, survival may increase significantly. Therefore, basic science research aimed at protecting the brain following CA-induced ischemia is of great importance and has the potential of increasing survival and quality of life for many people.

4.2. CA/CPR Model – Methods

The most frequently used animal model to study global cerebral ischemia induced by cardiac arrest is ventricular fibrillation followed by CPR in the dog. The pig has also been frequently used (for review *see* 28). These CA/CPR models in large animals are powerful in their ability to mimic the clinical condition; however, using large animals is labor intensive and expensive, and there are no well-characterized behavioral tests for assessing functional recovery in these species. In contrast, the mouse model of CA/CPR has many advantages – most importantly, the availability of transgenic and knockout strains and well-described behavioral tests.

To study the neurological effects of global ischemia in mice without damaging peripheral organs, a variety of alternative global cerebral ischemia models have been used that produce forebrain cerebral ischemia without affecting whole body circulation, i.e., 2-vessel and 4-vessel occlusion with or without hypotension. Recently, our group developed and characterized a new mouse model of CA/CPR that uses KCl to stop the heart and complete CPR methods to resuscitate, including chest compressions, ventilation, and epinephrine (29). This model exhibits many of the hallmarks of CA/CPR in humans, including histopathological damage to sensitive neuronal populations, cognitive deficits, and affective deficits, making it a valuable animal model to study the consequences of CA/CPR.

The details of this method are as follows: Anesthesia is induced with 3% isoflurane and maintained at 1–1.5%. Temperature probes are inserted into the left temporalis muscle and rectum. A catheter is inserted into the right internal jugular vein for drug administration. The animal is endotracheally intubated with a 22-ga IV catheter and connected to a mouse ventilator. EKG leads are connected. Cardiac arrest is induced by injecting 70 µL cold 0.5 M KCl via the jugular catheter. Cardiac arrest is confirmed by a flatline EKG. During CA, the head temperature is maintained at 37–38°C using a water-filled coil. The animal's body temperature may be cooled to 28–32°C during CA to induce peripheral hypothermia and protect peripheral organs from ischemic damage. CPR is initiated 8–10 min after arrest by injecting 16 µg/mL epinephrine, administering chest compressions (300/min), and ventilating with 100% oxygen. If return of spontaneous circulation is not achieved within 2.5 min of initiating CPR, resuscitation efforts are abandoned.

The surgical preparations, anesthetic exposure, and temperature modulation described above are similar for sham-operated animals, except that sham animals receive an injection of 0.5 µL isotonic saline instead of KCl and 0.5 mL isotonic saline instead of epinephrine. The sham animals do not experience CA/CPR and are not exposed to ischemia, epinephrine, or chest compressions. It has been determined that decreasing brain temperature to 27°C during the sham procedure does not affect physiological, histological, or behavioral outcomes (9). However, during the procedure, the brain temperature of half of the sham animals in each experimental group is maintained at 37–38°C, while the brain temperature of the other half is maintained at 28–32°C in order to mimic temperature control in experimental animals.

Assuming no statistical differences, the two sham groups are collapsed into one sham group for statistical analysis.

4.3. Behavioral Outcome Assessment – Methods

Many of the behavioral tests used with the cardiac arrest model are used in the MCAO model. Post-operatively, CA/CPR animals are kept on a warming blanket for 24 h; otherwise, post-injury care is identical. Some behavioral issues are unique to CA/CPR-injured animals and can influence the interpretation of the data obtained from the same tests used following MCAO. Cardiac arrest is a major insult with a mortality rate 20–40% greater than the 45–60-min MCAO model used to assess behavioral outcomes following focal cerebral ischemia. Again, a double-blind, randomized study design with targeted replacement is optimal.

4.3.1. Sensory-motor

Extensive sensory-motor deficits are common after stroke due to concentrated neuronal damage in the cortex and striatum but are less common after cardiac arrest because damage is limited primarily to the hippocampus. Indeed, during the first 4–5 days of recovery from CA/CPR, mice are hyperactive. The increase in locomotor activity likely reflects transient changes in glutamate transmission. It is important to allow enough time for locomotor activity to return to baseline levels before engaging in complex behavioral testing. Therefore, we perform neuroscore and open field (as described above) to assess recovery.

4.3.2. Cognitive

The hippocampus is a crucial region of the brain for performing spatial tasks, which makes the Morris water maze (MWM) an outstanding measure of cognitive performance after CA/CPR. The tank measures 1 m in diameter and is filled with opaque, white water that is maintained at 27°C. A platform is submerged 0.5 cm beneath the water surface. Each animal undergoes a single habituation trial (60 s) on the first day of testing with the platform removed. Beginning on the following day, each mouse undergoes a hidden platform acquisition trial 3 times a day for 10 days. For each trial, the mouse is placed in the pool at one of the four start positions around the edge of the pool and allowed to swim to the hidden platform. If the mouse does not find the platform after 60 s, the tester places it on the platform. After the tenth acquisition day, the platform is moved to a new location in a different quadrant of the pool (reversal training). For four additional days of training, 3 trials/day are conducted to determine how readily each mouse learns the new platform location. At the end of the acquisition training trials, a 30-s probe trial, for which the platform is removed, is conducted every other day. Time spent and distance traveled in the goal quadrant that normally contains

Morris Water Maze

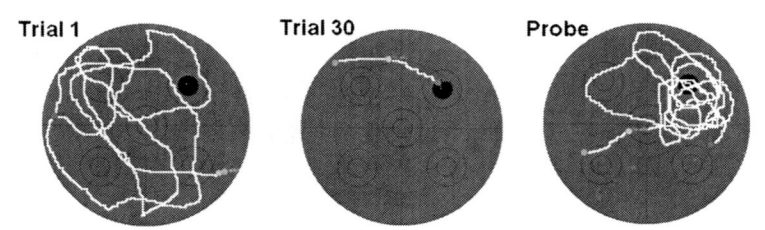

Fig. 12.3. On the first trial of the MWM, mice explore the entire tank searching for the hidden platform (the *white line* represents the path the mouse swam and the *black circle* represents the hidden platform). After 10 days of 3 trials/day, the typical mouse has learned the location of the platform and can swim directly to it from any place in the tank (Trial 30). During the probe task the platform is removed, and if the mouse has learned the location of the platform it will continue to swim overtop of the platform's previous location, as indicated in the *right-hand figure* (Probe). The *filled black circle* indicated the location of the platform. The other four *circles* are suggested locations for moving the platform during reversal testing and are not used in analysis of water maze data when the platform is located at a different point.

the platform are compared to the opposite quadrant. **Figure 12.3** illustrates a representative Morris water maze experiment performed on a control mouse. On the last day of water maze testing, a platform that is visible above the water surface is placed in a previously unused quadrant to assess gross visual ability and motivation to escape the water. Throughout the Morris water maze test, tracking software (HVS Image, San Diego, USA) is used to determine latency to reach the platform, distance swum, and mean swim speed for each trial. Because of the size of the MWM tank, it is not feasible to use new water with each trial, but the water's surface should be skimmed to remove any debris and the water mixed to disrupt any scent trails between each trial. Be sure that the surface is calm before starting each trial. Both novel object recognition and passive avoidance tasks described above are also suitable for assessing cognitive performance following CA/CPR.

4.3.3. Affective

Patients surviving cardiac arrest often report depression, anxiety, and social isolation. These changes in affective behaviors have also been documented in mice after CA/CPR (9). The same tests that assess affective changes following stroke, i.e., the elevated plus maze, open field, Porsolt swim test, and sucrose consumption (described above), can be used following CA/CPR but not until the hyperactivity has resolved. However, the tail suspension task is not recommended following CA/CPR because the mice tend to be more aggressive and will bite the tester, and their tails are more fragile.

5. Conclusions

Brain injury arising from focal or global ischemia remains a significant clinical problem. Furthermore, treatments for ischemic injury that have proved effective at the bench have also failed to translate well to patients. One key reason for translational failure is that many experiments fail to test functional efficacy of candidate treatments in animal models or do not use well-constructed, behavioral paradigms. Key aspects to enhance behavioral assessment in mouse models include consideration of natural mouse characteristics, proper attention to genetic strain, age, and sex effects, elimination of housing and breeding-related confounding variables, exquisite control of the testing environment, and selection of tests that are feasible for recovering animals.

Acknowledgments

We gratefully acknowledge Ms Kathy Gage for her editing expertise. Supported by NIH NS 49210.

References

1. Fisher M, Feuerstein G, Howells DW, Hurn PD, Kent TA, Savitz SI, Lo EH for the STAIR Group (2009) Update of the Stroke Therapy Academic Industry Roundtable preclinical recommendations. Stroke 40:2244–2250
2. Sudlow CL, Warlow CP (1997) Comparable studies of the incidence of stroke and its pathological types: results from an international collaboration international stroke incidence collaboration. Stroke 28:491–499
3. Cheng J, Hurn PD (2010) Sex shapes experimental ischemia brain injury. Steroids (Nov 2009) [epub ahead of print]
4. Majid A, He YY, Gidday JM, Kaplan SS, Gonzales ER, Park TS, Fenstermacher JD, Wei L, Choi DW, Hsu CY (2000) Differences in vulnerability to permanent focal cerebral ischemia among 3 common mouse strains. Stroke 31:2707–2714
5. Wellons JC, Sheng H, Laskowitz DT, Mackensen GB, Pearlstein RD, Warner DS (2000) A comparison of strain-related susceptibility in two murine recovery models of global cerebral ischemia. Brain Res 868:14–21
6. Beckmann N (2000) High resolution magnetic resonance angiography non-invasively reveals mouse strain differences in the cerebrovascular anatomy in vivo. Mag Reson Med 44:252–258
7. Schulte-Herbruggen L, Klehmet J, Quarcoo D, Meisel C, Meisel A (2006) Mouse strains differ in their susceptibility to post-stroke infections. Neuroimmunomodulation 13:13–18
8. Oliff HS, Coyle P, Weber E (1997) Rat strain and vendor differences in collateral anastomoses. J Cereb Blood Flow Metab 17:571–576
9. Neigh GN, Karelina K, Glasper ER, Bowers SL, Zhang N, Popovich PG, DeVries AC (2009) Anxiety after cardiac arrest/cardiopulmonary resuscitation: exacerbated by stress and prevented by minocycline. Stroke 40:3601–3607
10. Sugo N, Hurn PD, Morahan MB, Hattori K, Traystman RJ, DeVries AC (2002) Social stress exacerbates focal cerebral ischemia in mice. Stroke 33:1660–1664
11. Van Loo PL, Mol JA, Koolhaas JM, Van Zutphen BF, Baumans V (2001) Modulation of aggression in male mice: influence of group size and cage size. Physiol Behav 72:675–683

12. Crabbe JC, Wahlsten D, Dudek BC (1999) Genetics of mouse behavior: interactions with laboratory environment. Science 284:1670–1672
13. Weil ZM, Karelina K, Su AJ, Barker JM, Norman GJ, Zhang N, Devries AC, Nelson RJ (2009) Time-of-day determines neuronal damage and mortality after cardiac arrest. Neurobiol Dis 36:352–360
14. Weil ZM, Norman GJ, Barker JM, Su AJ, Nelson RJ, DeVries AC (2008) Social isolation potentiates cell death and inflammatory responses after global ischemia. Molr Psychiatry 13:913–915
15. Craft TKS, Glasper ER, McCullough L, Zhang N, Sugo N, Otsuka T, Hurn PD, DeVries AC (2005) Social interaction improves stroke outcome. Stroke 36:2006–2011
16. Karelina K, Norman GJ, Zhang N, Morris JS, Peng H, DeVries AC (2009) Social isolation alters neuroinflammatory response to stroke. Proc Natl Acad Sci (USA) 106:5895–5900
17. DeVries AC, Nelson RJ, Traystman RJ, Hurn PD (2001) Cognitive and behavioral assessment in experimental stroke research: will it prove useful? Neurosci Biobehav Rev 25:325–342
18. Quinn TJ, Dawson J, Walters MR, Lees KR (2009) Functional outcome measures in contemporary stroke trials. Int J Stroke 4:200–205
19. Sampei K, Goto S, Alkayed NJ, Crain BJ, Korach KS, Traystman RJ, Demas GE, Nelson RJ, Hurn PD (2000) Stroke in estrogen receptor-alpha-deficient mice. Stroke 31:738–743
20. Longa EZ, Weinstein PR, Carlson S, Cummins R (1989) Reversible middle cerebral artery occlusion without craniectomy in rats. Stroke 20:84–91
21. Hattori K, Lee H, Hurn PD, Crain BJ, Traystman RJ, DeVries AC (2000) Cognitive deficits after focal cerebral ischemia in mice. Stroke 31:1939–1944
22. Li X, Blizzard KK, Zeng Z, DeVries AC, Hurn PD, McCullough LD (2004) Chronic behavioral testing after focal ischemia in the mouse: functional recovery and the effects of gender. Exp Neurol 187:94–104
23. Uchida M, Palmateer J, Herson PS, DeVries AC, Hurn PD (2009) Dose-dependent effects of androgens on outcome following focal cerebral ischemia in adult male mice. J Cereb Blood Flow Metab 29:1454–1462
24. Craft TK, DeVries AC (2006) Role of IL-1 in poststroke depressive-like behavior in mice. Biol Psychiatry 60:812–818
25. Winter B, Juckel G, Viktorov I, Katchanov J, Gietz A, Sohr R, Balkaya M, Hörtnagl H, Endres M (2005) Anxious and hyperactive phenotype following brief ischemic episodes in mice. Biol Psychiatry 57:1166–1175
26. Moulaert VRMP, Verbunt JA, van Heugten CM, Wade DT (2009) Cognitive impairments in survivors of out-of-hospital cardiac arrest: a systemic review. Resuscitation 80:297–305
27. Arawwawala D, Brett SJ (2007) Clinical review: beyond immediate survival from resuscitation – long-term outcome considerations after cardiac arrest. Crit Care 11:235
28. Traystman RJ (2003) Animal models of focal and global cerebral ischemia. ILAR J 44:85–95
29. Kofler J, Hattori K, Sawada M, DeVries AC, Martin LJ, Hurn PD, Traystman RJ (2004) Histopathological and behavioral characterization of a novel model of cardiac arrest and cardiopulmonary resuscitation in mice. J Neurosci Methods 136:33–44

Chapter 13

A Comparative Analysis of Cellular Morphological Differentiation Within the Cerebral Cortex Using Diffusion Tensor Imaging

Lindsey A. Leigland and Christopher D. Kroenke

Abstract

Diffusion tensor imaging (DTI) is a magnetic resonance imaging (MRI) technique that provides information about cellular microstructure through measurements of water diffusion. Because inferences about neuroanatomy can be made from DTI, this methodology has been used to characterize cellular morphological changes associated with development of the cerebral cortex. Currently, however, the specific anatomical changes associated with DTI measurements directed at the cerebral cortex are incompletely characterized. Here, data collected in several laboratories, investigating five species (mouse, rat, ferret, baboon, and human), are compared to determine whether similarities in the trajectory of DTI measurements with development exist in the literature. Specifically, rates of change in fractional anisotropy (FA) of water diffusion were compared to rates of neuroanatomical development (based on the occurrence of specific neural events) in each species. In all species, decreases in FA with development were accurately approximated by fitting data to the same mathematical expression of exponential decay. Additionally, a high degree of correlation was found between rates of FA decay and rates of neuroanatomical development. This suggests that a common mechanism underlies decreases in FA with development across species. These results have two major implications. The ability of DTI to detect changes in neuroanatomy in the normal developing cerebral cortex introduces the potential for the use of this methodology in detecting cortical abnormalities associated with various developmental disorders. Additionally, the comparable patterns of neurodevelopment, and hence FA, across species imply that DTI methodology applied in non-human species can provide information about the human condition.

1. Introduction

Magnetic resonance imaging (MRI) is a non-invasive imaging technique that holds much potential for investigating brain disorders and brain development. A subclass of MRI experiments collectively termed diffusion tensor imaging (DTI) is particularly well suited to study the cellular-level bases of tissue changes associated with development and pathology. In DTI experiments, the image intensity recorded by MRI is rendered sensitive to diffusion-mediated displacement of water molecules over distances of approximately 10 μm in the tissue under study (1, 2). These measurements are extremely powerful because biological membranes impede water displacement due to diffusion (3), and thus the cellular-scale structure of tissue is reflected in water diffusion measurements. In tissue that is highly ordered on the cellular scale, water diffusion exhibits a directional dependence (4). Directional dependence in molecular diffusion is termed diffusion anisotropy and is most frequently measured within the context of MRI through application of the diffusion tensor formalism (5–7). The extent of diffusion anisotropy reflects the degree of cellular-scale order in tissue and is frequently quantified in terms of fractional anisotropy (FA), a parameter that ranges from 0 (isotropic diffusion, unstructured tissue) to 1 (extremely anisotropic diffusion, well-ordered tissue).

Several recent studies of brain development have provided evidence that DTI strategies are of potential utility for characterizing the development of the cerebral cortex. Water diffusion anisotropy decreases with maturation of the cerebral cortex, and it is believed that the observed changes in diffusion anisotropy reflect cellular morphological changes relevant to cortical development. In the immature state, at a time when cortical FA is high, neurons are undifferentiated and highly radially oriented (8, 9); subsequently, neurons undergo morphological differentiation, and it has been found that this event in cortical development is temporally associated with a decrease in diffusion anisotropy (10, 11).

Significant diffusion anisotropy in the immature cerebral cortex was first observed in studies of cats (12) and pigs (13). More recently, such measurements have been extended to several other species including humans. However, a systematic comparison of the rate of FA changes has yet to be referenced to species-specific rates of brain development, estimated in comparative analyses of timing of developmental events (14–16). A comparison of DTI studies to independent studies of neuroanatomical development could therefore provide support that a common underlying developmental process gives rise to patterns observed in DTI measurements. DTI-based studies of cortical development are reviewed in the following sections, and methodological issues associated with differences in acquisition and analysis procedures among laboratories are discussed. To obtain independent experimental confirmation that changes in cortical diffusion anisotropy occur at a rate that is consistent with the rate of morphological differentiation of the cerebral cortex across species, the comparative

DTI data reviewed here are referenced to species-specific rates of brain development estimated from comparative analyses of developmental event timings reported in classical anatomical studies (14–16).

2. The Phenomenon of Anisotropic Water Diffusion Within the Developing Cerebral Cortex

2.1. Water Diffusion Is Anisotropic Within Immature Cortical Gray Matter

Immediately following migration of pyramidal neurons from ventricular zones of the forebrain to the cortical plate, neurons exhibit simple morphology characterized by elongated cell somas and undifferentiated, radially oriented apical dendrites (9, 17). As the cerebral cortex matures, obliquely oriented collaterals of apical dendrites, basilar dendrites, and axons arborize to provide a scaffold for the formation of functional synapses (18). Changes in water diffusion anisotropy take place along with these cellular-level morphological transformations. In the immature cortex, preferential restriction of diffusion in directions parallel to the pial surface and relative lack of restriction in directions parallel to apical dendrites of pyramidal neurons (10–13, 19) are thought to be the primary causes of the prominent diffusion anisotropy seen at this time. As morphological differentiation occurs within the developing cerebral cortex, cortical FA decreases. Diffusion within mature cortex is nearly uniformly restricted in all directions, and diffusion anisotropy is measurable but subtle (20, 21). The temporally coincident evolution of cortical diffusion anisotropy with morphological development is schematized in **Fig. 13.1** (10).

2.2. Temporal, Laminar, and Regional Patterns in the Loss of Cortical Diffusion Anisotropy with Development Coincide with Morphological Differentiation of Neurons

Although there is a strong association between the loss of cortical diffusion anisotropy and morphological differentiation of the cerebral cortex, the ability to interpret FA values within the developing cerebral cortex in terms of the underlying anatomical properties of brain tissue will require a quantitative link between FA changes and specific cellular morphological changes to be established. The component of cerebral cortex termed the neuropil, which consists of axons, dendrites, and associated extracellular space, represents 70–80% of the cerebral cortical volume fraction at maturity (22, 23). Given the relative sizes (volume fractions) of other elements such as glial cells (3.6% (23)), vasculature (4.3% (23)), and neuron cell somas (22% (23)), it has been proposed that the loss of cortical diffusion anisotropy with brain maturation

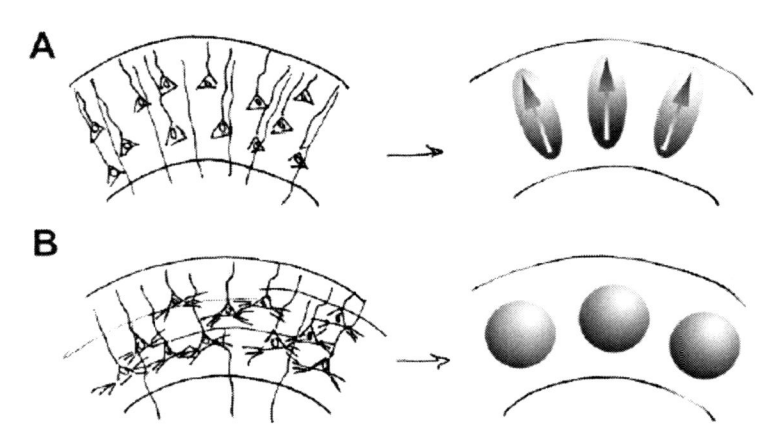

Fig. 13.1. Hypothesized relationship between cortical neuronal differentiation and cortical diffusion anisotropy, taken from (10) by permission of Oxford University Press. (a) The *left image* displays neurons in the immature cortex that are undifferentiated and radially organized. It is hypothesized that this neuroanatomical organization causes water diffusion in the immature cortex to be highly directionally dependent, as represented by the ellipsoidal shapes on the *right*. (b) The *left image* is an example of neuronal organization in the mature cortex, wherein neurons are highly differentiated and much of the radial organization found at earlier stages of development is lost. It is hypothesized that the high degree of differentiation of neural components in the mature cortex leads to highly isotropic water diffusion represented by the spherical shapes on the *right*.

can be attributed to morphological differentiation of the neuropil (**Fig. 13.1**) (10).

If cortical FA reflects the degree of differentiation of the neuropil rather than some other anatomical transformation, then FA would be expected to depend directly on the age of neurons within a given cortical region. Laminar (8) and regional (24, 25) patterns of neurogenesis have been extensively characterized in several species and are known to produce gradients in neuron age throughout the cerebral cortex. Previously, laminar and regional patterns of cortical FA have been examined at multiple stages of development following neurogenesis in an attempt to evaluate whether morphological properties of the neuropil determine cortical FA values in immature cortex.

Pyramidal neurons of the cerebral cortex are generated in an inside–out manner; neurons of deep cortical layers are born on earlier dates than neurons of more superficial layers (8). As a result, there is a laminar gradient in the age of these cells. Correspondingly, at early stages of cortical development, neurons of deep cortical cell layers have developed more extensive networks of obliquely oriented collateral branches of apical and basal dendrites than have neurons of superficial cell layers (17, 18). Based on these observations, a laminar gradient in cortical diffusion anisotropy would be expected, with superficial layers exhibiting higher anisotropy than deeper layers. Examination of the laminar dependence of cortical FA in fetal baboon brain provides evidence that a superficial-to-deep, high-to-low intracortical FA gradient exists (26, 27). Laminar gradients in FA have also been documented in rat cerebral cortex (28, 29).

A regional gradient in neurogenesis has also been described, in which neurons of a given cortical lamina are born earlier near the source of a transverse neurogenetic gradient (TNG (24, 25)) than at distal extremes of the cerebral cortical sheet. The neurodevelopmental mechanism giving rise to the TNG of cortical pyramidal neurons has not yet been elucidated, however the TNG source within the cortical sheet has been mapped onto models of the cortical surface for rodent species (24, 30), carnivores (ferrets (31, 32) and cats (33)), and primates (9). In ferrets and cats, the TNG gives rise to a 5-day age difference between neurons of a given lamina located at the TNG source and neurons located at the occipital pole (31–33). In the ferret (34), cortical FA values measured at ages P6 through adulthood were fitted to a model in which FA decreases exponentially with postnatal age following migration of pyramidal neurons to the cortical plate. From this analysis, it was found that the difference in FA between the TNG source and occipital pole corresponds to an age difference of 5 days, in agreement with previous histological estimates (31–33) and in close correspondence to differences in the ages of layer II neurons estimated by autoradiographic cell-birthdating studies of ferret somatosensory and visual cortical areas (35, 36). Similar rostral/lateral to caudal/dorsal cortical FA patterns corresponding to the TNG have also been observed in rat (28), baboon (27), and human (37).

Quantitative characterizations of laminar and regional patterns of cortical FA have thus established a correspondence with the age of pyramidal neurons within ferrets and baboons. This association is suggestive that a relationship exists between changes in cortical FA and morphological development of the neuropil. To provide further evidence of this potential relationship, a comparative analysis of data published from several laboratories is presented here to determine whether FA changes observed across species correspond to a common stage of brain development.

2.3. Application of Diffusion Anisotropy to Monitor Cerebral Cortical Development Is a Potential Strategy to Use DTI for Early Detection of Neurodevelopmental Disorders

A morphological abnormality common to individuals affected by a diverse array of neurodevelopmental disorders has been observed in which dendritic arbors of the early developing cerebral cortex are less elaborate than in age-matched control individuals (reviewed by Kaufmann and Moser (38) and in a special issue of *Cerebral Cortex* (39)). Simplified cerebral cortical neuronal morphology has been particularly well documented in analyses of Golgi-stained tissue from experimental animals exposed to alcohol during the fetal period (40–43) and in tissue from individuals with Rett syndrome (44) (as well as in MECP2-deficient mice (45), a genetic model of Rett syndrome). Given the potential relationship between cortical FA and morphological differentiation of the neuropil, it is possible that DTI measurements could be used to detect anatomical abnormalities within a class of neurodevelopmental disorders.

Current strategies that utilize DTI to characterize the neurobiological basis of several neurodevelopmental disorders have mostly focused on cerebral white matter (WM) of affected individuals at maturity or stages of development subsequent to the loss of cortical FA. Reduced WM FA relative to controls has been reported within contexts of disorders with genetic (e.g., Rett syndrome (46, 47), phenylketonuria (48)), environmental (e.g., fetal alcohol spectrum disorder (49–52), premature birth (53, 54)), or multifactorial (e.g., autism (55), schizophrenia (56–58)) origins. It is known that plasticity in the CNS decreases after critical/sensitive periods which end prior to the completion of myelination within brain WM (59–61). These critical periods are specific periods of time, during which particular functional systems can be formed. Abnormal development of, or coordination among, multiple neural components (i.e., neurons, dendrites, axons, synapses, sensory afferents) during this time could lead to abnormal maturation and permanent dysfunction. For example, deprivation of visual sensory experience early in brain development has been demonstrated to produce permanent effects on visual system processing (61–63). Thus, a DTI-based approach aimed at detecting morphological characteristics of disease within the developing cerebral cortex could potentially provide a new strategy that would extend the capabilities of DTI toward identification of anatomical abnormalities prior to the end of these critical developmental periods.

3. Methods for Quantifying Changes in Cortical FA with Brain Development

An objective of this review is to determine whether the high-to-low change in cortical FA with age for a given species occurs at a rate that would be predicted based on other comparative studies of brain development. In order to make such a comparison, cortical FA measurements reported by several laboratories, using a variety of experimental approaches, must be combined. It is therefore important to recognize methodological differences between the DTI experiments reviewed here so that potential sources of differences between studies may be appreciated. The following discussion summarizes the primary factors that lead to inter-laboratory differences in cortical FA values. As can be observed in **Fig. 13.2**, differences in the magnitude of FA

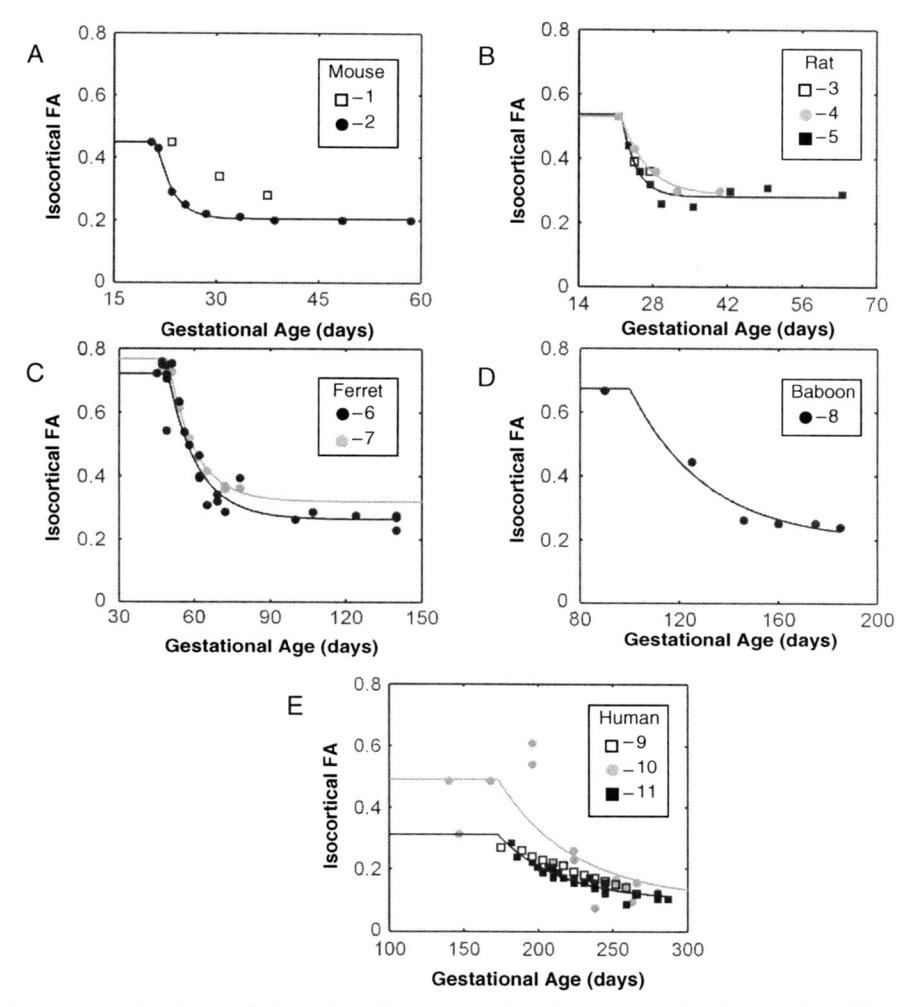

A Comparative Analysis of Cellular Morphological Differentiation   335

Fig. 13.2. Isocortical fractional anisotropy is plotted as a function of post-conceptional age in five different species. Data from 11 independent studies were fit to the empirical mathematical model expressed in Equation [5]. Fractional anisotropy across species shows an exponential decline with development over the age ranges pictured. Data in **a–e** are pictured for mouse, rat, ferret, baboon, and human species, respectively. Squares indicate data collected in vivo, while circles represent data collected from post-mortem samples. Open shapes indicate data that were insufficient for modeling a time constant for an exponential decline in FA (τ_{FA}); 1 = (89), 2 = (92), 3 = (29), 4 = (28), 5 = (75), 6 = (34), 7 = (93), 8 = (27), 9 = (37), 10 = (10), 11 = (76).

values exist between studies. However, an underlying assumption of the analysis presented here is that the rate of change in FA within each of the studies reviewed is, neglecting species-specific differences, consistent between studies despite differences in absolute FA values at a given age. Justification of this approximation is based on acknowledgment that experimental conditions are held fixed within each of the studies shown in **Table 13.1** with respect to the factors discussed below.

Table 13.1
Summary of DTI parameters and methodology employed in the reviewed articles

Species	Mouse		Rat			Ferret		Baboon	Human		
References	(92)	(89)	(29)	(75)	(28)	(93)	(34)	(27)	(10)	(76)	(37)
Tissue	Ex vivo	In vivo	In vivo	In vivo	Ex vivo	Ex vivo	Ex vivo	Ex vivo	In vivo	Ex vivo	In vivo
Measure	FA	FA	FA	FA	FA	RA	FA	RA	A_σ	FA	FA
N/point	~4	13–16	Unknown	~5	2	1–3	1–2	1–3	24 total	4–12	37 total
Time points (post-conceptional age)	20–59	23–73	24–28	21–64	21–41	45–140	47–241	90–185	182–287	140–371	175–266
Magnetic field strength (T)	9.4	9.4	4.7	7.0	9.4	4.7	4.7/11.7	4.7	1.5	1.5	1.5
Image resolution (RO/PE/slice) (mm)[a]	≤.125 × .125 × .125	.2 × .2 × .5	.125 × .125 × .5	.27 × .27 × .5	≤.14 × .14 × .14	≤.35 × .35 × .35	≤.35 × .35 × .35	≤.5 × .5 × .5	1.9 × 1.9 × 5	≤.94 × .94 × 3	1.4 × 1.4 × 3
TE (ms)	34	23.3	40	38	34	67	42–67	67	106	100	99.5
TR (ms)	800	2,000	1,500	4000	700	3,400–5,800	4,000–10,500	3,450–4,000	3,000	8,000	7,000
No. of dir.	6	6	6	21	6	25	22/25	25	7	10	6
b value (s/mm²)	1,758	~700	763	700	1,000	200–12,100	200–2,500	500–12,500	340–800	700	600
ROI location (or surface analysis)	L,R; mid-coronal	Mid-sagittal	Parietal cortex	L,R; mid-coronal	Assigned to functional cortical areas	Lateral to caudate nucleus	Surface	Surface	Parietal, occipital cortices	Parietal, occipital, frontal, temporal cortices	Precentral, post-central, sup. frontal, sup. occipital gyri

[a] RO, PE, and Slice refer to image voxel sizes along imaging readout, phase encode, and slice-selection directions, respectively.

3.1. DTI Measurements Have Been Performed on a Diverse Array of Experimental Subjects

3.1.1. In Vivo Versus Post-mortem

The use of post-mortem tissue permits long scanning times for imaging experiments, which can be used to obtain high-image spatial resolution with a high signal-to-noise ratio (SNR). As a result, MRI investigations of the CNS utilizing post-mortem tissue have seen increasing use over recent years (19, 26–28, 34, 64–71), and much progress has been made toward understanding the degree of correspondence between in vivo and ex vivo measurements (64, 65, 67, 70–74). Due to the sensitivity improvements that can be realized, many of the investigations of cortical FA changes with development listed in **Table 13.1** have been performed on post-mortem tissue.

One difference in water diffusion between in vivo and post-mortem tissue is that the apparent diffusion coefficient (ADC) of water decreases by a factor of ~2.7 (71) following death and tissue perfusion fixation. This decrease in the ADC likely arises from several differences between living and post-mortem tissue, including physiological (e.g., membrane permeability to water) and temperature (magnet bore vs. physiological temperature) variations (64, 65, 72, 73). A practical consequence of the low ADC value in post-mortem tissue is a requisite pulsed-field gradient system capable of producing strong (approximately 20 G/cm or greater) magnetic field gradients to achieve appropriate diffusion-sensitization settings ("b values") to characterize water diffusion without requiring excessively large echo time (TE) values. Such gradient capabilities are standard on small animal imaging MRI systems but are not included in human clinical instrumentation.

In spite of the large difference in ADC between living and post-mortem tissue, diffusion anisotropy differences between the two tissue states in adult GM and WM structures have been found to be less significant (64, 65, 72, 73) if present at all (67, 71). Specifically, within the context of the developing cerebral cortex, time courses of cortical FA changes with development in rat in vivo (29, 75) and post-mortem (28) are highly similar, as are comparisons between post-mortem human (68) or non-human (27) primates and in vivo human studies (10, 11, 37). The analysis presented below provides further evidence that similar conclusions with regard to development of the cerebral cortex can be drawn from in vivo and post-mortem DTI measurements.

With regard to comparisons between measurements performed on living and post-mortem tissue, an additional layer of complexity is introduced in studies of post-mortem tissue that has not undergone fixation with aldehydes. Empirical data show that

FA measurements obtained from post-mortem brain tissue vary with the amount of time that elapses between death and tissue fixation due to autolysis (64, 72). An implication of this finding is that measurements conducted on unfixed post-mortem tissue will depend on the post-mortem interval prior to data collection. For obvious reasons, this aspect of a DTI study of human subjects is not readily controlled (76, 77).

3.1.2. Measurements in Species with Gyrencephalic Versus Lissencephalic Brains

One striking difference between cerebral cortices of lower (mouse, rat) and higher (ferret, old world monkey, human) animal forms studied by DTI is the presence of sulcal and gyral fissures on the latter at maturity. Notably, the cerebral cortices of gyrencephalic species do not posses gyri and sulci at the developmental stage immediately following migration of pyramidal neurons to the cortex. The loss of cortical FA with development coincides with cortical folding in these species. Therefore, the question arises whether mechanical forces associated with cortical folding influence cortical FA measurements exclusively within gyrencephalic species and thus confound comparisons with rodents. The analysis below demonstrates that, despite significant differences in the extent of cortical folding between species, cortical FA changes occur over a remarkably consistent period of brain development. Further, the potential role of gyrus/sulcus formation on the loss of cortical FA was specifically investigated in a study of ferret brain development (34). It was observed that cortical FA was slightly larger within sulci than gyri at a given age, but that variation throughout the cortical sheet in FA due to position relative to gyri/sulci was modest relative to temporal, laminar, and regional patterns in FA (34).

3.2. DTI Measurements Have Been Performed Using Varying Experimental Settings

3.2.1. Image Resolution

In the studies summarized in **Table 13.1**, a range of image resolution settings were used for measuring diffusion anisotropy. As a result of variation in the thickness of the cerebral cortical wall with brain volume among species, the laminar gradient in cortical FA is usually expressed over a different distance scale for different species. Comparative studies of the dependence of mean cortical thickness upon cortical volume (78, 79) lead to estimates that rat cerebral cortex (mean cortical thickness = 1.05 mm) is 39% of that of human (2.72 mm). Given these dimensions, it should be recognized that cortical FA values typically reflect averaged morphological tissue properties over multiple cortical lamina. This is particularly true in the developing cerebral cortex, which is thinner than the cortex at maturity.

3.2.2. Number of Diffusion-Sensitization Directions

In DTI studies directed at WM, it has become recognized that water diffusion within volume elements that overlap multiple fiber tracts of different orientations can give rise to multimodal diffusion tensors (80). In order to infer complex WM fiber architecture, it is necessary to make use of measurement schemes in which several (e.g., 25 or more) diffusion-sensitization directions are sampled. Image acquisition procedures such as high-angular resolution diffusion imaging (81, 82), diffusion spectrum imaging (83), or q-ball (84) can be used to facilitate such sophisticated DTI analyses. In contrast to the multimodal fiber orientation distribution expected within some regions of WM, reports on the distribution of axonal and dendrite orientations inferred from DTI measurements of the developing cerebral cortex to date have revealed unimodal (85), radially oriented axial symmetry. The corresponding axial symmetry observed for water diffusion within the immature cerebral cortex enables data to be modeled with a less complex DTI expression than that typically used for WM (26, 85). Therefore, analyses of cortical diffusion anisotropy have, in many cases, made use of a modest number (less than 12) of diffusion-sensitization directions. Generally, however, in DTI investigations, the SNR increases with the amount of time devoted to acquiring data, and thus studies with additional diffusion-sensitization directions gain precision relative to DTI measurements that sample a coarser set of directions over a shorter period of time.

3.3. Multiple Post-acquisition Analyses Have Been Employed to Quantify Diffusion Anisotropy

In order to perform a quantitative analysis of cortical FA changes with age, a reliable method of extracting FA values from MRI data is needed. Specifically, it is important that an approach adopted to quantify cortical FA avoids potential covariation due to dependencies on laminar and regional position. A number of strategies have been described for accomplishing this. The array of analysis procedures may be broadly categorized into region of interest (ROI) and surface-based methods.

3.3.1. ROI Analyses

In the majority of studies summarized in **Table 13.1**, cortical FA is quantified within a number of individuals through manual delineation of a region of the cerebral cortex and reporting an average within the identified region. Such an approach can be used to produce highly reliable estimates of cortical FA because manual supervision can ensure that consistent regions are delineated across individuals and can verify the location of the ROI boundaries with respect to regional and laminar position. One drawback, however, is that ROI dimensions are large relative to individual cortical lamina and therefore averaged FA values include contributions from cortical tissue that has undergone varying extents of morphological differentiation. Lastly, due to the need for a rater to prescribe the location of ROI boundaries, this technique is susceptible to rater-induced bias.

Huang et al. (28) have described an ROI-based approach that addresses cortical FA variation due to laminar position. For each region-specific ROI, these investigators quantified FA within three zones defined relative to distance from the pial surface along a line normal to the local surface tangent plane. Laminar specificity is thus obtained from the authors' reports of cortical FA changes with age in outer, medial, and inner cortical zones (28).

3.3.2. Surface-Based Methods

An approach that can be used to address regional and laminar patterns in cortical FA utilizes surface models of the cortical sheet (86). For visualization purposes, Huang et al. (28) projected cortical FA values onto a cortical surface model by color-coding the model according to the FA within the voxel intersected by the surface. By incorporating surface registration procedures (86), it is possible to delineate the boundaries of an ROI on an atlas surface (to produce a "surface-based ROI" rather than a "volume-based ROI" as described above) and automatically project the ROI boundaries on surface models representing individual brains in order to facilitate quantitative analyses of cortical FA values. This approach was adopted in a study of baboon development (27). Surface-based methods also can facilitate an analysis that accommodates laminar variation in cortical FA. As an example, an approach similar to Huang et al. (28) was adopted in a study of ferret development, in which line segments normal to the local surface tangent plane were defined for every node on each animal's surface (34). To extract cortical FA values specifically from the immature superficial lamina of the developing cortex, the maximum cortical FA value intersected by each line segment was projected onto the surface models. Thus, surface-based methods can be used to generate automated analyses that account for potential variation in cortical FA values due to laminar and regional heterogeneities in temporal patterns of morphological differentiation.

4. Interspecies Comparison Between the Rate of Loss in Cortical Diffusion Anisotropy and the Rate in Which Developmental Milestones Are Surpassed

Indices of cortical diffusion anisotropy have been characterized throughout early development in five species: mouse (*Mus musculus*), rat (*Rattus norvegicus*), ferret (*Mustela putorius furo*), baboon (*Simia hamadryas*), and human (*Homo sapiens*) (**Table 13.1**). For the subset of (**Table 13.1**) studies suitable for the analysis described below, the species-dependent rate of reduction in cortical FA is quantified in terms of an exponential decay time constant, τ_{FA}. The estimated τ_{FA} values are compared to

4.1. Conventions Adopted to Facilitate Interspecies Comparisons

independently estimated rates of brain development for these species using a model developed by Finlay et al. (14–16).

It is well recognized that the length of gestation is highly variable relative to other aspects of development across species (14–16). As a result the relevant measure of age is relative to conception rather than to birth. In contrast, several DTI investigations report age in terms of postnatal age. Gestation lengths of 18.5, 21.5, 41, 185, and 270 for mouse, rat, ferret, baboon, and human, respectively (15), have been assumed for purposes of converting postnatal to post-conceptional age.

An additional convention adopted herein is that the term cerebral cortex is used synonymously with the terms "isocortex" or "neocortex." The inclusion of isocortical regions was a criterion for inclusion of published studies in this analysis because allocortical and deep gray matter structures do not undergo developmental trajectories of migration, followed by morphological differentiation, similar to the isocortex (9). Importantly, non-isocortical gray matter structures also do not exhibit high-to-low changes in FA with development (27, 34).

4.2. Extracting Species-Specific Developmental Time Constants from Investigations of Cortical FA

4.2.1. Accumulation of Data

Table 13.1 summarizes 11 published DTI investigations of cerebral cortical development. Criteria for inclusion in Table 13.1 were that cortical FA values during the fetal and/or perinatal periods are reported in graphical or tabular format, on healthy (or experimental control) individuals, and that no redundancy in the set of individuals may occur between data sets. Diffusion anisotropy values were obtained from these articles either directly from the authors, from data values listed in the published paper in a tabular format, or by estimating data values from published graphs. In the latter case, grids were placed over high-resolution images of graphs to obtain diffusion anisotropy values and/or ages, at which data were collected. Many of the reports included in Table 13.1 were directed at different regions of interest within the isocortex. If the entire isocortex was not studied (e.g., in a surface-based analysis), all regions of interest studied were averaged at each developmental time point for the analysis presented here to provide an approximation of overall isocortical FA. Additionally, where possible, analysis presented here focuses on superficial layers of the cerebral cortex. As described above, mean FA values incorporate several cortical lamina, and thus represent a heterogeneous set of FA values across the cortical depth. In cases where values were reported for different layers, data were collected only from superficial layers.

For one of the studies conducted on human subjects (37), it was not feasible to extract FA values from the published graphs. However, in this case, FA changes were modeled as a linear decrease with age within four distinct cortical areas. Cortical FA values were thus obtained by averaging the four reported slope and intercept values and interpolating FA values along a line described by the mean slope and intercept values over the age range reported for individuals in this study.

4.2.2. Conversion of Anisotropy Parameters

As different measures of cortical diffusion anisotropy were used among the studies listed in **Table 13.1**, all measures were converted to FA.

$$\mathrm{FA} = \sqrt{\frac{3}{2}} \sqrt{\frac{(\lambda_1 - \mathrm{ADC})^2 + (\lambda_2 - \mathrm{ADC})^2 + (\lambda_3 - \mathrm{ADC})^2}{\lambda_1^2 + \lambda_2^2 + \lambda_3^2}} \quad [1]$$

$$\mathrm{ADC} = (\lambda_1 + \lambda_2 + \lambda_3)/3 \quad [2]$$

In Equations [1] and [2], ADC is the apparent diffusion coefficient, and λ_1, λ_2, and λ_3 are eigenvalues of the diffusion tensor (6, 7). Two alternative rotationally invariant measures of diffusion anisotropy are A_σ and relative anisotropy (RA). The conversion of A_σ to FA is

$$\mathrm{FA} = A_\sigma \sqrt{3/\left(2A_\sigma^2 + 1\right)} \quad [3]$$

and the conversion of RA to FA is (87, 88)

$$\mathrm{FA} = \frac{\mathrm{RA}}{\sqrt{2}} \sqrt{3/\left(\mathrm{RA}^2 + 1\right)} \quad [4]$$

Figure 13.2 shows cortical FA values versus postconceptional age for the 11 studies listed in **Table 13.1**.

4.2.3. Modeling Temporal Changes as an Exponential Decay Toward a Positive Offset

It has been previously proposed that, for a single position with respect to the regional cortical FA gradient, cortical FA decreases exponentially from a maximal value ($\mathrm{FA_{max}}$) observed immediately following migration of immature neurons to the cortical plate, toward an asymptotic value characterizing cortical FA at maturity ($\mathrm{FA_{min}}$) at a rate that can be characterized by the time constant τ_{FA} (34)

$$FA = \begin{cases} FA_{max} \text{ if } t < t_{init} \\ (FA_{max} - FA_{min}) \exp\left(-(t - t_{init})/\tau_{FA}\right) + FA_{min} \text{ if } t \geq t_{init} \end{cases} \cdot [5]$$

In Equation [5], t_{init} is the post-conceptional age at which FA begins to decrease and is associated with the initiation of morphological differentiation within a given cortical area.

The majority of studies listed in **Table 13.1** do not provide an analytical expression to model FA changes with age. Exceptions are the ferret study in which the Equation [5] expression was proposed (34), a study of human subjects in which FA changes were considered to decrease linearly with age (37), and a study of human post-mortem brains in which cortical FA was approximated using a quadratic function of age (76, 77). A drawback to linear and quadratic models of cortical FA changes with age is they do not asymptotically converge to a value that is reflective of differentiated cortex. Thus, for reports listed in **Table 13.1** that produced data suitable for analysis using Equation [5], τ_{FA} values were calculated to provide comparisons to independently measured species-specific rates of brain development.

Three of the studies included in **Table 13.1** could not be analyzed using Equation [5]. For one of these cases, cortical FA was reported at two ages (29) and an additional study reported cortical FA at four ages (89). For these two studies, an analysis using Equation [5] is underdetermined because there are as many (or more) adjustable parameters in the model as there are measured data values. The study of Deipolyi et al. (37) was also not used for the Equation [5] analysis because extraction of FA values required reference to a model in which FA linearly decays with age and this step in the data analysis would introduce systematic error in subsequent fitting using the exponential function in Equation [5]. Data from studies not analyzed using Equation [5] are plotted as *open symbols* in **Fig. 13.2**.

Data reported in the remaining eight studies listed in **Table 13.1** were fitted using the Equation [5] model and the nonlinear least squares optimization routine implemented in Matlab (The Mathworks, Boston, MA). Results of these analyses are summarized in **Table 13.2** and are represented as *solid curves* in **Fig. 13.2**. For species in which more than one study is available (rat, ferret, and human), analyses were performed such that t_{init} values were constrained to a common species-specific value. Due to the lack of data at early and/or late ages for a subset of species, it was necessary to impose educated guesses as constraints in some optimization calculations. For the primate species, unconstrained optimization yielded t_{init} values of 125 and 195 days for baboon and human, respectively. Such values are considered unrealistically

Table 13.2
Parameters utilized in the comparison between time constants for exponential decrease in FA (τ_{FA}) and neurodevelopmental events (τ_{Event})

Species	References	t_{init} (days)	FA_{max}	FA_{min}	τ_{FA} (days)	τ_{Event} (days)
Mouse	(92)	20	0.53	0.20	3.3	12.02
Rat	(75)	22	0.54	0.28	2.7	14.54
	(28)	22	0.53	0.29	5	14.54
Ferret	(34)	49	0.77	0.32	10.7	32.74
	(93)	49	0.75	0.26	11.9	32.74
Baboon	(27)	100[a]	0.67	0.19	31.9	80.54[d]
Human	(77)	173[b]	0.49	0.10[c]	51.2	91.75
	(10)	173[b]	0.32	0.10[c]	39.8	91.75

[a] Baboon t_{init} constrained to be ≤ 100 days.
[b] Human t_{init} constrained to be ≤ 175 days.
[c] Human FA_{min} constrained to be ≥ 0.10.
[d] Baboon τ_{Event} score converted from macaque score based on gestation lengths of 185 and 165 days for baboon and macaque, respectively. See text for details.

high because cortical FA is known to decrease in baboon following gestational day 100 (unpublished observations), and data from human studies indicate cortical FA decreases over the period from 175 to 195 days gestation (11, 37, 68, 90). Thus, optimizations were performed by constraining t_{init} values to 100 and 175 days in baboon and human species, respectively. Additionally, optimizations performed, in which FA_{min} was not constrained produced unrealistically small values for this parameter of approximately 0.05 in human studies. Although a realistic estimate of FA_{min} will depend on experimental conditions as discussed above, previous estimates of cortical FA at maturity are on the order of 0.2 (21). Thus, FA_{min} values for the studies of human subjects were constrained to be 0.1 or greater.

4.3. Comparisons of Cortical Diffusion Anisotropy Data to Species Scores

A translational model of brain development (14–16) was used to obtain a quantitative measure of the rate of neuroanatomical development in each species that has the same physical units as τ_{FA}. The translating time model (http://www.translatingtime.net/) relates post-conceptional age to a species score, an event score, an interaction term, and a constant factor of 4.34 through the expression

$$\text{Post conceptional age} = \exp(\text{species score} + \text{event score} + \text{interaction term}) + 4.34. \quad [6]$$

The species score reflects the rate of development of particular species; species that develop relatively slowly will have higher scores. The event score reflects the timing of specific neural events (e.g., the appearance of the external capsule) with events occurring later in development having higher scores. The interaction term reflects a slower rate of cortical development in primates; it equals 0.249 for events occurring within the cerebral cortex in primate species and equals 0 otherwise. The constant factor represents the amount of time over which extremely early developmental events (i.e., blastulation) occur.

For each of the five species included in **Table 13.1**, the period of time between the end of neurogenesis in cortical layers II/III (characterized by an event score of 1.929) and eye opening (characterized by an event score of 2.546) was calculated. This time is defined here as τ_{Event} and is expressed in units of days. Numerical values for τ_{Event} are given for mouse, rat, ferret, and human in **Table 13.2**. To obtain a τ_{Event} for baboon (a species not explicitly included in the translating time model (14–16)), a correction factor, which is the ratio of the lengths of gestation for baboon divided by that for macaque (185/165 days) was multiplied by τ_{Event} for macaque to obtain the value given in **Table 13.2**. The baboon τ_{Event} derived is in qualitative agreement with observations reported in a histological study of baboon brain development (91).

Figure 13.3 provides a graphical comparison of τ_{Event} and τ_{FA} values for the five species. Generally, variation in τ_{Event} is matched by a proportional variation in τ_{FA}, which reflects agreement between cortical FA changes and other developmental events in estimating the rate of brain development. The line intersecting the origin, fit to the **Fig. 13.3** data (*dashed line*), has a slope of 0.454, indicating that the time constant associated with cortical FA decay with age is 45.4% the length of the time elapsed between cortical layers II/III neurogenesis and eye opening.

The most notable potential deviation from a proportional relationship between τ_{Event} and τ_{FA} in **Fig. 13.3** is observed for human species, in which the τ_{FA} values for both studies are larger than predicted based on τ_{Event}. Given the small number of studies investigated, it is conceivable that random error or inaccuracies in the measurement and/or modeling strategies employed for these studies are the source of this discrepancy. Two additional potential sources of difference between data for human and other species could also be considered. First, the difficulty of obtaining measurements of "normal" human individuals should be acknowledged. The in vivo studies (10, 37) utilized prematurely delivered human infants that showed no overt signs of neurological disorder as subjects. However, it could be argued that premature birth in itself predisposes individuals to heightened risk of abnormal brain development, and the inflated τ_{FA} values observed here reflect

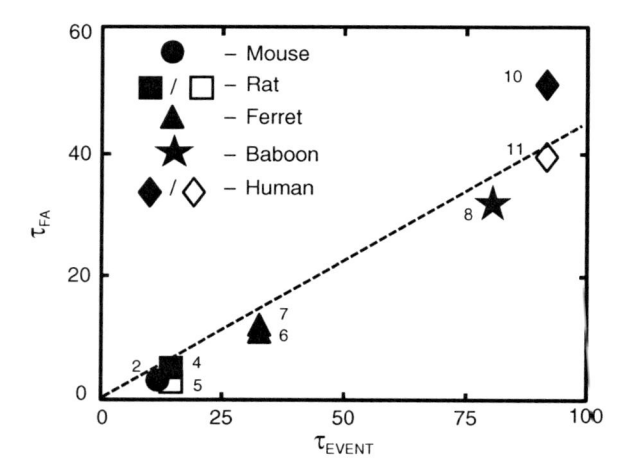

Fig. 13.3. Time constants for the exponential decrease in fractional anisotropy (τ_{FA}) are plotted as a function of neuroanatomical development times (τ_{Event}) for five different species. Time constants for decreases in fractional anisotropy were calculated based on the empirical mathematical model expressed in Equation [5]. Neurodevelopmental event times were calculated based on the days between the end of neurogenesis in cortical layer II/III and eye opening for each species (14–16). A high degree of correlation was found between time constants among and across species, suggesting that neuroanatomical development in the isocortex underlies patterns in diffusion anisotropy seen with age. The *dashed line* was obtained by fitting a line intersecting the origin to the eight data values. The slope of this line is 0.454. *Open symbols* indicate data collected in vivo, while *closed symbols* represent data collected from post-mortem samples. References are numbered as in **Fig. 13.2**.

reduced rates of development relative to control individuals in this population. A similar argument could be made that the brain tissue obtained in the post-mortem analysis (76) is not representative of "normal" brain. A second potential source of discrepancy between human τ_{Event} and τ_{FA} values is a species-specific mechanistic difference between biophysical determinants of cortical FA and development milestones used to build the translating time model. Additional study is necessary to discern whether true differences are being observed between humans and other species in τ_{Event} and τ_{FA} comparisons.

5. Summary of Findings and Conclusion

Despite high variability in many of the parameters employed in the studies reviewed here, patterns of cortical diffusion anisotropy during early developmental stages, found among studies both within and across species, are strikingly similar. Changes in FA can be accurately approximated using an empirical mathematical expression (Equation [5]), and the time periods over which the decreases in FA occur correspond to similar developmental events. This suggests that there is a common mechanism underlying DTI measurements among these studies.

The time periods reviewed correspond to periods of neuronal differentiation, specifically the differentiation of neurons or periods of increasing complexity of neuritic arbors. Therefore, inferences about cortical microanatomy should be possible based on animal studies employing DTI methodology. The comparisons between time constants calculated from exponential decreases in FA and neuroanatomical events across five different species support this concept (**Fig. 13.3**). In this comparison, there was a significant correlation between τ_{FA} and τ_{Event} ($r = 0.996$, $p < 0.0001$; statistics implemented in Statview [SAS, Cary, NC]). This direct correlation suggests that rates of reduction in FA correspond to overall rates of neural development for these species. In addition to suggesting that DTI methodology can infer neuroanatomical differentiation in the cerebral cortex, this comparison suggests that utilizing DTI methodology in animal species should allow for results to be applicable to humans.

Acknowledgments

This work was supported by National Institutes of Health Grant T32 AA007468-22 and the OHSU 11.7 T MRI system is supported by the Keck Foundation. The authors would like to thank Jaime F. Olavarria for helpful discussion during the preparation of this manuscript.

References

1. Le Bihan D (2003) Looking into the functional architecture of the brain with diffusion MRI. Nat Rev Neurosci 4:469–480
2. Mori S, Zhang J (2006) Principles of diffusion tensor imaging and its applications to basic neuroscience research. Neuron 51:527–539
3. Beaulieu C (2002) The basis of anisotropic water diffusion in the nervous system – a technical review. NMR Biomed 15:435–455
4. Moseley ME, Cohen Y, Mintorovitch J, Chileuitt L, Shimizu H, Kucharczyk J, Wendland MF, Weinstein PR (1990) Early detection of regional cerebral ischemia in cats: comparison of diffusion- and T2-weighted MRI and spectroscopy. Magn Reson Med 14:330–346
5. Basser PJ, Pierpaoli C (1996) Microstructural and physiological features of tissues elucidated by quantitative-diffusion-tensor MRI. J Mag Reson Ser B 111:209–219
6. Basser PJ, Mattiello J, LeBihan D (1994) Estimation of the effective self-diffusion tensor from the NMR spin echo. J Magn Reson B 103:247–254
7. Basser PJ, Mattiello J, LeBihan D (1994) MR diffusion tensor spectroscopy and imaging. Biophys J 66:259–267
8. Rakic P (1995) A small step for the cell, a giant leap for mankind: a hypothesis of neocortical expansion during evolution. Trends Neurosci 18:383–388
9. Sidman RL, Rakic P (1982) Development of the human central nervous system. In: Haymaker W, Adams RD (eds) Histology and histopathology of the nervous system. Charles C. Thomas, Springfield, IL, pp 3–145

10. McKinstry RC, Mathur A, Miller JP, Ozcan AO, Snyder AZ, Schefft GL, Almli CR, Shiran SI, Conturo TE, Neil JJ (2002) Radial organization of developing human cerebral cortex revealed by non-invasive water diffusion anisotropy MRI. Cereb Cortex 12:1237–1243
11. Neil JJ, Shiran SI, McKinstry RC, Schefft GL, Snyder AZ, Almli CR, Akbudak E, Aaronovitz JA, Miller JP, Lee BCP, Conturo TE (1998) Normal brain in human newborns: apparent diffusion coefficient and diffusion anisotropy measured using diffusion tensor imaging. Radiology 209:57–66
12. Baratti C, Barnett A, Pierpaoli C (1997) Comparative MRI study of brain maturation using T1, T2, and the diffusion tensor. Proceedings of the ISMRM, Vancouver, CA
13. Thornton JS, Ordidge RJ, Penrice J, Cady EB, Amess PN, Punwani S, Clemence M, Wyatt JS (1997) Anisotropic water diffusion in white and gray matter of the neonatal piglet brain before and after transient hypoxia–ischaemia. Magn Reson Imaging 15:433–440
14. Clancy B, Kersh B, Hyde J, Darlington RB, Anand KJ, Finlay BL (2007) Web-based method for translating neurodevelopment from laboratory species to humans. Neuroinformatics 5:79–94
15. Darlington RB, Dunlop SA, Finlay BL (1999) Neural development in metatherian and eutherian mammals: variation and constraint. J Comp Neurol 411:359–368
16. Finlay BL, Darlington RB (1995) Linked regularities in the development and evolution of mammalian brains. Science 268:1578–1584
17. Juraska JM, Fifkova E (1979) A Golgi study of the early postnatal development of the visual cortex of the hooded rat. J Comp Neurol 183:247–256
18. Conel JL (1939) The postnatal development of the human cerebral cortex, vol 1. Harvard University Press, Cambridge, MA
19. Mori S, Itoh R, Zhang J, Kaufmann WE, van Zijl PC, Solaiyappan M, Yarowsky P (2001) Diffusion tensor imaging of the developing mouse brain. Magn Reson Med 46:18–23
20. McNab JA, Jbabdi S, Deoni SC, Douaud G, Behrens TE, Miller KL (2009) High resolution diffusion-weighted imaging in fixed human brain using diffusion-weighted steady state free precession. Neuroimage 46:775–785
21. Bhagat YA, Beaulieu C (2004) Diffusion anisotropy in subcortical white matter and cortical gray matter: changes with aging and the role of CSF-suppression. J Magn Reson Imaging 20:216–227
22. Granger B, Tekaia F, Le Sourd AM, Rakic P, Bourgeois JP (1995) Tempo of neurogenesis and synaptogenesis in the primate cingulate mesocortex: comparison with the neocortex. J Comp Neurol 360:363–376
23. Miller MW, Potempa G (1990) Numbers of neurons and glia in mature rat somatosensory cortex: effects of prenatal exposure to ethanol. J Comp Neurol 293:92–102
24. Bayer SA, Altman J (1991) Neocortical development. Raven Press, New York, NY
25. Caviness VS Jr, Goto T, Tarui T, Takahashi T, Bhide PG, Nowakowski RS (2003) Cell output, cell cycle duration and neuronal specification: a model of integrated mechanisms of the neocortical proliferative process. Cereb Cortex 13:592–598
26. Kroenke CD, Bretthorst GL, Inder TE, Neil JJ (2005) Diffusion MR imaging characteristics of the developing primate brain. Neuroimage 25:1205–1213
27. Kroenke CD, Van Essen DC, Inder TE, Rees S, Bretthorst GL, Neil JJ (2007) Microstructural changes of the baboon cerebral cortex during gestational development reflected in magnetic resonance imaging diffusion anisotropy. J Neurosci 27:12506–12515
28. Huang H, Yamamoto A, Hossain MA, Younes L, Mori S (2008) Quantitative cortical mapping of fractional anisotropy in developing rat brains. J Neurosci 28:1427–1433
29. Sizonenko SV, Camm EJ, Garbow JR, Maier SE, Inder TE, Williams CE, Neil JJ, Huppi PS (2007) Developmental changes and injury induced disruption of the radial organization of the cortex in the immature rat brain revealed by in vivo diffusion tensor MRI. Cereb Cortex 17:2609–2617
30. Smart IHM (1983) Three dimensional growth of the mouse isocortex. J Anat 137:683–694
31. McSherry GM (1984) Mapping of cortical histogenesis in the ferret. J Embryol Exp Morphol 81:239–252
32. McSherry GM, Smart IH (1986) Cell production gradients in the developing ferret isocortex. J Anat 144:1–14
33. Marin-Padilla M (1978) Dual origin of the mammalian neocortex and evolution of the cortical plate. Anat Embryol (Berl) 152:109–126
34. Kroenke CD, Taber EN, Leigland LA, Knutsen AK, Bayly PV (2009) Regional patterns of cerebral cortical differentiation determined by diffusion tensor MRI. Cereb Cortex 19:2916–2929

35. Jackson CA, Peduzzi JD, Hickey TL (1989) Visual cortex development in the ferret. I. Genesis and migration of visual cortical neurons. J Neurosci 9:1242–1253
36. Noctor SC, Scholnicoff NJ, Juliano SL (1997) Histogenesis of ferret somatosensory cortex. J Comp Neurol 387:179–193
37. Deipolyi AR, Mukherjee P, Gill K, Henry RG, Partridge SC, Veeraraghavan S, Jin H, Lu Y, Miller SP, Ferriero DM, Vigneron DB,, Barkovich AJ (2005) Comparing microstructural and macrostructural development of the cerebral cortex in premature newborns: diffusion tensor imaging versus cortical gyration. NeuroImage 27:579–586
38. Kaufmann WE, Moser HW (2000) Dendritic anomalies in disorders associated with mental retardation. Cereb Cortex 10:981–991
39. Nitkin RM (2000) Dendritic mechanisms in brain function and developmental disabilities. Cereb Cortex 10(Issue 10):925–926
40. Hammer RP Jr (1986) Alcohol effects on developing neuronal structure. In: West JR (ed) Alcohol and brain development. Oxford, New York, NY, pp 184–203
41. Fabregues I, Ferrer I, Gairi JM, Cahuana A, Giner P (1985) Effects of prenatal exposure to ethanol on the maturation of the pyramidal neurons in the cerebral cortex of the guinea-pig: a quantitative Golgi study. Neuropathol Appl Neurobiol 11:291–298
42. Hammer RP Jr, Scheibel AB (1981) Morphologic evidence for a delay of neuronal maturation in fetal alcohol exposure. Exp Neurol 74:587–596
43. Granato A, Van Pelt J (2003) Effects of early ethanol exposure on dendrite growth of cortical pyramidal neurons: inferences from a computational model. Brain Res Dev Brain Res 142:223–227
44. Armstrong DD, Dunn K, Antalffy B (1998) Decreased dendritic branching in frontal, motor and limbic cortex in Rett syndrome compared with trisomy 21. J Neuropathol Exp Neurol 57:1013–1017
45. Kishi N, Macklis JD (2004) MECP2 is progressively expressed in post-migratory neurons and is involved in neuronal maturation rather than cell fate decisions. Mol Cell Neurosci 27:306–321
46. Izbudak I, Farage L, Bonekamp D, Zhang W, Bibat G, Mori S, Naidu S, Horska A (2009) Diffusion tensor imaging findings in Rett syndrome patients. Proceedings of the 17th ISMRM Scientific Meeting, Honolulu, HI
47. Naidu S, Kaufmann WE, Abrams MT, Pearlson GD, Lanham DC, Fredericksen KA, Barker PB, Horska A, Golay X, Mori S, Wong DF, Yablonski M, Moser HW, Johnston MV (2001) Neuroimaging studies in Rett syndrome. Brain Dev 23(Suppl 1): S62–S71
48. Leuzzi V, Tosetti M, Montanaro D, Carducci C, Artiola C, Carducci C, Antonozzi I, Burroni M, Carnevale F, Chiarotti F, Popolizio T, Giannatempo GM, D'Alesio V, Scarabino T (2007) The pathogenesis of the white matter abnormalities in phenylketonuria: a multimodal 3.0 Tesla MRI and magnetic resonance spectroscopy (1H MRS) study. J Inherit Metab Dis 30:209–216
49. Lebel C, Rasmussen C, Wyper K, Walker L, Andrew G, Yager J, Beaulieu C (2008) Brain diffusion abnormalities in children with fetal alcohol spectrum disorder. Alcohol Clin Exp Res 32:1732–1740
50. Ma X, Coles CD, Lynch ME, Laconte SM, Zurkiya O, Wang D, Hu X (2005) Evaluation of corpus callosum anisotropy in young adults with fetal alcohol syndrome according to diffusion tensor imaging. Alcohol Clin Exp Res 29:1214–1222
51. Sowell ER, Johnson A, Kan E, Lu LH, Van Horn JD, Toga AW, O'Connor MJ, Bookheimer SY (2008) Mapping white matter integrity and neurobehavioral correlates in children with fetal alcohol spectrum disorders. J Neurosci 28:1313–1319
52. Wozniak JR, Mueller BA, Chang PN, Muetzel RL, Caros L, Lim KO (2006) Diffusion tensor imaging in children with fetal alcohol spectrum disorders. Alcohol Clin Exp Res 30:1799–1806
53. Anjari M, Srinivasan L, Allsop JM, Hajnal JV, Rutherford MA, Edwards AD, Counsell SJ (2007) Diffusion tensor imaging with tract-based spatial statistics reveals local white matter abnormalities in preterm infants. Neuroimage 35:1021–1027
54. Counsell SJ, Edwards AD, Chew AT, Anjari M, Dyet LE, Srinivasan L, Boardman JP, Allsop JM, Hajnal JV, Rutherford MA, Cowan FM (2008) Specific relations between neurodevelopmental abilities and white matter microstructure in children born preterm. Brain 131:3201–3208
55. Alexander AL, Lee JE, Lazar M, Boudos R, DuBray MB, Oakes TR, Miller JN, Lu J, Jeong EK, McMahon WM, Bigler ED, Lainhart JE (2007) Diffusion tensor imaging of the corpus callosum in Autism. Neuroimage 34:61–73
56. Kubicki M, McCarley R, Westin CF, Park HJ, Maier S, Kikinis R, Jolesz FA, Shenton ME (2007) A review of diffusion tensor imaging studies in schizophrenia. J Psychiatr Res 41:15–30

57. Kubicki M, Park H, Westin CF, Nestor PG, Mulkern RV, Maier SE, Niznikiewicz M, Connor EE, Levitt JJ, Frumin M, Kikinis R, Jolesz FA, McCarley RW, Shenton ME (2005) DTI and MTR abnormalities in schizophrenia: analysis of white matter integrity. Neuroimage 26:1109–1118
58. Kumra S, Ashtari M, Cervellione KL, Henderson I, Kester H, Roofeh D, Wu J, Clarke T, Thaden E, Kane JM, Rhinewine J, Lencz T, Diamond A, Ardekani BA, Szeszko PR (2005) White matter abnormalities in early-onset schizophrenia: a voxel-based diffusion tensor imaging study. J Am Acad Child Adolesc Psychiatry 44:934–941
59. Fields RD (2008) White matter in learning, cognition and psychiatric disorders. Trends Neurosci 31:361–370
60. Innocenti GM, Price DJ (2005) Exuberance in the development of cortical networks. Nat Rev Neurosci 6:955–965
61. Katz LC, Crowley JC (2002) Development of cortical circuits: lessons from ocular dominance columns. Nat Rev Neurosci 3:34–42
62. Hubel DH (1982) Exploration of the primary visual cortex, 1955–78. Nature 299:515–524
63. Wiesel TN (1982) Postnatal development of the visual cortex and the influence of environment. Nature 299:583–591
64. D'Arceuil H, de Crespigny A (2007) The effects of brain tissue decomposition on diffusion tensor imaging and tractography. Neuroimage 36:64–68
65. D'Arceuil HE, Westmoreland S, de Crespigny AJ (2007) An approach to high resolution diffusion tensor imaging in fixed primate brain. Neuroimage 35:553–565
66. Flynn SW, Lang DJ, Mackay AL, Goghari V, Vavasour IM, Whittall KP, Smith GN, Arango V, Mann JJ, Dwork AJ, Falkai P, Honer WG (2003) Abnormalities of myelination in schizophrenia detected in vivo with MRI, and post-mortem with analysis of oligodendrocyte proteins. Mol Psychiatry 8:811–820
67. Guilfoyle DN, Helpern JA, Lim KO (2003) Diffusion tensor imaging in fixed brain tissue at 7.0 T. NMR Biomed 16:77–81
68. Huang H, Xue R, Zhang J, Ren T, Richards LJ, Yarowsky P, Miller MI, Mori S (2009) Anatomical characterization of human fetal brain development with diffusion tensor magnetic resonance imaging. J Neurosci 29:4263–4273
69. Pfefferbaum A, Sullivan EV, Adalsteinsson E, Garrick T, Harper C (2004) Postmortem MR imaging of formalin-fixed human brain. Neuroimage 21:1585–1595
70. Sun SW, Neil JJ, Liang HF, He YY, Schmidt RE, Hsu CY, Song SK (2005) Formalin fixation alters water diffusion coefficient magnitude but not anisotropy in infarcted brain. Magn Reson Med 53:1447–1451
71. Sun SW, Neil JJ, Song SK (2003) Relative indices of water diffusion anisotropy are equivalent in live and formalin-fixed mouse brains. Magn Reson Med 50:743–748
72. Shepherd TM, Flint JJ, Thelwall PE, Stanisz GJ, Mareci TH, Yachnis AT, Blackband SJ (2009) Postmortem interval alters the water relaxation and diffusion properties of rat nervous tissue – implications for MRI studies of human autopsy samples. NeuroImage 44:820–826
73. Shepherd TM, Thelwall PE, Stanisz GJ, Blackband SJ (2009) Aldehyde fixative solutions alter the water relaxation and diffusion properties of nervous tissue. Magn Reson Med 62:26–34
74. Thelwall PE, Shepherd TM, Stanisz GJ, Blackband SJ (2006) Effects of temperature and aldehyde fixation on tissue water diffusion properties, studied in an erythrocyte ghost tissue model. Magn Reson Med 56:282–289
75. Bockhorst KH, Narayana PA, Liu R, Ahobila-Vijjula P, Ramu J, Kamel M, Wosik J, Bockhorst T, Hahn K, Hasan KM, Perez-Polo JR (2008) Early postnatal development of rat brain: in vivo diffusion tensor imaging. J Neurosci Res 86:1520–1528
76. Gupta RK, Hasan KM, Trivedi R, Pradhan M, Das V, Parikh NA, Narayana PA (2005) Diffusion tensor imaging of the developing human cerebrum. J Neurosci Res 81:172–178
77. Trivedi R, Gupta RK, Husain N, Rathore RK, Saksena S, Srivastava S, Malik GK, Das V, Pradhan M, Sarma MK, Pandey CM, Narayana PA (2009) Region-specific maturation of cerebral cortex in human fetal brain: diffusion tensor imaging and histology. Neuroradiology 51:567–576
78. Hofman MA (1988) Size and shape of the cerebral cortex in mammals. II. The cortical volume. Brain Behav Evol 32:17–26
79. Hofman MA (1985) Size and shape of the cerebral cortex in mammals. I. The cortical surface. Brain Behav Evol 27:28–40
80. Behrens TE, Berg HJ, Jbabdi S, Rushworth MF, Woolrich MW (2007) Probabilistic diffusion tractography with multiple fibre orientations: what can we gain? NeuroImage 34:144–155
81. Frank LR (2001) Anisotropy in high angular resolution diffusion-weighted MRI. Magn Reson Med 45:935–939

82. Tuch DS, Reese TG, Wiegell MR, Makris N, Belliveau JW, Wedeen VJ (2002) High angular resolution diffusion imaging reveals intravoxel white matter fiber heterogeneity. Magn Reson Med 48:577–582
83. Wedeen VJ, Hagmann P, Tseng WY, Reese TG, Weisskoff RM (2005) Mapping complex tissue architecture with diffusion spectrum magnetic resonance imaging. Magn Reson Med 54:1377–1386
84. Tuch DS (2004) Q-ball imaging. Magn Reson Med 52:1358–1372
85. Kroenke CD, Bretthorst GL, Inder TE, Neil JJ (2006) Modeling water diffusion anisotropy within fixed newborn primate brain using Bayesian probability theory. Magn Reson Med 55:187–197
86. Van Essen DC, Dickson J, Harwell J, Hanlon D, Anderson CH, Drury HA (2001) An integrated software system for surface-based analysis of cerebral cortex. J Am Med Inf Assoc 41:1359–1378
87. Kingsley PB (2005) Introduction to diffusion tensor imaging mathematics: part II. anisotropy, diffusion-weighting factors, and gradient echo schemes. Concepts Magn Reson Part A 28A:123–154
88. Le Bihan D, Mangin JF, Poupon C, Clark CA, Pappata S, Molko N, Chabriat H (2001) Diffusion tensor imaging: concepts and applications. J Magn Reson Imaging 13:534–546
89. Larvaron P, Boespflug-Tanguy O, Renou JP, Bonny JM (2007) In vivo analysis of the postnatal development of normal mouse brain by DTI. NMR Biomed 20:413–421
90. Mukherjee P, McKinstry RC (2006) Diffusion tensor imaging and tractography of human brain development. Neuroimag Clin N Am 16:19–43
91. Dieni S, Inder T, Yoder B, Briscoe T, Camm E, Egan G, Denton D, Rees S (2004) The pattern of cerebral injury in a primate model of preterm birth and neonatal intensive care. J Neuropathol Exp Neurol 63:1297–1309
92. Baloch S, Verma R, Huang H, Khurd P, Clark S, Yarowsky P, Abel T, Mori S, Davatzikos C (2009) Quantification of brain maturation and growth patterns in C57BL/6J mice via computational neuroanatomy of diffusion tensor images. Cereb Cortex 19:675–687
93. Barnette AR, Neil JJ, Kroenke CD, Griffith JL, Epstein AA, Bayly PV, Knutsen AK, Inder TE (2009) Characterization of brain development in the ferret via MRI. Pediatr Res 66:80–84

INDEX

A

Acuity................................31, 46–49, 51–52
Addiction........................148, 269–273, 281, 295
Allocentric..92
Alzheimer's disease (AD)............2–3, 8–14, 17–18, 94
Ambulation.............................56–57, 249, 257
Anxiety....80, 99, 127–128, 148, 165, 168, 193–194, 243, 256–257, 259, 269, 275, 312, 314, 318–320, 325
Associative conditioning............................267
Autism.....................................153–185, 334
Autism spectrum disorders (ASD)......154–162, 171–174

B

Behavioral assessment................155, 308, 314, 326
Behavioral sensitization..........................267–295
Bradykinesia....................55–57, 59–60, 62–63, 85

C

Cardiac arrest (CA)....................308, 312, 322–325
Cardiopulmonary resuscitation (CPR)......312, 322–325
Cascade.............................161, 237–238, 246
Cerebellum..............................1, 4–5, 9, 11–14
Cerebral ischemia.....................307–308, 310–325
Cholesterol..................................12–15, 17
Circadian.................................217–233, 311
Conditioned place preference (CPP)...133–149, 166–168, 172, 181–182
Conditioned stimulus (CS)...2–3, 7–8, 11, 13–18, 21–22, 112–129, 134, 138, 173–174, 182–185, 274
Confocal laser scanning microscope....................36
Contrast sensitivity..............................48–50
Cortisol............218, 221, 227–229, 231–232, 275, 279

D

Depression...174, 176, 178, 194, 208, 243, 256, 271, 313, 319–320, 325
Development...2–5, 42, 50, 128, 157–158, 162, 165, 247, 268–270, 273, 285, 291, 294–295, 309–310, 329–347
DHEAS........................221, 227–229, 231–232
Diffusion tensor imaging (DTI)............244, 329–347
Drug abuse........145, 148, 269–270, 279, 282, 284, 294

E

Egocentric...92
Electroretinogram (ERG)....................32–33, 36–45
Epidemiology............................238, 247, 307
Excitotoxic cascades.................................238

F

Fear conditioning....................111–129, 173–174, 182–184
Fear extinction.................................124, 126
Fear potentiated startle..........................113, 118
Fetal alcohol syndrome (FAS).......................2–8
Fractional anisotropy (FA).....................330–347

G

Gene microarrays....................................232

H

Hierarchical...196
Hippocampus....1–2, 8–9, 11–12, 91, 94–95, 97–98, 116, 246, 254, 268, 324

I

Immunohistochemistry........................31–32, 76
International Classification of Functioning, Disability, and Health (ICF).........................56, 59

L

Locomotor...66, 70, 77–84, 140–141, 148–149, 177, 180, 269, 271–272, 286–287, 289–290, 293, 308, 311, 318, 320, 324

M

Magnetic resonance imaging (MRI).........159, 213, 284, 329–330, 337
Memory Island.........................98–102, 104–106
Mouse..10, 32–33, 36–44, 51, 66, 77–78, 81–83, 93, 100, 106, 141, 153–185, 203, 257, 286, 291–292, 308–312, 314–326, 335–336, 338, 340–341, 344–345

N

Neuroadaptations...............................272, 294
Neuropathology......................................12
Nonhuman primate..........92, 94–96, 98, 106, 217–233
Novel Image Novel Location (NINL).........98, 102–106

P

Parkinson's disease (PD)....................9, 55–85, 283
Photoperiod sleep.....................217, 223, 229–233
Physiology..29–52, 92, 113, 119, 121, 156, 166, 168–169, 200, 217–233, 268, 276, 279, 283, 309–311, 323, 337

J. Raber (ed.), *Animal Models of Behavioral Analysis*, Neuromethods 50,
DOI 10.1007/978-1-60761-883-6, © Springer Science+Business Media, LLC 2011

ANIMAL MODELS OF BEHAVIORAL ANALYSIS
Index

R

Rabbit 2–3, 6, 8, 10–15, 17–18
Rat .. 3–7, 10–11, 15–16, 18–22, 46–47, 49, 51–52, 65–77, 83, 117, 125, 134, 136, 139–141, 148, 165, 168–170, 182, 202, 208, 246, 248–249, 251–258, 269, 274, 308, 313–314, 321, 333, 335–338, 340–341, 343–345
Reinstatement 112, 118, 125, 135, 139–141, 147–149
Relapse 140, 149, 268–269, 272–273, 295
Retina 29–31, 34–36, 39, 45–46, 50–52

S

Sensory .. 4–5, 11, 51, 57, 85, 116, 154, 194, 196, 200–204, 243, 245, 248–249, 252, 255, 258, 309, 313, 334
Sequelae 237, 240, 243–244, 246, 258
Social interaction 148, 153–185, 205, 256–257, 259
Spatial Foodport Maze 95–96, 106
Spatial learning 91–106, 247, 253
Stroke 52, 60–61, 307–309, 312–322, 324–325
Substantia nigra pars compacta 55, 65, 67, 77

T

Traumatic brain injury (TBI) 237–259

U

Uncertainty 126, 193–214
Unconditioned stimulus (US) 2–3, 7–8, 13–18, 21–22, 112–120, 124–127, 129, 134, 139, 173

V

Virtual Optomotor System (VOS) 31, 49–52
Visual ... 4, 16, 31, 33, 44, 46–52, 79, 92, 95, 97, 114, 117, 120, 129, 144, 155, 160, 162, 169, 171, 194, 204, 221–222, 229–230, 232, 275, 277, 281, 317, 325, 333–334, 340

W

Water maze .. 92–94, 96–97, 100, 106, 246, 252–253, 258, 311–312, 315, 324–325
Withdrawal .. 140, 171, 269, 275–276, 278, 280–281, 284, 291, 293